Recent Advances in
OCCUPATIONAL HEALTH

EDITED BY

J. M. HARRINGTON

NUMBER THREE

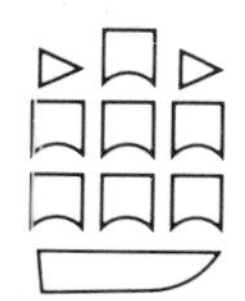

CHURCHILL LIVINGSTONE
EDINBURGH LONDON MELBOURNE AND NEW YORK 1987

CHURCHILL LIVINGSTONE
Medical Division of Longman Group UK limited

Distributed in the United States of America by
Churchill Livingstone Inc., 1560 Broadway, New York,
N.Y. 10036, and by associated companies, branches
and representatives throughout the world.

First Published 1987

ISBN 0443 03577 6

British Library Cataloguing in Publication Data
Recent advances in occupational health. –
No. 31
1. Industrial hygiene – Periodicals
613.6′2 RC967

Library of Congress Cataloging in Publication Data
Recent advances in occupational health. – No. 1- –
Edinburgh; New York: Churchill Livingstone, 1981-
v.: ill; 24 cm
Irregular.
Editor: 1981, J. C. McDonald.
ISSN 0261-1449 = Recent advances in occupational health.
1. Industrial hygiene—Periodicals. 2. Occupational diseases — Periodicals.
I. Mcdonald, J. C. (John Corbett)
[DNLM: 1. Occupational Medicine — periodicals. W1 RE105VNM]
RC967.R42 613.6′2′05 — dc19 85-644853

Printed in Great Britain at The Bath Press, Avon

SBHMC BOOK
54009000004479

Recent Advances in
OCCUPATIONAL HEALTH

J. M. HARRINGTON BSc MSc MD FRCP FFOM
Professor of Occupational Health, University of Birmingham, UK

R. G. PARRISH II MD
Division of Environmental Hazards and Health Effects, Center for Environmental Health, US Department of Health and Human Services, Atlanta, Georgia 30333, USA

A. J. RAWLINGS MSc BA
Ergonomics Group, Department of Engineering Production, Birmingham University, UK

J. RUTENFRANZ MD PhD
Professor, Institut fur Arbeitphysiologies, Dortmund, Germany

P. A. SCHULTE PhD
Chief, Screening and Notification Activity, Industrywide Studies Branch, National Institute for Occupational Safety and Health, Cincinnati, Ohio, USA

D. M. SMITH MBChB MFOM DIH
Employment Medical Adviser, Health and Safety Executive, Worcester, UK

C. A. SOUTAR MD MRCP
Head of Medical Branch, Institute of Occupational Medicine, Edinburgh, UK

J. TAYLOR MB BS FFOM DIH
Medical Adviser, Department of Transport, Swansea, UK

D. THOMPSON PhD MSc
Ergonomics Group, Department of Engineering Production, Birmingham University, UK

Contents

first sight, the production of oxygen in this way appears to demand a violation of the Second Law of Thermodynamics!

The solution is to use two (or better three) beds, each of such a size that it can generate far more oxygen than is needed by the aircrew. They can then be made to operate in a push-pull fashion by means of a solenoid valve that switches the flow of incoming gas between them (Miller et al, 1980). The pressure in the spent bed is reduced by connecting it to the outside atmosphere, to facilitate desorbtion, and the excess product from the active bed is passed through it in the reverse direction; when the scrubbing is complete, the forward flow of air is again diverted by the valve to the now clean bed.

The molecular sieve oxygen concentrator is mechanically simple, but its inherent efficiency is low. The theoretical maximum of its performance is, of course, the extraction of 21 litres of oxygen from 100 litres of air, but the need for back-flushing means that a large fraction of the gas supplied to it is eventually voided back into the ambient atmosphere. In a typical system, an input of about 300 litres per minute is needed to supply breathing gas at a rate of 15 litres per minute, but because the source of supply is air bled from the aircraft engine, the extravagance is unimportant. Electrical power is needed to operate the valves and heaters, and a back-up source of gaseous oxygen must be available for use in cases of engine failure, but emergency supplies are also necessary to guard against failure in more conventional oxygen systems.

The gas produced by a molecular sieve is not pure oxygen. The 0.93% of argon naturally present in the atmosphere is not adsorbed by the zeolite, and it is concentrated to the same extent as the oxygen. Thus, the product gas can contain, at most, 95% of oxygen. Argon has no specific adverse physiological effects, and acts as a simple diluent. Its presence *does* reduce the maximum safe altitude at which the system can be used. The accepted hypoxia threshold is 40 000 feet when pure oxygen is breathed, and the presence of 5% of argon reduces this by about 1000 feet; from a practical standpoint this matters not at all. In fact, although it is generally preferable (in aviation but *not* in clinical medicine) to give too much oxygen than too little, there are significant objections to the use of high concentrations save at very great altitudes. The complexity of conventional oxygen regulators is increased by the requirement to add judicious amounts of air to the breathing mixture, but that need brings the hidden benefit of conserving the supply of bottled or liquid gas. With OBOGS, that economy is unnecessary, and it is actually necessary deliberately to degrade the efficiency of the oxygen concentrator so that, at low altitudes, the breathing mixture contains about 40% of inert gas (Ernsting, 1984).

It can be argued that systems based upon compressed or liquid supplies are perfectly satisfactory, and that the incidence of failure or contamination is too small to justify the development of alternative techniques. However, in terms of cost of ownership, of safety, and of ease of handling, the on-board generator scores heavily. Nor should the spin-off to medical practice be forgotten, for the advantages of molecular sieve oxygen systems are gaining practical recognition in hospitals and homes, as well as in Harriers.

HELICOPTER SAFETY

The greatest scope for improvement in flight safety undoubtedly lies in the domain of the helicopter. Rotary wing aircraft have been, to some extent, the Cinderella of aviation, and their accident record has been significantly poorer, in terms both of numbers of accidents and of fatalities, than that of their conventional cousins (Gilbert & MacMahon, 1982). Overall, the accident rate has significantly diminished as engines, rotors and airframes have become more reliable. The fatality rate has not declined to the same degree, because the envelope in which helicopters fly has expanded (Vyrnwy-Jones, 1985). The maximum capability of the modern helicopter is, in fact, not far removed from that at which ejection seats are judged to be necessary in fixed wing aircraft, and a logical and positive step would be to introduce similar escape systems into helicopters. The key to the problem is to avoid the rotor disc, or by jettisoning the rotor and gearbox as the first stage of an upward ejection sequence. Expense and complexity preclude such solutions; what is practicable is not always reasonable.

It is possible to escape from a crippled helicopter in flight by baling out, but only if the aircraft is stable and at sufficient altitude, and if the occupants are equipped with parachutes. The only other hope of survival lies in the skilful use of autorotation, in which the rotor free-wheels but is still, to some extent, under the control of the pilot. A crash in autorotation is to the helicopter what a forced landing is to an aircraft; in ideal circumstances it leads to a normal but unpowered arrival, but in the worst case it results in a destructive impact. The force of such a heavy landing is applied mainly in the vertical axis of the body, in which the tolerance of impact is low, because little support is received from the seat back or the restraint harness. The probability of spinal injury is high, and American figures show an incidence of fracture of about 20% (Laananen, 1980). Other data indicate that the risk of vertebral fracture rises from about 5% when the peak force is 15 G to more than 60% at 25 G (Kazarian & Greaves, 1977; Howard, 1982). In any crash, the decelerative force depends only on the velocity at impact and the stopping distance. It is easy to show that, for a crash involving a vertical speed of 40 feet per second, the extreme value of 25 G will be exceeded if the stopping distance is less than 14 inches. If 15 G is taken as a more realistic criterion for the threshold of acceptable injury, that distance must be increased to 22 inches. If the vertical speed of the aircraft when it strikes the ground is 75 feet per second, a decelerating distance of 70 inches is required to limit the force to 15 G. Some absorption of the energy is afforded by the undercarriage and by deformation of the airframe, but small helicopters have little capacity to cushion the shock.

What is needed is a device that will come into operation when the impact force reaches a predetermined level, and will then dissipate the remaining energy by controlled collapse (Desjardins et al, 1982). The crew seat is the obvious site for such a mechanism, but the limiting factor is the space available within the shell of the aircraft. Any protection that can be given by an energy absorbing seat must be achieved by adding no more than about 1 foot to the vertical travel of the occupant between the moment of impact and the final arrest of the seat, but even such a small increase in stopping distance would greatly improve the survivability in a majority of crashes.

Some elegant and ingenious approaches to energy absorption have been tested, based on a variety of physical principles (Svoboda & Warrick, 1981; Schulman, 1982). These include the crushing of a thick metal honeycomb, and the pumping of hydraulic fluid through narrow orifices. In one of the more successful seats, developed in the UK, tapered rods attached to the rails are forced through a series of bushes mounted on the seat itself. The stroke begins when the applied acceleration reaches 15 G, and it has a maximum travel of about 12 inches; the exact distance depends upon the weight of the occupant. This system, which is now being brought into service, provides very good attentuation. In tests it transformed a triangular pulse of 56 G applied to the vehicle into a fairly level plateau of less than 25 G on the man. No device of this type can be wholly effective, because all crashes involve multi-directional forces. Although any reduction in the vertical component of an impact is to be welcomed, the ideal must be to provide energy absorption in all three axes. The associated engineering problems are formidable, but the day of the truly crashworthy helicopter seat cannot be far distant.

SURVIVAL IN COLD WATER

At first thought, the problems of survival after immersion in cold water are more the concern of the sailor than of the aviator. Nonetheless, a great deal of military flying takes place above the sea, and the use of helicopters to ferry large numbers of workers from the mainland to offshore oilfields continues to increase. Emergencies leading to ejection or to ditching are not unknown, and the risk of being immersed is not negligible. Thus, there is a moral (if not a legal) obligation to provide safeguard for aircrew and for passengers. That requirement has spawned a wide variety of immersion garments, and sparked considerable controversy about the degree of protection needed (Goldman, 1980).

The fundamental question is how much thermal insulation should be incorporated into the survival suit. It can be light in weight and comfortable in wear, or it can be an effective barrier against the cold and wet, but it cannot be both (Hayward et al, 1978). The *minimum* requirement is set by a balance between the factors that govern the probable time of survival in the water, and those that determine the likely delay before rescue can be effected. The latter depend upon weather, and the distance and availability of resources, while the former include the state and temperature of the sea; none of these lies within the scope of aviation medicine. It is possible, however, to specify in round terms the probability of dangerous hypothermia as a function of the temperature, the exposure time, and the thermal properties of the clothing. (It must be remembered that the impairment of physical capacity caused by cold can lead to drowning with a body temperature well above the clinical definition of hypothermia, and the threat to survival becomes serious at a core temperature of about 34°C.)

In March, the highest water temperature in the North Sea is 7°C, and the lowest can be 2 or 3°C. Few measurements have been made in such conditions, but after immersion at 5°C an average man will have a 50% chance of surviving for one hour before he succumbs to hypothermia (White & Roth, 1979). The *safe* limit must clearly be set well below this LD_{50}. By measuring the rate at which heat is lost from the body during brief periods of immersion, it has been possible to construct complex mathematical models for predicting the time taken for the deep body temperature to fall to the critical hypothermic threshold, and to assess the efficiency of different forms of thermal insulation. Worn under an impervious immersion garment, light summer clothing increases the survival time at 5°C only to 90 minutes (Nunneley & Wissler, 1980). The pile fabric bunny suit undergarment that is *de rigeur* for military aviators in winter raises it to about 11 hours. Even with allowance for the difficulties of search and rescue in bad weather and at night, this is a very generous margin of safety and, in order to achieve it, the wearer must be exposed to a not inconsiderable heat load during normal operations. The value of the model lies not in the demonstration of such over-protection, but in the ability to specify physiological requirements in more objective and operational terms. For example, a realistic specification might be that the clothing assembly must provide an immersed subject with an average insulation of 0.5 clo for a minimum period of 3 hours (Allan, 1983). This simple statement contains two recent advances. It recognises that external compression by the surrounding water can make a nonsense of insulation measurements quoted by manufacturers, and that some leakage (and hence loss of thermal protection) is almost inevitable in any practical immersion garment (Allan et al, 1985).

The recent development of light-weight fabrics that allow the escape of water vapour but not the ingress of water, and which have improved thermal properties, offers the prospect of immersion garments that are both protective and comfortable. The new techniques for the accurate assessment of their efficacy give hope that the hazard of hypothermia after ejection or ditching can be avoided (Pidcock, 1984). All that will then remain will be to persuade those most at risk that they should wear the suits.

FLYING AFTER DIVING

In aviation, the syndromes of decompression sickness are of mercifully rare occurrence, and strict adherence to well documented schedules should also spare divers from the condition. The popularity of scuba diving as a pastime first drew attention to the importance of symptomless or sub-clinical decompression sickness, made manifest in flight (Balldin, 1978). Holiday makers who enjoy one last dive before flying home a few hours later may experience the malaise or aching joints of mild dysbarism, but their symptoms will commonly be attributed to overindulgence in other local delights.

For professional divers (engaged, for example, in off-shore oil operations) the risk is higher. The depths at which they work are greater and their sojourn is longer; the rate at which they return to the surface is often perilously close to or beyond the recommended limit; the incentive to return to civilisation with all possible speed is strong. Time and thought are not spared for the slow process of denitrogenation of the tissues after the dive, and exposure to even a moderate altitude can then precipitate severe decompression sickness. The rules laid down by various authorities for the time that must elapse between return to the surface and flying are less than helpful. After a shallow dive, the recommended resting interval ranges from 2 hours to 24 hours; following a saturation dive, a wait of as little as 24 hours or as long as 1 week may be decreed.

In an attempt to bring science and sense to this confusion, a Workshop was held in 1982 at which international representatives from the aviation and diving medical communities tried to establish rational and acceptable guidelines. They considered two types of diving; those in which air or a nitrogen/oxygen mixture was breathed, and those employing mixed gases, including helium/oxygen. Two subsequent flight

could be a limiting factor in spaceflight may yet be realised by the osteoporotic vertebrae of an ageing astronomer.

REFERENCES

Allan J R 1983 Survival after helicopter ditching. A technical guide for policy makers. International Journal of Aviation Safety 1: 291–296

Allan J R, Higenbottam C, Redman P J 1985 The effect of leakage on the insulation provided by immersion protection clothing. Aviation, Space and Environmental Medicine 56(11): 1107–1109

Anon 1983 CHIRP — New human factors reporting scheme. International Journal of Aviation Safety 1(14): 482–484

Arthur D C, Margules R A 1982 The pathophysiology, presentation and triage of altitude-related decompression sickness associated with hypobaric chamber operation. Aviation, Space and Environmental Medicine 53(5): 489–494

Bagshaw M, Stott J R R 1985 Current techniques for the treatment of chronic motion sickness in aircrew. IAM Report No 640, RAF Institute of Aviation Medicine, Farnborough

Balldin U I 1978 Intercranial gas bubbles and decompression sickness while flying at 9000 m within 12–24 hours of diving. Aviation, Space and Environmental Medicine 49(11): 1314–1318

Benson A J 1978 Motion sickness. In: Dhenin G (ed) Aviation medicine, Vol 1. Tri Med-Books, London, ch 22, p 468–496

Breck D W 1974 Zeolite molecular sieves — structure, chemistry and use. John Wiley & Sons, New York

Bungo M W, Charles J B, Johnson P C 1985 Cardiovascular deconditioning during spaceflight and the use of saline as a countermeasure to orthostatic tolerance. Aviation, Space and Environmental Medicine 56(10): 985–990

Calvin M, Gazenko O G (eds) 1975 Foundations of space biology and medicine. National Aeronautics and Space Administration, Washington

Civil Aviation Authority 1985 Accidents to aircraft on the British Register, 1984. CAP Report 506, Civil Aviation Authority

Collins W E, Schroder D J, Elam G W 1982 A comparison of some effects of three anti-motion sickness drugs on nystagmic responses to angular accelerations and to optokinetic stimuli. Aviation, Space and Environmental Medicine 53(12): 1182–1189

Cooke J N C 1980 Use of beta-blockade in the treatment of aircrew with hypertension. AGARD Conference Proceedings No 310. Advisory Group on Aerospace Research and Development, Paris

DeHart R L (ed) 1985 Fundamentals of aerospace medicine. Lea and Febiger, Philadelphia

DeHart R L, Beers K N 1985 Aircraft accidents, survival, and rescue In: DeHart R L (ed) Fundamentals of aerospace medicine. Lea and Febiger, Philadelphia, ch 30, p 862–887

Desjardins S P, Coltman J W, Laananen D H 1982 Development of improved criteria for energy-absorbing aircraft seats. AGARD Conference Proceedings No 322. Advisory Group on Aerospace Research and Development, Paris

Dhenin G (ed) 1978 Aviation medicine. Tri-Books, London

Diving Medical Advisory Committee 1982 Recommendations for flying after diving. Summary report of International Workshop, London, January 1982

Dobie T G 1972 Aeromedical Handbook for Aircrew. AGARDograph No 154, Advisory Group for Aerospace Research and Development, Paris

Ernsting J 1984 Molecular sieve on board oxygen generating systems for high performance aircraft. AGARD Report No 697, Advisory Group on Aerospace Research and Development, Paris

Gazenko O G, Shumakov V I, Kakurin L I, Katkov V E, Chestukin V V, Nikolayendo E M, Gvozdev S V, Rumyantsev V V, Vasilyev Y K 1982 Effects of various countermeasures against the adverse effects of weightlessness on central circulation in healthy man. Aviation, Space and Environmental Medicine 53(6): 523-530

Gilbert G, MacMahon B 1982 How safe are helicopters? Business and Commercial Aviation 61–66

Glaister D H 1981 Effects of beta-blockers on psychomotor performance: A review. Aviation, Space and Environmental Medicine 52(IIii): S23–S30

Goldman R F 1980 Immersion survival — The key factors. AGARD Conference Proceedings No 286, Advisory Group on Aerospace Research and Development, Paris

Guedry F E, Holtzman G L, Lentz J M, O'Connell P F 1981 Airsickness during naval flight officer training. Report No NAMRL-1275, Naval Aerospace Medical Research Laboratory, Pensacola, Florida

Harding R M, Mills F J 1983 Aviation medicine. British Medical Journal, London

Hargreaves J 1982 The prophylaxis of seasickness — a comparison of cinnarizine with hyoscine. Practitioner 226: 160

Hayward J S, Lisson P A, Collis M L, Eckerson J D 1978 Survival suits for accidental immersion in cold water. University of Victoria
Howard P (ed) 1982 Physiopathology and pathology of spinal injuries in aerospace medicine. AGARDograph No 250, Advisory Group for Aerospace Research and Development, Paris
Hull D H, Wolthuis R A, Triebwasser J H, McAfoose D A 1978 Treatment of hypertension in aircrew: a clincical trial with aldactazine. Aviation, Space and Environmental Medicine 49(3): 503–511
Ikels K G, Thesis C F 1985 The effects of moisture on molecular sieve oxygen concentrators. Aviation, Space and Environmental Medicine 56(1): 33–36
Johnson R L, Hoffler G W, Nicogossian A E, Bergman S A, Jackson M M 1977 Lower body negative pressure: third manned Skylab mission. In: Johnson R S, Dietlein L F (eds) Biomedical results from Skylab. National Aeronautics and Space Administration, Washington
Johnson R S 1977 Skylab medical program overview. In: Johnson R S, Dietlein L F (eds) Biomedical results from Skylab. National Aeronautics and Space Administration, Washington
Kazarian L E. Greaves G A 1977 Compressive strength characteristics of the human vertebral column. Spine 2(1): 1–14
Laananen D H 1980 Aircraft crash environment and human tolerance. In: Aircraft crash survival design guide. Vol 2; 33–38
Leach C S 1981 An overview of the endocrine and metabolic changes in manned space flight. Acta Astronautica 8: 977–986
Levy R A, Jones D R, Carlson H 1981 Biofeedback rehabilitation of airsick aircrew. Aviation, Space and Environmental Medicine 53(5): 449–453
Miller R L, Ikels K G, Lamb M J, Boscola E J, Ferguson R H 1980 Molecular sieve generation of aviator's oxygen: performance of a prototype system under simulated flight conditions. Aviation, Space and Environmental Medicine 51(7): 665–673
Neil-Dwyer G 1981 Clinical importance of lipid solubility in beta-blockers. Aviation, Space and Environmental Medicine 53(IIii): S19–S22
Nicholson A N 1979a Effects of the anti-histamines bromophenamine maleate and tripolidine hydrochloride on performance in man. British Journal of Clinical Pharmacology 8: 321–324
Nicholson A N 1979b Hypnotics to-day. Practitioner 223: 479–484
Nicholson A N 1984 Long-range air capability and the South Atlantic campaign. Aviation, Space and Environmental Medicine 55(4): 269–270
Nicholson A N, Stone B M 1982a Performance studies with the H1-histamine receptor antagonists, astemizole and terfenadine. British Journal of Clinical Pharmacology 13: 199–202
Nicholson A N, Stone B M 1982b Sleep and wakefulness: A handbook for flight medical officers. AGARDograph No 182, Advisory Group on Aerospace Research and Development, Paris
Nicholson A N, Roth T, Stone B M 1985 Hypnotics and aircrew. Aviation, Space and Environmental Medicine 56(1): 33–36
Nicogossian A E 1985 Biomedical challenges of spaceflight. In: DeHart R L (ed) Fundamentals of aerospace medicine. Lea & Febiger, Philadelphia
Nunneley S A, Wissler E H 1980 Prediction of immersion hypothermia in men wearing anti-exposure suits and/or using life rafts. AGARD Conference Proceedings No 286, Advisory Group on Aerospace Research and Development, Paris
Parker J F, West V R 1984 Bioastronautics data book. National Aeronautics and Space Administration, Houston, Texas
Pidcock F 1984 The maintenance and testing of survival suits. International Journal of Aviation Safety 1(4): 482–484
Rambaut P C, Goode A W 1985 Skeletal changes during space flight. Lancet ii: 1050–1052
Rambaut P C, Johnson R S 1979 Prolonged weightlessness and calcium loss in man. Acta Astronautica 6: 1313–1322
Rayman R B 1982 Clinical aviation medicine. Vantage Press, New York
Reason J T, Brand J J 1975 Motion sickness. Academic Press, London
Robertshaw D 1977 Environmental physiology II. International review of physiology Vol 15. University Park Press, London
Royal L, Jessen B, Wilkins M 1984 Motion sickness susceptibility in student navigators. Aviation, Space and Environmental Medicine 55(4): 277–280
Schulman M 1982 The US Navy approach to crashworthy seating systems. AGARD Conference Proceedings No 322, Advisory Group on Aerospcace Research and Development, Paris
Stott J R, Bagshaw M 1984 The current status of the RAF programme of desensitisation for motion sick aircrew. AGARD Conference Proceedings No 372, Advisory Group on Aerospace Research and Development, Paris
Stupakov G P, Kazaykin V X, Kozlovsky A O, Korolev V V 1984 Evaluation of changes in human axial skeletal bone structures during long-term spaceflight. Kosmicheskaya Biologiya i Aviakosmicheskaya Meditsina 18: 33–37

Svoboda C M, Warrick J C 1981 Design and development of variable load energy absorbers. Report No NADC 80257–60, Naval Air Development Center, Philadelphia
Tilton F E, Degioanni J J C, Schneider V S 1980 Long-term follow-up of Skylab bone demineralization. Aviation, Space and Environmental Medicine 51(11): 1209–1213
Tissandier G 1877 Histoire de mes ascensions. Maurice Dreyfus, Paris
Toscano W B, Cowings P S 1982 Reducing motion sickness: a comparison of autogenic-feedback training and an alternative cognitive task. Aviation, Space and Environmental Medicine 52(2): 118–121
Vyrnwy-Jones P 1985 A review of Army Air Corps helicopter accidents, 1971–1982. Aviation, Space and Environmental Medicine 56(5): 403–409
White G R, Roth N J 1979 Cold water survival suits for aircrew. Aviation, Space and Environmental Medicine 50(10): 1040–1045

2. The offshore oil and gas industry

D. Elliott A. M. Grieve

THE DIMENSIONS OF THE OFFSHORE INDUSTRY

During the 20 years that have elapsed since oil and gas were discovered in recoverable quantities under the sea bed around the shores of the UK a major industry has developed in these seas which occupies thousands of men and creates tremendous revenue for the oil producing countries. The British section is bounded to the east by an artificial line which has been drawn midway down the North Sea between the UK and its continental neighbours. The principle areas from which development and subsequently production have been mounted are East Anglia and the North East of Scotland but the actual worksite varies from between 60 to 300 miles out to sea from the operating base.

The first exploratory and drilling rigs began work in 1964 and since then there has been a rapid escalation in the numbers of people working offshore and in the complexity of the tasks which they perform.

In 1985 there were 41 major operating oil companies owning assets in the UK sector of the North Sea with 71 production platforms and all the attendant vessels in support. A total of 31 500 workers are involved regularly in the offshore part of this industry whereas many more employees of oil companies and specialist technicians travel sporadically to work in the oil field. For every person employed either full-time or part-time offshore there are many more people involved on the UK mainland either in administration, logistic backup or engineering preparation and support.

Travel to and from the work place is invariably by helicopter and in 1985 over 1.5 million passengers flew to and from Aberdeen airport in the course of their duties offshore, making it one of the busiest airports in the country and indeed in terms of numbers of departures and arrivals one of the busiest airports worldwide.

Platforms and vessels

A whole variety of structures exist offshore to facilitate the extraction of gas and oil. In the stable state which we see today much interest devolves upon the production platforms. These are large structures up to 700 ft in total height from the sea bed and 300 ft above the surface of the water. The dimensions of the platform may be up to 200 m × 200 m and four or five deck levels may be present to accommodate anything from 200 to 500 people.

In order to place these massive structures in position however, a whole armada of other vessels is required before speculation can metamorphose through exploration to the point where the production platform is conceived, constructed and finally floated out to be mounted on the sea bed to within an accuracy of a few centimetres in its positioning.

Amongst these specialist vessels are the following.

Seismic survey vessels
Relatively conventional monohull boats which provide the reconnaissance capability and, with geologists and scientific equipment aboard, go out to discover areas where oil and gas might be found.

Drilling rigs (usually semi-submersibles)
These are marine craft in their own right. Mounted on two large pontoons which float below sea level, the working platform is supported by six or eight legs. They have a capacity, using anchors or dynamic positioning to hover directly over a part of the sea bed while exploration drilling takes place. They may accommodate between 50 and 100 persons while working. Mono hull drilling vessels do still operate in some areas, and in shallow water, 'Jack Up' drilling rigs can be used, raising themselves off the sea bed on giant legs.

Accommodation vessels (Flotels)
Because of the large numbers of staff required to build some of these platforms, or commission them once they are constructed, peak increases in the offshore population are catered for by the use of floating accommodation units. Originally these were drilling barges which had been converted to produce a rudimentary hotel type environment but more recently custom-built accommodation units have transformed the living environment for the casual worker offshore. They may accommodate between 400 and 800 'guests' at any one time.

More specialist vessels are necessary to create the scenario in which the production platforms can be brought to life. They are variously referred to as vessels or barges but each specialist craft will cope with either trenching, pipe laying or rock dumping. In addition to these smaller vessels heavy cranes with lifting capacity of up to 6000 tons also operate autonomously mounted on twin hull vessels.

THE OFFSHORE WORKFORCE

One reason for describing the various hardware is that there are many dimensions into which the offshore workforce can be polarised. Depending on the part of this large group to which an individual belongs, he or she may expect widely differing terms and conditions of service, salary, security of employment and welfare provisions. Work patterns and occupational hazards also vary enormously amongst the different groups and a whole spectrum exists between the well established secure employee of a large company, through the self-employed specialist who may only work occasionally but can attract a substantial reward to the casual contract labourer. Peaks in construction, commissioning or refurbishment lead to widely fluctuating demands in the number of people working offshore.

Work cycles vary considerably depending on the type of operation being undertaken and the type of vessel on which an offshore worker is employed. The operation continues night and day 24 hours a day for every day of the year. Some companies prefer to send men offshore for 7 days at a time and others for 14. Custom and practice is that for every period of time spent offshore, a field break or equivalent period of leave will be given. Consequently men will work 7 days on, 7 days off, throughout the year and this carries its own benefits in terms of regular time off but carries its own penalties in terms of a variety of stresses which obtain in the offshore working environment.

Not all of the people working offshore have chosen to move their family either to East Anglia or to North East Scotland. A substantial commuting population exists where individuals travel each week from as far afield as Plymouth, Leeds and Birmingham to Aberdeen in order to work offshore. Commuting of this sort is at the individual's own expense and of course the time taken for such travel erodes the period of field break. Where weather and flying delays can occur, there can be considerable frustrations and additional stresses for the worker.

In what appears to be a massive and perhaps cosmopolitan workforce there is therefore considerable stratification into much smaller groups according to employing company, nature of contract and the type of specialist work which individuals undertake. Some enjoy great affluence, whereas others, who might be providing domestic services, retained by a catering contractor in intense competition with others, may not enjoy more than a basic salary. Such relative deprivation can assume undue proportions because of the close proximity in which people have to live with each other and the very interdependent nature of their jobs.

RELEVANT LEGISLATION CODES OF PRACTICE

When considering the problems of health and safety in the offshore workforce one must first realise that the legislative requirements differ from those of a normal industry onshore.

The provisions of the National Health Service in Britain do not extend below the low water mark around the shores. All routine medical care during the offshore work period, and emergency medical services must be provided by the industry itself and the final responsibility to ensure that these provisions are made is that of the operating oil compnay which holds the concession for exploration or production in the appropriate part of the North Sea.

The Health and Safety at Work Act of 1974 did not originally apply offshore but it was extended to encompass all working within the UK Continental Shelf area by an Order in Council of 1977. Specialist legislation with respect to lead, radiation, asbestos and other matters does also apply offshore but not all onshore legislative requirements translate directly to the offshore environment. (It is therefore prudent for the doctor who is not regularly involved in dealing with an offshore population to contact the Medical Advisor to the operating oil company involved when in any doubt over legislative matters.)

The Offshore Installations (Operational Safety, Health and Welfare) Regulations 1976 (Statutory Instruments 1976) is currently still in force and new legislation is about to emerge as The Health and Safety (Offshore Installations and Pipeline Works) First Aid Regulations 198–. Whilst each of these specifies requirements for sick bays, drugs and medical equipment, and the qualification of any medical attendant required, the latter will prescribe increased requirements for pre-service and in-service training of the Medical Attendant offshore. A distinction is drawn however between those vessels which are accountable to the Department of Trade (e.g. a drilling rig in transit) and those which are accountable to the Department of Energy (e.g. a drilling rig whilst drilling). A vessel of foreign registration (perhaps flying a flag of convenience) will also have to comply with the regulations of the country in which it is registered, e.g. USA or Norway, and these may not be compatible with the UK legislation in detail. For any one type of vessel the medical provisions are generally scaled in proportion to the number of passengers onboard at any one time. In general as other aspects of the business offshore have developed and become more refined medical provisions have followed suite and legislation relating to medical and first aid provisions offshore has improved from the point whereby 'written instructions should be provided on an installation to specify what arrangements are made for liaison with medical practitioners' to the point where an operator should make arrangements with a medical practitioner who has suitable knowledge and experience of the offshore environment and can give advice to the offshore staff, and where necessary mobilise to the offshore installation to give care. There is however no proposal that any list of approved doctors be created in respect of general medical provisions offshore.

The United Kingdom Offshore Operators Association (UKOOA) which includes representatives from every operating company in the UK sector of the North Sea, has also taken a responsible attitude towards the questions of health and safety offshore and, through its Medical Advisory Committee, has produced recommended general medical standards of fitness for designated offshore employees (UKOOA 1986). These provide advice to doctors on the generalities of examining persons for their fitness to work offshore, outline the range of conditions which might make an individual unsuitable for such work, and recommend a frequency of examination. An important principle of that guidance is that

> 'the operating company should reserve the right for its own Medical Advisors wherever necessary to make a definitive decision about the medical fitness of any individual who is to work in its operation'.

In parallel with these recommendations on standards of fitness, UKOOA also produces environmental health guidelines, thus acknowledging the importance of well controlled living conditions in a very cramped working and living environment (UKOOA 19—). Both of these sets of guidelines are under consant review to ensure that they remain appropriate to the needs of the industry.

STANDARDS OF FITNESS

A number of factors, either singly or in combination, may render an otherwise apparently innocuous medical condition to be a bar to an individual working offshore. In particular there is the problem of geographical isolation. Production may be taking place out to sea 300 miles from a base hospital. Travel by helicopter is essential in order to complete the journey in reasonable comfort and time. Such transport however is at the mercy of unpredictable delays which can happen for many reasons such as mechanical problems with the aircraft, severe storm or fog or the need to re-route. Delays disrupt meal schedules and anyone taking regular medication (in particular insulin) might well be considerably disadvantaged. Even a routine helicopter flight can change its complexion completely given a mechanical emergency and the need to land on the sea.

Against this backdrop there can of course only be limited medical provisions made offshore. For the most part the structures and vessels carry comprehensive sick bays and have a competent nurse or paramedical person (rig medic) onboard. Such a person, working with limited equipment, can only be expected to undertake simple routine treatments (e.g. foreign body in the eye), and in the event of a serious emergency, stabilise someone who might be badly injured and prepare him for transport onshore. It would be irresponsible to allow persons to go to work offshore with medical conditions which may have sudden acute complications (e.g. active duodenal ulcer, insulin dependent diabetes, epilepsy or a previous history of coronary heart disease). Whilst all such cases must be judged on their merits, a small but finite range of medical conditions such as these must regrettably prohibit individuals from being included in the offshore workforce.

Safety considerations dictate that anyone should be able to look after himself and fit enough to evacuate himself from a platform or a ditched aircraft, posing no additional risk to others in the event of either a rig abandonment or a helicopter emergency. Training is given to all members of the offshore workforce in safety, fire-fighting, first aid and personal survival. Further to this pre-employment training, regular drills are carrried out offshore as required by statute (SI 1976) but are generally performed more frequently in accordance with individual company health and safety policies.

The physical demands upon the offshore worker can be considerable: not only ladders between decks and from decks to superstructure but also, for some, ladders 100 m down the inside of the platform legs. Some workers may have relatively sedentary jobs whilst offshore, but all must be sufficiently fit to undertake the initial and periodical emergency and survival at sea training.

Regular offshore workers, whether on a 7 + 7 day or 14 + 14 day work-rest cycle, generally fall into two types of shift pattern. The majority of production technicians and maintenance personnel work a 12-hour shift with a 12-hour rest period and work one shift per day during their time offshore. Where required exceptionally, overtime may be worked over and above the 12 hours, but must be recorded. Excessive overtime is actively discouraged by all responsible operating companies and not least for reasons of safe working practice.

Members of staff such as the rig medical attendant, may work in what is referred to as a single responsibility position. The implication of this phrase is that the individual is the only one of his type on the platform and therefore is on call throughout the period of time which he spends offshore. Such a role obviously calls for self-discipline on the part of the individual so that he paces himself properly between work and rest, and a degree of constraint on the part of his colleagues so that they do not call him excessively or frivolously whilst he is resting.

In addition to ensuring that people with unacceptable physical ill health are not allowed to be exposed to the risks of working offshore there is a more nebulous requirement for people of the correct temperamental suitability. Not everyone is able to cope with intermittent isolation from the rest of society, travel by helicopter, the intensity of the work cycle or the claustrophobic nature of living cheek by jowl with all ones contemporaries during the work period. Most frequently where people find themselves incompatible with work offshore they tend to select themselves out and move onto other jobs elsewhere but there are occasions when the manager along with the medical department and the career development advisors have to relocate individuals who, in their experience, are not compatible with this unusual working situation.

OCCUPATIONAL HAZARDS

General

Occupational hazards in the offshore oil industry start at the heliport with the risks (albeit statistically extremely small) of helicopter travel and the possibility of immersion in cold water (Allan, 1983a,b). Whilst working offshore the individual is in an industrial complex which contains elements of light/medium engineering, chemical processing, electricity generation, and in which there are liquid hydrocarbons and gasses within a closed pipework system often at pressures of up to 6000 lbs per square inch. The risks of leakage and fire/explosion are ever present but the situation seems to exemplify the concept that the higher the hazard, the less the risk, as all offshore workers are acutely aware of the potential for conflagration which exists around them. There is a well rehearsed system of alarms to detect leakage or gas, whereupon emergency procedures are put in hand. Only very limited parts of the installation outside the accommodation modules are designated as smoking areas. The accommodation, and some control rooms are kept at higher than atmospheric pressure to prevent the ingress of gas (which has no smell) and smoking is permissable in these areas.

All in all therefore the processes on an oil production platform very closely parallel those of an oil refinery or chemical plant which might normally occupy several square miles but offshore have to be shoe-horned into a space 200 m × 200 m and possibly four floors high.

Noise and vibration

Given the very close nature of this working environment and the fact that electricity generation and gas compression are principal activities, then noise is unavoidable. On older generation platforms, in the process areas, ambient noise levels of up to 110 dB are not uncommon. It would be impossible to substitute a significantly quieter process in these areas and acoustic cladding, whilst helpful, would not be capable of reducing the noise level to the point where hearing protection would be unnecessary, without creating intolerable penalties for the engineers in terms of either bulk or weight. The technical workforce offshore must be well educated on the subject of noise-induced hearing loss and the need to be scrupulous in their use of personal hearing protection when working in the well labelled noisy areas. Many companies operate comprehensive hearing conservation programmes consisting of noise measurement, information/education, the provision of personal hearing protection and periodical audiometry, although there is no specific legal requirement in many of these areas. When setting hygiene standards for the amount of noise which employees should be exposed the equal energy concept is generally adopted. In an offshore situation however, the assumption of this concept may not be applicable as an operator who is exposed to high noise levels during his work period may not be subject to sufficiently low noise levels during his time off to allow any temporary threshold shift in his hearing to recover. This simplistic interpretation of personal noise dose is further complicated by the fact that the individual will be exposed to noise levels quite outwith the control of the employer during the 7- or 14-day field break between work periods.

Drilling muds

During the drilling operation a fluid known as drilling mud is circulated down the drill string to the drill bit. The exact constituents of drilling mud are a jealously guarded commercial secret as the quality and efficacy of such muds can have a profound impact on the success of the drilling operation. The purpose of the mud is to control the oil reservoir conditions round about the drilling operation and also to cool the bit as it is cutting.

Some of these muds are water based whereas others are oil based and a number of factors have to be balanced in order to achieve the optimum effect. These include: the aromatic ingredients of the mud, the proportion of light to heavy aromatics and whether straight chain or branched paraffines are to be used. The potential toxicological effects of these muds to man are still poorly understood.

Sour gas

In the North Sea very few of the oil reservoirs are sour i.e. there is a very low sulphur content in the oil and very little hydrogen sulphide results from the drilling process. This is to be contrasted with the Dutch section of the North Sea, where, when drilling for gas, sometimes perforation must take place through a cap of hydrogen sulphide (H_2S). During such operations breathing apparatus and eye protection is worn against the risk of intoxication. The principle source of H_2S in the UK sector of the North Sea is from the action of sulphate reducing bacteria in parts of the platforms where an oil/water mixture may exist (e.g. down the platform leg). The potential risk that inhalation of H_2S might take place through a tympanic perforation has now been discounted, however, (Ronk & White, 1985) and workers with such a perforation need no longer be excluded.

Radioactivity

Initially most of the North Sea oil reservoirs gave freely of their resources although in many cases they yielded an effervescent oil from which gas had to be separated as part of the process. Now that the reservoirs are ageing, more energy needs to be put into the recovery of the oil and one technique which is now well developed in several areas is that of water injection. Desalinated and deoxygenated sea water is injected at high pressure down into the reservoir through one drill string to drive the oil towards another drill string up which it can be extracted. A consequence of this process which is becoming increasingly apparent is that barium and radium salts naturally occurring in the rock formation can be precipitated as sulphate in the oil which is extracted. Within the tubulars along which the oil is lifted, and at valves and flanges in the process, deposits of low specific activity radioactive scale can be found. The vast proportion of this activity is from beta emitters and is entirely harmless to man, whilst the pipework remains intact and production proceeds. Now that the potential hazard of radioactive scale has been recognised, procedures are in place with all operators whereby during a period of production shutdown for refurbishment, measurements are made, and appropriate operating practices are adopted for the descaling of the pipework in a safe manner.

Other agents

In addition to varying the constituents of drilling mud the petroleum engineers also use a variety of chemicals to either inhibit or stimulate production wells depending on a wide variety of factors. Various techniques are used but some of the chemicals involved (e.g. hydrogen fluoride) could be extremely harmful to the operator. Procedures must be scrupulously adhered to in order to prevent personal exposure. Personal protective clothing, where necessary, must be properly used and the medical facilities must be prepared to cope with any emeregency which might arise as the result of a mishap when handling these chemicals.

As well as water injection a number of other techniques have been tried to enhance oil recovery from reservoirs which are becoming depleted. Various surfactants are in use and gases such as carbon dioxide and nitrogen have been suggested. Practical difficulties arise with all of these possible solutions (e.g. if the nitrogen is not completely oxygen free when compressed to high pressure for injection then a risk of explosion will exist). The significance of gasses such as carbon dioxide and nitrogen for the operator are less troublesome as these have no caustic or narcotic affect at normal temperature and pressure but where a vessel has been used to contain these substances anyone proposing to enter that container must check for oxygen content before doing so because of the risk of hypoxic asphyxia.

Asbestos

There is still some asbestos present on offshore installations. The Asbestos Licencing Regulations apply offshore and operators are generally very well aware of where the asbestos exists, in what quantity, and the condition of the asbestos containing material. For the most part it exists in thermal or acoustic insulating materials and where these are in good condition (and are very often sealed) no hazard exists for the worker. As a general principle however these are being removed wherever possible with the emphasis being placed on areas (e.g. around valves and flanges) which might have to be disturbed periodically for maintenance).

THE ROLE OF OCCUPATIONAL HEALTH PERSONNEL

In all of the above areas there is need for vigilance by operators where potential occupational health problems exist. People working there need to be aware of the process in which they are involved and the substances which they are handling or to which they are to be exposed. Company safety and medical departments provide input by occupational hygienists and occupational physicians in the compilation of safe working practices and operating procedure guides which outline, according to well established principles, ways in which these various substances can safely be used to achieve the industrial purpose. In addition appropriate medical screening procedures can be adopted.

Because of the severe limitations on space available, and the maximum weight with which offshore structures can safely be loaded, it is sometimes impossible within the state of the art or within reasonable parameters of cost to produce a change to a completely safe or harmless process. Under these circumstances identification of the hazard, awareness amongst the workforce and the provision of suitable personal protection is the route of choice.

SPECIFIC OCCUPATIONS

Catering staff

UKOOA Medical Guidelines exceed the normal onshore legislative requirements of the Food Hygiene (General) Regulations. In many operating companies twice yearly routine stool sampling is practised and it is surprising how often this routine process yields positive results revealing intestinal pathogens which one would not normally expect to find in Europe.

Other special groups

Those working with processes where legal requirements obtain must also be examined periodically in accordance with the appropriate legislation (e.g. asbestos, radiation, lead and mercury). Helicopter pilots, whilst not technically part of the offshore workforce do in some cases live and work offshore and in these cases aviation medical examiners should bear in mind the UKOOA Medical Guidelines when deciding suitability for this particular type of work.

Divers

There are additional environmental hazards for one particular section of offshore workers: the divers. These workers will continue to be needed offshore at least until remotely-controlled robots become totally reliable and also are capable of undertaking all the tasks currently performed by man underwater. This is not likely to happen in the near future. As diverless systems develop there will be less demand for divers at the more expensive exreme depths but the use of divers, particularly compressed-air divers in shallower waters, will continue, possibly indefinitely. Divers will be required not only for the offshore oil and gas industry but for military and police activities and also, for instance, for the inspection and maintenance of ships' hulls, hydro-electric dams, docks and bridges. The applied physiology and occupational medicine of diving are dealt with in full elsewhere, for instance Bennett & Elliott (1983) and Shilling et al (1984). As divers numerically form but a small percentage of the workforce offshore, only some highlights of the recent advances will be given space in the following outline.

Applied physiology

The laws of physics are immutable as, therefore, are the consequences of raised environmental pressure upon man. But slowly improvements in our understanding of the basic physiological mechanisms are being made.

One factor in the recent reduction in the output of relevant research worldwide has been the almost complete withdrawal of military naval sponsorship from civilian-sector research consequent upon the limitation of most naval operational diving to around 300 m. The expense of deep diving research is greater than can be borne by any one company alone and even sponsorship by an association within the international oil and diving industries has recently demonstrated its inability to fill this gap. Current research programmes in Europe, North America and Japan are therefore limited, focussing upon the practical aspects of diving while fundamental physiology tends to be neglected. In spite of this trend there have been a number of relevant reviews which each consolidate progress in a specialist research area. Notable among them, and useful for being bound in one volume, are the papers of a Royal Society Discussion Meeting (Paton et al, 1984).

Underwater physiology is a subject in which physical principles have guided research more effectively than is usual in biology and the value of physico-chemical arguments can be illustrated: particularly nitrogen narcosis (Smith, 1984). This topic was reviewed by a workshop (Undersea Medical Society, 1985b) with special emphasis on adaptation to narcosis, especially in saturation diving, and varying individual susceptibility.

The effects of pressure on the molecular structure and the function of cell membranes (Macdonald, 1984) influence passive permeability, active transport, membrane excitability and synaptic transmission. This one paper illusrates better than many others the basic complexity underlying some of the divers' intrinsic occupational hazards and, while the paper is concerned predominantly with extremely high pressures, it provides an essential foundation to any attempted understanding of the mechanisms of phenomena encountered by man at pressure.

Interest in pulmonary and neurological oxygen toxicity, the cellular mechanisms and, in particular, any possible extension of man's tolerance to increased partial pressure of oxygen is not confined to the diving industry. Thus a review of oxygen toxicity (Undersea Medical Society, 1983) is necessarily multidisciplinary and also refers the reader to relevant studies of a biochemical and clinical nature. Subsequently the US Navy has conducted several hundred man dives using closed-circuit 100%-oxygen scuba and, while recommending that safety lines connect two divers in a swim pair, proposed new oxygen limits for swimming, such as 240 min at 25 fsw (8 m), significantly longer than previous US Navy limits (Butler & Thalmann, 1985).

The diver can continue to work within the known thresholds of nitrogen narcosis and oxygen toxicity, but the physiological limits to deep diving have yet to be defined. Among the many experimental dives of the last decade, the pressure chamber dive 'Atlantis III' at Duke University still provides the benchmark for man's achievement in the laboratory and thus a target for his potential in the undersea workplace. Three men spent more than 11 days at an environmental pressure greater than 600 m (2000 ft) including 24 hours at 686 m (2500 ft) (Bennett & McLeod, 1984). During this time, psychometric performance was such that arterial catheters could be inserted and used to demonstrate the relative normality of Pa CO_2 during physical workloads of up to 240 watts for 5 min.

In the course of many recent deep dives in Europe, North America and Japan there have been extensive studies of other aspects of applied physiology. Some examples include psychometric performance (Naquet et al, 1984), cardio-pulmonary function (Salzano et al, 1981; Stolp et al, 1985), thermal balance with particular reference to the lost bell and to respiratory heat loss (Virr, 1984; Undersea Medical Society, 1985a), liver function (Doran et al, 1985) and haematological changes (Shiraki et al, 1983).

The dives at Duke University were undertaken to investigate the optimal percentage of nitrogen to be added during compression to the conventional oxy-helium mixture in order to minimise the adverse effects of the high pressure neurological syndrome (hpns), as was a French dive to 610 m (Gardette & Carlioz, 1984). The theoretical basis of hpns, and of the contrasting pressure reversal of anaesthesia and narcosis, has been the subject of many reviews over recent years (for instance Bennett & Elliott, 1983). The Royal Society discussion included Brauer (1984), Smith (1984), Bennett & McLeod (1984) and Naquet et al (1984). Not all the laboratories agree with the trimix approach and Hempleman et al (1984), from the many experimental dives conducted in the UK by the Ministry of Defence are among several who consider that the addition of the nitrogen in deep dives seems to offer little to offset its inherent disadvantages such as narcosis.

In terms of occupational hazard for the offshore diver such studies contribute to a better understanding of the complex phenomena associated with exposure to raised environmental pressure, but it needs to be emphasised that most commercial deep diving is at less than half depth achieved in laboratory research. Recent advances in deep diving have therefore been not in physiology but in those areas, such as bio-engineering, that currently inhibit the further development of man as an underwater worker, for example the design of breathing apparatus, the control of inspiratory gas temperature, the duration of the emergency bale-out bottle and the provision of reliable communication systems.

The physiology of compressed air diving, limited because of nitrogen narcosis to 50 m in the UK Sector of the North Sea, has not been studied so extensively as deep diving in recent years. Besides the oxygen and nitrogen reviews already quoted, there have been a number of mathematical studies of gas elimination leading to the development of new decompression tables. The development of Canadian tables (Lauckner & Nishi, 1984) based upon a pneumatic analogue and the more pragmatic approach of the US Navy, in proposing tables based on empirical probabilistic models to choose an acceptable risk of 10 or 5% (Weatherby et al, 1985), have not yet led to proposals for evaluation in the North Sea. Of greater relevance has been the recalculation Arntzen & Eidsvick (1984) of surface decompression tables culminating in 1600 logged Norwegian dives of significant depths and durations with only one case of decompression sickness.

Fitness

Medical standards of fitness for divers are necessarily strict and the Health and Safety Executive Guidance Notes (1981) are currently being rewritten. An analysis of weight measurements of 1520 divers whose records were in the MRC Decompression Sickness Registry suggested that divers are significantly heavier than other populations on whom the height/weight tables have been based (McCallum & Petrie, 1984). More than 10% of divers might be rejected as overweight using conventional American-derived tables. The underlying principles for medical approval of fitness have been reviewed recently by Hickey (1984) for sports, commercial and naval diving.

For the UK sector only approved doctors may sign the annual certificates of fitness to dive. Many general practitioners in the UK and overseas have HSE approval to do so, usually after attending a 1 week introductory course such as that run regularly by the Royal Navy. A particularly difficult decision for approved doctors is that of whether a diver can return to diving after illness or injury. In the absence of any available factual evidence, a guidance note on fitness to return to diving after decompression illness has been prepared (Diving Medical Advisory Committee, 1983). The Royal Navy policy to cease exposure to raised environmental pressure after pulmonary barotrauma (Leitch & Green, 1985b) is challenged by the proposal that some may be fit to return to diving (Calder, 1985). That some have done so successfully does not, of course, mean that it is safe for all to do so.

Decompression sickness

Other than the introduction of revisions excursion (bounce) and saturation decompression tables both for shallow diving (air and oxy-nitrogen) deep (oxy-helium and other mixed gases), little has been developed in the preventive area of decompression sickness. Bayne et al (1985) have concluded that the use of Doppler ultrasonic techniques has no diagnostic value. Guidance notes have been prepared on the limitations upon flying after diving but these are based on pragmatism rather than scientific fact (Diving Medical Advising Committee, 1982).

While the collection and discussion of unusual presentations of decompression sickness has enlarged our limited knowledge, the extensive and continuing studies on the pathogenesis of decompression sickness (Lynch et al, 1985; Sykes & Yaffe, 1985) have yet to clarify factors affecting individual susceptibility and prognosis.

The treatment of deeply occurring decompression sickness though sometimes difficult seems relatively satisfactory in contrast to that of decompression sickness arising at the surface from compressed-air dives. Neurological decompression sickness is more common and, unless treated promptly and effectively, is more refractory and more likely to relapse. A review of 20 years of Royal Navy records concluded from the 14 relevant cases that further recompression to 50 m rarely altered the recovery of cases which were not already responding at 18 m. The only case with motor deficits to recover at 50 m later relapsed (Leitch & Green, 1985a). They conclude that the only real justification for compression to 50 m is rapid deterioration at 18 m, even though this will not necessarily be effective.

Among options available for treatment, the US Navy table 5 (RN equivalent table 61) for pain-only decompression sickness is no longer recommended for the North Sea. Notwithstanding its effectiveness in suitable cases, it was felt that its use could be potentially dangerous by non-medically qualified personnel who might miss the subtle manifestations of a more serious incident that required more vigorous recompression.

For many years it has been accepted that another option is recompression to 30 m using tables, such as Comex 30, based on those of the French Navy and using 50% oxygen : 50% nitrogen breathing mixtures with occasional air intervals. More recently (James, 1983) it has been proposed that 50% oxygen 50% helium at 30 m would provide a more effective treatment for compressed air decompression sickness. While others have concluded only that use of helium : oxygen as a breathing gas does not cause worsening of severe decompression sickness in an animal model (Catron et al, 1985), the consensus view of a group of diving doctors has been that the helium option should remain available to those doctors who prefer it, but no general recommendation could be made without substantial further experience.

Another option for difficult cases of decompression sickness is oxy-nitrogen saturation. With a PO_2 of 0.4 bar there is no threat of oxygen toxicity to hasten a premature decompression in life-threatening cases and continuing experience shows that the recommended saturation-decompression (Miller et al, 1978) is usually effective.

The basis for drug and adjuvant therapy in the treatment of decompression sickness has been reviewed Bove (1982) and Catron & Flynn (1982).

Long-term effects
Environmental pressure is distributed throughout the body and so, therefore, is the potential for long-term health effects. While chronic changes to other systems such as pulmonary and hearing have been described, there are three areas of current study.

The prevalence of osteonecrosis in divers has been studied thoroughly in the past decade. The radiographs of more than 6000 divers reveal 274 (4%) with one or more sites of damage but only 72 (1.1%) with juxta-articular lesions of whom 11 (0.2%) had some disability (McCallum, 1984). Contributory factors and possible pathogenetic mechanisms are reviewed but further epidemiological studies will be required.

Long-term damage to the central nervous system is a natural consequence of acute decompression sickness, particularly in severe cases or where treatment has been delayed or was inadequate. Although a functional recovery may be apparent, extensive cord damage may be present with very few abnormal signs on neurological examination (Palmer et al, 1981). Separately, the possibility has been raised that hpns, in which decrements of psychomotor performance may be encountered, is associated with long-term changes in cns function. A workshop in Norway concluded that while some transient changes showed that concern was justified, there was no evidence from the many dives reviewed of any long term neurological sequelae (Shields et al, 1983). More data is required before any authoritative statements can be justified.

Chromosome aberrations in divers form the third area of concern. Six out of 153 (2 air, 4 heliox) had an abnormally high number of structural alterations but concentrated in only one or two of the lymphocytes scored (Fox et al, 1984). These aberrations could not be statistically associated with any known causative factors though were typical of those induced by ionising radiation. The damage these abnormal cells contained was so extreme that most would be likely to die at mitosis. There is no direct evidence to support any consequential health effects.

These several papers individually contain no evidence of serious health risk to the diving population but collectively should eliminate any complacency. The next decade should see a further extension of knowledge in this important aspect of occupational health.

COLD WATER IMMERSION

All offshore and maritime workers share a common and ever present hazard, that of sudden immersion in a cold sea. A seminar, sponsored by industry, was held in 1983 and reached a number of conclusions by consensus (Allan 1983a). From this and reviews published elsewhere (Adam, 1981; Golden, 1982; Allan, 1983b; Keatinge, 1984), the basic problems potentially confronting each offshore worker can be gleaned. A number of helicopter flights in the offshore oil field have a duration of only a few minutes. Provided rescue services are organised to be able to rescue every survivor quickly, there will be no danger of life-threatening body cooling. In these circumstances a Neoprene jacket, not unlike the upper part of a diver's wet suit, has been developed which will fit snugly over the passenger's clothing. The life-jacket worn over this is provided with an integral splash-guard or spray-hood which can be deployed when the survivor is on the surface of the sea. Such a survival jacket will provide good protection against cold shock and significant insulation which may prolong survival time (Grieve, 1985; Allan et al, in press). Where a significant number of survivors may need to be cared for after immersion, active therapy may not be practical, and indeed it has been suggested that passive rewarming may be the method of choice (Golden, 1983). The problems of immersion in cold water are also addressed in Chapter 1.

CONCLUSION

The offshore oil and gas industry has grown up around our shores with remarkable rapidity but in a piecemeal fashion over the last 20 years. It has generated a fascinating pot-pourri of physical and psychological problems for the occupational physician, some of which are traditional and well documented, eg noise, but many of which have yet to be acknowledged or clearly defined far less addressed, eg the effects of prolonged intermittent absence from home. There therefore remains a large number of avenues for worthwhile research amongst what is currently a poorly documented population of healthy workers. Because of the multiplicity of different types of work, different work cycles, and different conditions of service leading to enormous variations in the mobility and turnover of the workforce it will not be easy to identify the more subtle trends which may be present in this community.

REFERENCES

Adam J M (ed) 1981 Hypothermia ashore and afloat, proceedings of the 3rd international action for disaster conference, Aberdeen, Aberdeen University Press
Allan J R 1983a Survival after helicopter ditching, a technical guide for policy makers. International Journal of Aviation Safety 1: 291–296
Allan J R 1983b Survival after helicopter ditching. United Kingdom Offshore Operators Association, London
Allan J R, Elliott D H, Hayes P A The thermal performance of partial coverage wet suits, Aviation Space and Environmental Medicine. In press
Arntzen A J, Eidsvick S 1984 Norwegian diving tables and procedures for diving with air or nitrox as breathing gas. In press
Bayne C G, Hunt W S, Johnson D C, Flynn E T, Wethersby P K 1985 Doppler bubble detection and decompression sickness, a prospective trial. Undersea Biomedical Research 12: 327–332
Bennett P B, Elliott D H (eds) 1983 The physiology and medicine of diving. 3rd edn. Ballière Tindall, London
Bennett P B, McLeod M 1984 Probing the limit of human deep diving. In: Paton Sir William, Elliott D H, Smith E B (eds), Diving and life at high pressures. The Royal Society, London, p 105–116
Bove A A 1982 The basis for drug therapy in decompression sickness. Undersea Biomedical Research 9: 91–112
Brauer R W 1984 Hydrostatic pressure effects on the central nervous system. Perspectives and outlooks. In: Paton Sir William, Elliott D H, Smith E B (eds), Diving and life at high pressures. The Royal Society, London, p 17–29
Butler F K, Thalmann E D 1985 Establishment of new exposure limits for the US Navy closed circuit oxygen scuba divers. Undersea Biomedical Research 12 (suppl): 20
Calder I M 1985 Autopsy and experimental observations on factors leading to barotrauma in man. Undersea Biomedical Research 12: 165–182
Catron P W, Flynn E T 1982 Adjuvant drug therapy for decompression sickness, a review. Undersea Biomedical Research 9: 161–174
Catron P W, Flynn E T, Thomas L B 1985 Effect of ventilation with helium — oxygen in experimental decompression sickness following air dives. Undersea Biomedical Research 12 (suppl): 7–18
Cox R A F (ed) 1982 Offshore medicine. Springer Verlag, Berlin, New York
Diving Medical Advisory Committee 1982 Recommendations for flying after diving. AODC, London
Diving Medical Advisory Committee 1983 Guidance on assessing fitness to return to diving. AODC, London

Doran G R, Chaudry C, Brubakk A O, Garard M P 1985 Hyperbaric liver disfunction in saturation divers. Undersea Biomedical Research 12: 151–164

Fox D P, Robertson F W, Brown T, Whitehead A R, Douglas J D M 1984 Chromosome aberations in divers. Undersea Biomedical Research 11: 193–204

Gardette B, Carlioz M 1984 Entex 9—Plongées en immersion jusqu'à 610 metres Medsubhyp 3: 41–50

Golden F St C 1982 Hypothermia and survival research. In: Health and hazards in the changing oil scene. John Wiley, New York

Golden F St C 1983 Hypothermia exposure, rescue and treatment. In: Proceedings of safety and health in the oil and gas extractive industries. Graham and Trottman, London

Grieve A M 1985 The shuttle jacket concept. In: Proceedings of emersion suits. A seminar for policy and decision makers. Robert Gordon's Institute of Technology, Aberdeen

Health and Safety Commission 198– Draft health and safety (offshore installation and pipeline works) first aid regulations. HMSO, London

Health and Safety Executive, Employment Medical Advisory Service 1981 MA1. The medical examination of divers. Employment Medical Advisory Service, London

Hempleman H B, Florio H T, Garrard M P, Harris D J, Hayes P A, Hennis A T R, Nicol G, Torok Z, Winsborough M M 1984 UK deep diving trials. In: Paton Sir William, Elliott D H, Smith E B (eds) Diving and life at high pressures. The Royal Society, London, p 119–140

Hickey D D 1984 Outline of medical standards for divers. Undersea Biomedical Research 11: 407–432

James P B 1983 Revision of the AODC/EUBS Round Table Document on the treatment of the decompression-related illnesses offshore. Annual Meeting of the Association of Diving Contractors, AODC, London

Keatinge W R 1984 Thermal balance in cold water immersion in ice diving. Stockholm, p 25–42

Lauckner G R, Nishi R Y 1984 Decompression tables and procedures for compressed air diving based on the DCIEM 1983 model. Downs U Ontario, Defense and Civil Institute of Environmental Medicine, Report 84-R-74

Leitch D R, Green R D 1985a Additional pressurisation for treating non-responding cases of serious air decompression sickness. Alverstoke, Institute of Naval Medicine, Report 4/85

Leitch D R, Green R D 1985b Recurrent pulmonary barotrauma. Alverstoke, Institute of Naval Medicine, Report 5/85

Lynch E R, Brigham M, Tuma R, Wiedeman M P 1985 Origin and time course of gas bubbles following rapid decompression in the hamster. Undersea Biomedical Research 12: 105–114

MacDonald A G 1984 The effects of pressure on the molecular structure and physiological functions of cell membranes. In: Paton Sir William, Elliott D H, Smith E B (eds) Diving and life at high pressures. The Royal Society, London, p 47–68

McCallum R I 1984 Bone narcrosis due to decompression. In: Paton Sir William, Elliott D H, Smith E B (eds) Diving and life at high pressures. The Royal Society, London, p 185–191

McCallum R I, Petrie A 1984 Optimum weights for commercial diver. British Journal of Industrial Medicine 41: 275–278

Miller J N, Fagraeus L, Bennett P B, Elliott D H, Shields T G, Grimstad J 1978 Nitrogen-oxygen saturation therapy in serious cases of compressed air decompression sickness, Lancet i: 169–171

Naquet R, Lemair C, Rostain J-C 1984 High pressure nervous syndrome. Psychometric and clinical electrophysiological correlations. In: Paton Sir William, Elliott D H, Smith E B (eds) Diving and life at high pressures. The Royal Society, London, p 95–101

Palmer A C, Calder I M, McCallum R I, Mastaglia F L 1981 Spinal cord degeneration in a case of recovered spinal decompression sickness. British Medical Journal 283: 88

Paton Sir William, Elliott D H, Smith E B (eds) 1984 Diving and life at high pressures. The Royal Society, London

Ronk R, White M K 1985 Hydrogen sulphide and the possibilities of 'inhalation' through a tympanic membrane defect. Journal of Occupational Medicine 27: 337–340

Salzano J V, Stolp B W, Moon R E, Camporesi E M 1981 Exercise at 47 and 66 ATA. In: Bachrach A J, Mantzen M M (eds) Underwater physiology VII. Undersea Medical Society, Bethesda MD, p 181–196

Shields T G, Minsaas B, Elliott D H, McCallum R I (eds) 1983 Long term nurological consequences of deep diving. Norwegian Petroleim Directorate and European Undersea Biomedical Society, Stavanger

Shilling C W, Carlston C B, Mathias R A (eds) 1984 The physicians guide to diving medicine. Kleenum, New York, London

Shiraki K, Sagawa S, Konda N, Nakayama H, Mstsuda M 1983 Properties of red blood cells after multiday exposure to 31 ATA. Undersea Biomedical Research 10: 349–358

Smith E B 1984 The biological effects of high pressures. Underlying principles. In: Paton Sir William, Elliott D H, Smith E B (eds) Diving and life at high pressures. The Royal Society, London, p 5–16

Statutory Instruments. The offshore installations (operational safety, health and welfare) regulations 1976, No 1019, HMSO London
Stolp B W, Moon R E, Camporesi E M 1985 2,3 DPG levels during saturation diving at 650 msw. Undersea Biomedical Research 12 (suppl): 21
Sykes J J W, Yaffe L J 1985 Spinal cord decompression sickness, shades of severity or more than one mechanism? Undersea Biomedical Research 12 (suppl): 17
Undersea Medical Society 1983 Oxygen. An indepth study of its pathophysiology. Undersea Medical Society, Bethesda MD
Undersea Medical Society 1985a Cold and the diver, prevention, protection and performance. A Bibliography of Informative Abstracts. Undersea Medical Society, Bethesda MD
Undersea Medical Society 1985b Nitrogen narcosis. Undersea Medical Society, Bethesda MD
Underwater Engineering Group 1985 Tables for saturation and excursion diving of nitrogen-oxygen mixtures UEG, Construction Industry Research and Information Association. London
United Kingdom Offshore Operators Association 19— Environmental health guidelines for offshore installations, United Kingdom continental shelf. UKOOA, London
United Kingdom Offshore Operators Association 1986 Recommended general medical standards of fitness for designated offshore employees. UKOOA, London
Virr L E 1984 Thermal protection equipment, design and operational considerations. Underwater Technology 10: 23–32
Weathersby P K, Hayes J, Randol J, Survanshi S S, Woolmer L D, Art B C, Flynn E T, Bradley M E 1985 Statistically based decompression tables II. Equal risk air diving decompression schedules. United States Naval Medical Research Institute, Bethesda MD, Report 85–17

3. The electricity supply industry

J. A. Bonnell

Electricity is the corner stone of the power industry and is essential for the maintenance of the standard of living of the developed nations and also for the efficient operation of industry in those countries.

The first public electricity supply in the world was inaugurated in 1882 at Godalming in Surrey, the first steam generating plant to supply public lighting was introduced in the City of London a year or so later. In the 100 years since then, there has been a phenomenal growth in the demand for electricity. (Table 3.1.)

Table 3.1 UK generating plant capacity by type of plant

Date	Power (MW)	Coal	Oil	Nuclear	Hydro	Pumped storage	Other
1925	3 917	(almost all due to coal and Hydro)					
1955	18 845	18 516	266	—	53	—	
1965	36 905	33 280	924	2341	118	360	
1975	58 677	40 773	12 016	3462	112	360	1954
1985	51 127	32 634	9 166	4499	112	2088	2630

Currently in England and Wales there is an interconnected network comprising 81 power generating stations with an installed capacity of 57 375 Megawatts (MW) and 9531 circuit kilometres of 400 kV and 4315 circuit kilometres of 275 kV transmission lines. Power is generated from coal, oil and nuclear power sources with a small amount of hydro-electric generation. There are in addition pump storage schemes in North Wales. Scotland has its own electricity supply industry. Two utilities, the South of Scotland Electricity Board (SSEB) and the North of Scotland Hydro Electric Board (NSHEB) own and operate all the power stations, transmission system and distribution network. There are close links at Board and working level between the organisation of electricity supply in Scotland with England and Wales and the transmission system is interconnected on both the East and West Coast ensuring an integrated system in the UK as a whole.

There is in addition a Direct Current cross channel link between the Central Electricity Generating Board (CEGB) and Electricite de France allowing for the two way transfer of 2000 MW of electricity between the British and French system. This was commissioned in 1986. Since France exchanges electricity with all its neighbours the UK is thus linked into the European network.

THE UNITED KINGDOM ELECTRICITY SUPPLY INDUSTRY

As in many countries the electricity supply industry in the UK is publicly owned and in money terms is one of the very largest businesses in the UK. Under the terms of the Electricity Acts 1948 and 1958 the Electricity Supply Industry is responsible for the generation, transmission and distribution of electricity efficiently, economically (Fig. 3.1) and safely. The Acts require that there should be adequate and proper provision of health, safety and welfare facilities and that particular attention must be paid to environmental factors.

The Industry is established on the basis of an Electricity Council for England and Wales and two boards, the SSEB and the NSHEB in Scotland. The Electricity Council consists of a small centralised organisation co-ordinating the activities of one large generating board (CEGB) and 12 distribution Area Boards, all 13 Boards having autonomous management responsibilities. The CEGB is responsible for the design, con-

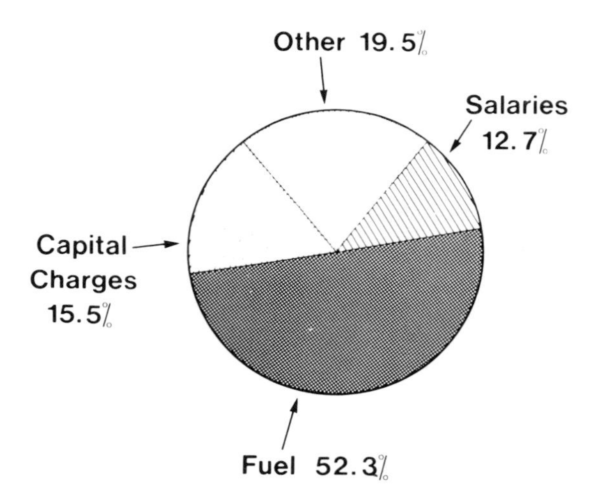

Fig. 3.1 A demonstration of the overwhelming importance of fuel charges for electricity production compared with salaries (courtesy of the Electricity Council).

struction and operation of power plant and for the bulk transmission system. The CEGB sells its electricity to the Area Boards and to a number of organisations operating on a national scale, such as British Rail. The transmission of electricity is at 400 kV or 275 kV, distribution is by the area boards and is at 132 kV and below, stepped down eventually to the domestic voltage of 240 volts.

The Electricity Council's responsibilities are for central financial arrangements with central government for various commercial and financial policies. It negotiates national agreements for the industry with trade unions and through its various joint consultation machinery provides the means for discussion of all aspects of health and safety. The National Health and Safety Advisory committee (HESAC) is reflected through regional, area and local HESACs to the shop floor. At all levels professional health and safety staff attend the various HESAC committee and sub-committee meetings.

The industrial relations within the industry is good, joint consultative committees have been well established since nationalisation in 1948. There is a clear distinction between the agreements concerning salaries, wages and conditions of employment and those concerned with health safety and welfare. The organisation of these various joint committees at national level are reflected throughout the industry by equivalent committees at Regional and Area Board level and at all management locations. Medical and safety staff attend all the committees concerned with health and safety and they attend as advisers to the committee not as representatives of management. Joint training committees including the joint training of safety representatives have been well established.

Electricity generation

In the UK electricity is produced by the three generating Boards: the CEGB in England and Wales and the SSEB and the NSHEB in Scotland. The two Scottish Boards also operate the distribution networks in Scotland.

The European electrical utilities vary depending upon the political framework. France has an almost identical system to the UK, Electricite de France being a large centralised organisation responsible for the construction, generation, transmission and distribution of over 95% of the electricity produced. There are in addition small private undertakings. Italy, Spain and Portugal are similar all having an integrated interconnected system. In Federal Germany each state has its own electricity system, mostly privately owned and operated but integrated nationally as in other European countries. Additionally, all European Countries are interconnected one with another.

In principle the United States and Canada are established on a similar footing. In these countries and in the USSR because of the long distances over which electricity is transmitted the system voltages have been stepped up to 500 kV and 765 kV, experimental lines have been constructed with transmission voltages of 1000 kV.

In order that integrated electrical networks such as these can operate efficiently it is necessary for the operation of all the power stations and the transmission system to be co-ordinated centrally. In order to achieve this, for example, in the UK there is a system operations department, who operate the network, and they maintain a merit order list of power stations based on efficiency and availability which are brought into operation according to the electrical demand. The more modern larger power stations and the nuclear stations are operated as base-load stations. The electricity demand can fluctuate enormously from hour to hour and minute to minute according to the season of the year and even to the popularity of TV programmes (Fig. 3.2).

OCCUPATIONAL HEALTH AND SAFETY

Because the primary 'product' of the Industry (electricity) is potentially lethal, safety procedures are well established. Comprehensive safety rules are drafted by the management and consultation takes place with the trade unions. Final approval having been obtained by HESAC the various safety rules become mandatory within the industry. Deliberate breach of the rules result in disciplinary action including dismissal. The

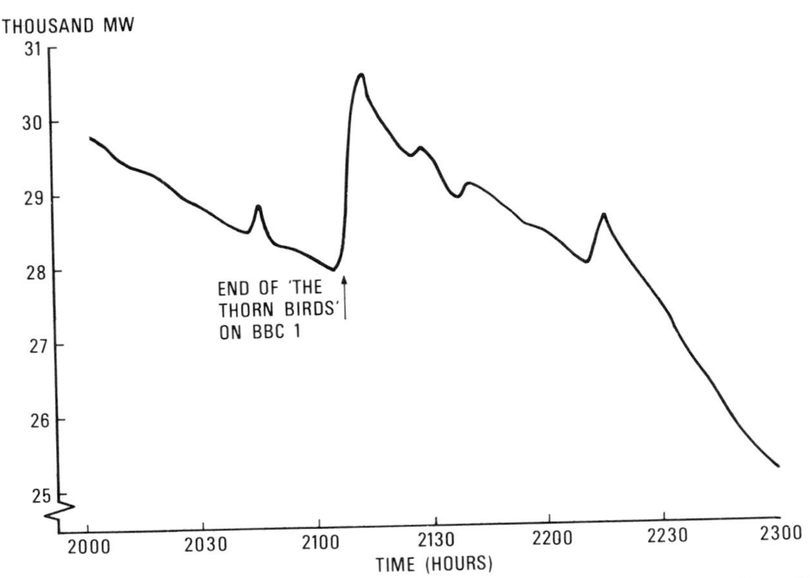

Fig. 3.2 Demand curve for the evening of Sunday 22 January 1984 when the end of The Thorn Birds on BBC 1 caused a demand of power increase of 2600 MW at 21.07 h.

three volumes of the ESI Safety Rules are Electrical and Mechanical Rules, Radiological Safety Rules and the Electricity Distribution Rules. All are dependent on a clearly defined system of permits to work, with only senior operational staff being authorised to sign the permits. Codes of practice are widely used to regulate safe working methods in relation to physical and chemical hazards, e.g. ionising radiation, noise and toxic substances.

Electrical accidents give rise to serious injury, in particular burns (which are full thickness) and electrocution. It is, therefore, vital that adequate precautions are taken to prevent accidents. The Industry's safety organisation is two fold: the Electricity Council has a Chief Safety Officer who with a staff of regional safety officers provide advice and guidance to safety engineers employed by the area boards; the CEGB has its own comprehensive health and safety organisation. Accident rates including fatalities are shown in Table 3.2.

The CEGB has a Director of Health and Safety directly responsibile to the executive of the Board for all health and safety considerations and with no other responsibility. It is constituted in five branches: medical, nuclear safety (operations), nuclear safety (development), industrial safety, and safety strategy. The Chief Medical Officer is head of the medical branch and has direct access to the executive in clinical medical matters.

This structure parallels the organisation of the Health and Safety Executive and is the channel of communication on technical matters between the Board and the Executive.

Occupational health is the responsibility of the medical branch which has a staff of medical officers, nurses and first aiders backed up by occupational hygiene and health physics support, which is available in the Health and Safety Directorate.

First aid training is on a voluntary basis, the Electricity First Aid Centre is affiliated to the St John's Ambulance Association. Training is to the level of the First Aid

Table 3.2 Accidents to employees in the electricity supply industry in England and Wales during 1984 and 1985

Classification	Number of accidents Non-fatal (1)	Fatal (2)	Sum of columns (1) and (2) expressed as percentages of all accidents
Handling object (by hand or hand truck)	650		23.5
Persons falling	649		23.4
Stepping on or striking against objects	424		15.3
Hand tools	223		8.1
Machinery (hand and power operated)	154		5.6
Transport (including mechanical trucks)	124	1	4.5
Objects falling	117		4.2
Accidents directly associated with electricity*	115	2	4.2
Hot substances	78		2.8
Explosions, fires and boiler backdraughts	11		0.4
Poisonous and corrosive substances and occupational diseases (including gassing)	10		0.3
Miscellaneous	213		7.7
Total	2768	3	100.0

* Accidents directly associated with electricity.

	Non-fatal	Fatal	Percentage of total
CEGB	3	0	0.1
Area boards	112	2	4.1

at Work Certificate, with additional training in the CEGB for an Occupational First Aid Certificate to a syllabus approved by the Health and Safety Executive.

The manager of a power station is responsible for the maintenance and operation of his plant. He has available to him various specialist advisors including those relating to health and safety in order to achieve his objectives. Occupational health advice to each location is provided by a senior occupational health physician operating on an area or regional basis with the assistance of part-time physicians, nursing staff and other auxiliary and first aid staff.

In summary the CEGB's organisation consists of headquarters departments, including health and safety, five operating regions and three divisions. The five regions are responsible for the operation and maintenance of the power stations and transmission system in the different geographical areas; the construction and development functions associated with power generation and transmission are consolidated into two divisions, Generation Construction and Development Division and the Transmission and Technical Services Division. The Technology Planning and Research Division operate three large central laboratories and in addition each region has its own scientific services department. Currently there are plans to reorganise the management structure of the Board to reflect the reduction in the number of operational units.

SOCIOLOGICAL ASPECTS

Electricity consumption per head of population is an accurate index of the standard of living in any particular country (Fig. 3.3).

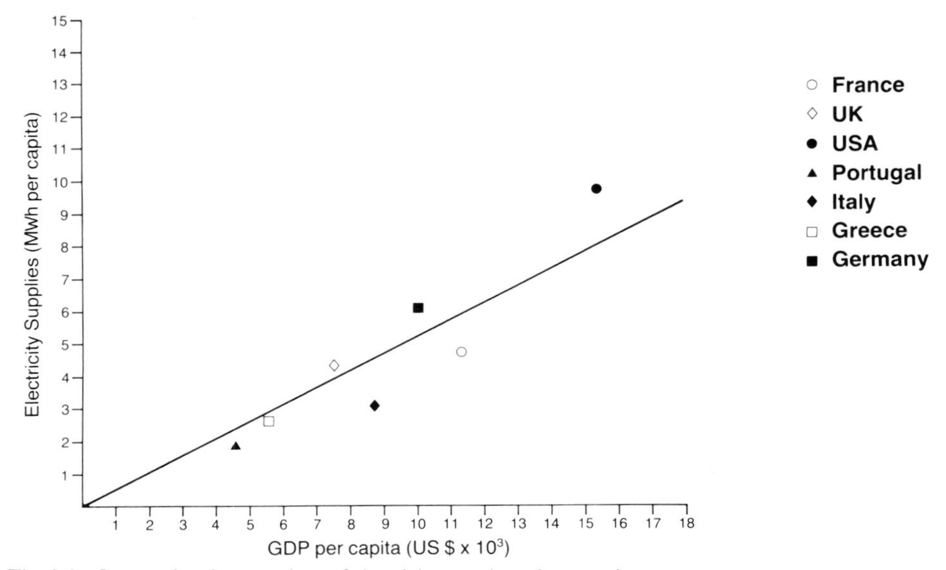

Fig. 3.3 International comparison of electricity supply and prosperity.

The construction of a large modern power station and its associated transmission network may be regarded by environmental groups as an intrusion into the environment and frequently gives rise to objections from some members of the general public, resulting in extensive public inquiries. The siting of a large power plant present a number of problems. From the operational point of view a coal fired power station ideally needs to be sited near to a coal field, an oil fired power station in the vicinity of an oil refinery. There is a need for an adequate source of water for cooling purposes, suitable geological strata for good foundation conditions and at the present time nuclear power stations have to be sited in relatively unpopulated areas. Clearly, sites meeting all these requirements are hard to come by in the small area of the UK. On the other hand the construction of a large industrial plant brings, in its turn increasing employment opportunities on its own account and also attracts other industries.

Sociological studies have been carried out demonstrating the beneficial effects of power station development in deprived areas. The County Planning Authority in North Wales commissioned a series of studies which were carried out jointly by the county councils and the Generating Board, (Chadwick & French, 1977) to study the economic and social impact of a power station in a rural community in North Wales. The immediate short term benefits included markedly increased employment during the construction stage, and the increase in the amount of money spent within the area. Longer term there was some increase in employment prospects, infrastructure improvements (roads etc) increased turnover to sections of the local economy, additional tourist attraction and increase in the rateable value (i.e. local taxation). These would have to be balanced against possible damage to the local ecology of lakes and rivers, visual amenity impact and disturbances from construction activities. Generally the benefits out weighed the disbenefits. In a later report (Lewis, 1985) the Economic impact of the closure without replacement of a nuclear power station in the same area was studied by the Institute of Economic Research, University College of North

Wales. It was apparent that the closure would have a significant impact on unemployment rates. These could be summarised as (a) loss of work for a proportion of the station staff (b) loss of indirectly created jobs (c) loss of future opportunities and (d) reduction in potential migration into the area.

Fears and anxieties about the possible health related effects of power station operation also feature as a major source of objection. These objections relate to effluent discharges of sulphur oxides and nitrogen oxides from coal and oil fired plants and radioactive waste discharges from nuclear plants. Considerable research efforts are expended in the study of these problems and the reports have been widely disseminated in the medical and scientific literature (see References). More recently, anxiety has been expressed about the possible problems associated with the long range dispersion of acidic effluents the so called acid rain problem. Scandinavian countries have suggested that the acidity of lakes resulting in fish death, and also damage to forests have been due to the long range drift of acid effluents from Northern European and UK industrial areas with particular reference to fossil fuel power stations (Howells & Kallend, 1983; Blank, 1985). This problem is the subject of major studies mounted under the auspices of the Royal Society. International electrical organisations are also involved in a cooperative study of the problem, notably in North America where Canada has expressed concern about the long range drift of pollution from the industrialised areas of the USA. Whilst these problems are not a direct hazard to the employees or to the immediate locality they present serious problems which the industry is seeking to resolve.

Hazards of the industry

The potential hazards of the industry can be categorised as those which could affect the general environment and those which present an occupational health problem.

Environmental problems

These associated with the combustion of large quantities of fossil fuels (coal and oil). Additionally large hydro electric installations can have a marked effect on the environment and on the ecology of a region. The construction of large dams are themselves obtrusive and the effects of damming rivers with the production of large artifical lakes can have an overriding effect. There is also the remote possibility of dam rupture with a resulting devastating effect on the country down river. Happily such events are rare. In the UK, coal and oil fired power stations are constrained by legislation in that their effluent discharges are subject to the requirements of the Alkali Acts and the Clean Air Act. The intake and discharge of water used for cooling purposes have been studied as a potential source of disturbance to the ecology of rivers and estuaries by producing unacceptable changes in the water temperature. The CEGB has a large research programme which is carried out in association with government departments and international institutions to study these problems. At coal and oil fired power stations the British Tall Stack Policy for the discharge of gaseous effluents is a direct result of these investigations (Annual Report of the Chief Alkali Inspector, 1974). The cost of meeting the stringent requirements of the Alkali Acts for the discharge of grit and dust and acidic gases represents a major item in the cost of power station construction.

The electrostatic precipitators used to collect dust and grit are required to be demonstrably 99.7% efficient. The controversy regarding the contribution (if any) of the discharge of acidic gases to the general environment and whether this plays any part in the acidification of lakes in Scandinavia and the consequent effect on fish life, together with reports of damage to forest land in Northern Europe is also the subject of international research and cooperation. It is clearly a complex subject and requires careful assessment.

In the domestic environment the distribution Area Boards have been concerned (in common with the gas industry) over the possible problem of asbestos used for heat insulation purposes in domestic electrical storage heaters. The studies carried out indicate that there is no hazard to members of the general public in the inspection and maintenance of these units.

Occupational health hazards

In the industry, these are more specifically associated with power station and transmission line construction and operation. The construction of power stations is carried out by major contractors and their subcontractors. During the construction phase the hazards are those of the construction industry. From an agreed date, usually when the electrical plant is commissioned, the industry's safety rules become mandatory on site. The generation development and construction division provide the specification for new power stations and are responsible for ensuring that the terms of the contract are complied with. Specifications include architectural constraints, ergonomic requirements, maximum noise levels and the specification of materials used. An example of the latter activity is the banning of asbestos; on the advice of the medical staff no asbestos-containing lagging materials have been used in any power stations for which the contract was placed after 1968.

At nuclear power stations the specification includes requirements to limit radiation doses to the operators in the form of dose targets which currently are 20% of the annual dose limit recommended by ICRP. There are also requirements to ensure the limited release of radioactive materials in the event of a nuclear emergency. These measures were discussed at length at a recent public enquiry for a nuclear power station in east Suffolk.

There are a number of potential health hazards in the operation of power stations, these vary depending on the type of power plant e.g. coal, oil or nuclear plants. Accidents at power stations are largely due to lifting and handling or accidents related to falls, either persons falling or falling objects (see Table 3.3). Electrical accidents are infrequent. At coal fired power plants a source of serious accidents are those associated with the coal handling plant. Coal is delivered in 'merry-go-round' trains carrying 18 000 tonnes of coal on continuous journeys from coal mine to power station. The coal is delivered automatically as the train passes slowly through the power station. The coal is either deposited in large storage areas or transferred on a conveyor belt system to hoppers, then converted to pulverised coal in large mills from whence it is treated as a 'fluid' and burnt through jets to heat the boilers. The hazards during this process are first, the possible emission of carbon monoxide due to slow spontaneous combustion of the coal in the hoppers: next the finely ground coal (average diameter 5–10 μm) forms an explosive mixture in air and a number of incidents of mill explosions have occurred over the years. Great care is taken to prevent such a mishap by careful design of the mills and in the care and maintenance of the plant. Coal dust in this form is a potential cause of pneumonocosis but no cases have been confirmed in men exclusively exposed to coal dust in power stations.

Power station coal is of relatively poor quality and after combustion produces 20% of ash (pulverised fuel ash). This material has many uses both in the building trade and the construction industry and its uses and potential toxic properties are discussed in a later section (see below).

The accident experience at oil fired power stations is different due to the absence of coal handling plant. Heavy duty residual fuel oil contains a small proportion of vanadium which during combustion sublimates in those parts of the boiler known as the economiser where the circulation of the hot gases is diverted to extract the maximum heat from the combustion process. This presents a potential problem to men employed as boiler cleaners in this type of power station.

Noise is a serious occupational health problem at all power stations. Noise levels may reach 110–120 dBA in some parts of older power stations, but these are not areas normally occupied by workers. A hearing conservation programme has been in operation for some years. Recent studies carried out to assess the noise hazard and the efficiency of hearing conservation measures over a 5-year period confirm that provided the programme as specified in the CEGB Code of Practice is carried out, no hazard to hearing exists. Leq surveys confirm that a small proportion of workers would be exposed to noise equivalent levels in excess of 95 dBA if hearing protection devices were not worn. Whilst efforts are made to reduce noise at source it is extremely difficult to secure the movement of heavy plant at 3000 rev/min without making a noise. Many large modern power stations only require man access to the turbine halls for brief periods each shift and this of itself limits the exposure of personnel to noise and adequate hearing protection devices can be worn during this time.

From time to time the noise levels are enhanced by the occurrence of steam leaks which give rise to high intensity high frequency noise. Efforts are made to isolate sections of plant responsible for the leakage of steam in as short a time as possible: but for operational reasons it may not be possible to do so immediately.

Even though asbestos has not been present in thermal insulating materials for nearly 20 years there are still many power stations in the CEGB which do contain vast amounts of asbestos. The Board's policy, introduced in 1968, has been to remove asbestos where it is insecure or loose and to seal-in the asbestos in those areas of plant which are not disturbed during routine maintenance activities. Since it was frequently found that crocidolite or amosite was usually present in the asbestos mixes the policy has been to work to the more restrictive level of 0.2 fibres/ml rather than try to demonstrate that only chrysotile was present. Despite these precautions cases of asbestos related disease have occurred (Bonnell et al, 1970), none to date in persons employed solely in the industry since 1968. It is of interest that since the temperature conditions at power stations are not excessive as compared for example with the steel or glass industry, the various man made mineral fibres (MMMF) such as glass wool and rockwool were found to be perfectly adequate as thermal insulating materials. Because of the large quantities of these materials required during the enormous power station construction programme of the 1960s and 1970s the electrical generating industry provided a sufficiently large market for the development of the MMMF manufacturing capability for the replacement of asbestos in many and various industrial and domestic situations.

NUCLEAR POWER GENERATION

In 1955 the Government published a White Paper making recommendations for the development of nuclear power as a source of electricity generation. The first phase was the construction of a series of nuclear power stations in which the reactors were of the natural uranium type. The first of these were officially opened in 1962. The second phase was the advanced gas cooled reactor (AGR) in which the fuel was enriched uranium oxide in stainless steel and again using carbon dioxide coolant. Announced in 1965 the first AGR stations were commissioned in 1980. In 1983 the CEGB gave notice of its intention to build a pressurised water reactor PWR at Sizewell in Suffolk. This resulted in the longest public inquiry in UK history lasting over 2 years. The result is still awaited.

At the present time about 15% of the electricity generated by the CEGB is from nuclear power. The industry's record in radiation protection is outstanding. Since 1962, there have been around 16 000 persons classed as radiation workers and only on three or four occasions have any individuals exceeded the recommended annual dose limit of 5 rem and then only a marginal excess. These classified workers receive annual medical and blood examinations and detailed records are kept of their radiation doses (Table 3.3). Those employed as from January 1976 are included in the National

Table 3.3 Personal radiation doses (rems) 1984 CEGB and contractors personnel

	<0.5	0.5–1.5	1.5–2.5	2.5–5	>5
Berkeley	1335	79	1	0	0
Bradwell	709	50	12	2	
Dungeness A and B	2290	6	0	0	0
Hinkley Pt A and B	1962	112	9	2	0
Oldbury	933	11	1	0	0
Sizewell	878	22	1	0	0
Trawsfyndd	892	40	1	0	0
Wylfa	728	2	0	0	0
Berkeley Nuclear Laboratories	573	13	1	2	0

Heysham and Hartlepool Power Stations not included in 1984 (only partially commissioned).

Register for Radiation Workers (NRRW) maintained by the National Radiological Protection Board (NRPB) and those employed between 1962 and 1976 have been identified and flagged at the Central Registry of the OPCS. These data together with those of employees of the United Kingdom Atomic Energy Authority (UKAEA) and British Nuclear Fuels Limited (BNFL) form the basic population for the long term study of the biological effects of exposure to low levels of ionising radiation.

Radioactive waste management

One aspect of nuclear power generation which causes a considerable amount of public concern is the management of radioactive waste arising from the operation of the power station. There are basically three types of waste matter. First the long lived fission products which result from the chemical processing of the spent radioactive fuel; secondly the radioactive materials induced in equipment and instrumentation which has formed part of the reactor and also the small quantities of fission products which have leached out of the fuel elements during their short stay on site and thirdly the low level radioactive waste matter which arises during the day to day operation of the plant. The only material which is of concern is that which is associated with the fission materials in the fuel. There is no chemical treatment of the fuel at nuclear power stations. The public discussions are associated with the preferred method of disposal namely glassification and disposal at sea (sea dumping) and land burial. The problem of disposal has yet to be resolved but one feature of the problem which is frequently overlooked is the small bulk of the material to be disposed.

Nuclear emergency plans

All nuclear establishments are required statutorily to prepare detailed plans to deal with emergencies arising from accidents or maloperation of the plant. The plans include a contingency for assessing and dealing with any hazard to members of the public. These plans are discussed in detail with appropriate officers of the police and local authorities who include nuclear emergencies among their civil emergency plans. In the event of a release of fresh fission products from a nuclear reactor it is likely that the iodine isotopes would be preferentially released. They are present in greatest abundance in the uranium fuel, they are volatile and are selectively absorbed and retained in the thyroid. The plans include the distribution of stable iodine in the form of potassium iodate to the public living in the 30° sector downwind and out to a specified previously agreed distance; usually about 1–1½ miles from the plant. (Adams & Bonnell, 1962). Temporary evacuation of the public in this sector may be necessary. These plans are discussed publicly through local liaison committees and copies of the plans are available in public libraries in the areas concerned.

Transport of nuclear fuel

The reactors are designed for refuelling during normal on-load operation. The spent fuel is transferred to cooling ponds on site where they are stored for a period of approximately 3 months. This allows for the radioactive decay of the shorter lived nuclides. The fuel is then transported by road and rail from the nuclear power station to the chemical processing plant operated by BNFL in Cumberland. The containers (or flasks) are specially designed to carry about 2 tonnes of fuel, the flasks weigh about 50 tonnes and have been demonstrated by exhaustive tests to be extremely robust.

In a recent demonstration accompanied by great publicity a flask remained intact having been struck by a locomotive drawing three coaches at 100 mph. The locomotive was totally destroyed. It was considered that the cost was worth the reassurance provided to the public.

CHEMICAL HAZARDS

Other various chemical substances have been the subject of study namely polychlorinated biphenyls (PCBs) in transformers and capacitators and sulphur hexafluoride in 400 kV circuit breakers. Orthotolidine was used for a period during the 1950s and 1960s in the chemical analysis of water purity. Several hundred chemists and their staff have been listed and flagged by the Office of Population, Censuses and Surveys (OPCS) and the men concerned regularly provide urine specimens for cytology, since the material is a potential bladder carcinogen.

The policy for dealing with problems such as those listed above is to produce safety rules or codes of practice which define the problem and specify the actions required to ensure safe practice. Health monitoring of the persons concerned may be instituted as part of the procedure for the protection of the individual.

TRANSMISSION OF ELECTRICITY

This is by means of an integrated transmission network operating at 400 or 275 kV. Interconnections are made at substations (switching stations) to allow transfer of power from one part of the country to another and to the distribution Area Board network, the voltage being stepped down to 132 kV. The CEGB operates one of the largest power systems under single integrated control in the world. The year 1985 marked the 50th anniversary of the initial grid system which entered full commercial operation in 1935.

Apart from electricity itself the transmission system presents few occupational health problems. Repair and maintenance of the system is carried out by small teams of overhead linemen. On occasions their work may be harrowing particularly during severe adverse weather conditions. One problem which presents itself to the occupational physician is the physical requirements for climbing towers which may be difficult in men aged 55 years and over.

Visual inspection of the lines is carried out from aircraft, at one time helicopters were used but more recently it has been found that fixed wing aircraft are more effective. On occasions work is required on a transmission line which cannot be taken out of service. For this purpose special protective suits made of conducting material are available for working on live high voltage lines, the suits operate as a mobile Faraday cage and allows contact with live conductors at 400 kV (Looms, 1971). This is rarely required in the UK because of the multiple interconnections provided by the grid which allows for almost any section of line to be taken out of service for routine maintenance work by transferring the power along alternative routes.

Recent suggestions have been made regarding the possibility of long term adverse health effects of electro magnetic fields. The evidence for these suggestions is not convincing and will be discussed in a later section (see p. 50).

PROTECTIVE CLOTHING

In many instances working procedures have been facilitated by the development of protective clothing specifically designed for the purpose. Transmission department personnel are frequently required to carry out repair and maintenance on overhead lines in severe weather conditions. Suitable clothing has been designed allowing mobility, weather protection, and for work on energised lines (Looms, 1971). Such suits are now widely used in North America and Europe but to a lesser extent in the UK. Protective clothing at nuclear establishments is used to allow a workman to enter areas of high radioactive contamination by taking with him his own microenvironment, these suits have been widely used for many years. They present some unfortunate problems due to the accumulation of water from sweat unless specially designed to have adequate air throughput. An additional problem arises at nuclear power stations as man access to the boilers (or heat exchangers) may be required for inspection and maintenance purposes. These boilers, mainly at AGR but also some magnox stations, take weeks or months to cool to 30°C, for economic reasons man access is required in the shortest possible time after shut-down. For this purpose air cooled suits utilising vortex tubes for cooling have been designed, so that work can be carried out in temperatures of up to 80°C with heat stress equivalent to 30°C. Simulators have been built to allow training for work in these suits at operational temperatures.

The upper limit for work of this kind has been set at 60°C with special dispensation in some situations of up to 70°C. The development of protective clothing has been brought about by the coordination of the industry's health and safety and research divisions.

PULVERISED FUEL ASH

Whilst the Electricity Supply Industry was established to provide a reliable and safe electrical system for its consumers it is also permitted to sell its major waste product namely pulverised fuel ash, which it does under the trade name PFA. It is necessary, because of international semantic problems to define what is meant by the term. There is a heavy material called furnace bottom ash or clinker which is used for in fill and hard core; precipitator ash is the material trapped in the electrostatic precipitators and is the material sold as PFA. There is also a fine globular material which is called stack ash (fly ash in the USA) and is the material which is not collected by the EPs and is discharged to the atmosphere.

PFA has many uses as a building and construction material based on its pozzalanic properties for which it has considerable commercial advantages. (A pozzalan is a material which without being itself a cement has the property of setting hard in the presence of lime and water.) Suggestions have been made about its possible toxicity and carcinogenic activity. Since it is sold by the CEGB to various organisations in the building industry it is incumbent upon the Board under Section 6 of the Health and Safety at Work Act to provide evidence as to its toxicity and to make recommendations for its safe handling.

The possible association between PFA exposure and pneumoconiosis in workers has been studied in the industry since 1950 (Bonnell et al, 1975). Men employed in dusty jobs in the Midlands and South Wales were examined clinically and radiologically at intervals between 1950 and 1961 and again between 1974 and 1977. These studies only revealed the presence of pneumoconiosis in men previously employed in the coal mining industry and the extent of their disease was directly related to the time they had worked in that industry. From 1974 to 1977, four men out of a total of 246 with category '01' pneumoconiosis were identified who had not been employed in dusty jobs elsewhere. This finding was not considered significant (Bonnell et al, 1980). This work was confirmed by further extensive radiological studies in power station operatives in the North East Region.

Published reports (Chrisp et al, 1978, 1979) from the USA suggested that stack ash (described as fly ash) gave positive Ames tests. As a result extensive cytotoxicity studies have been carried out (Reid, 1984) which have failed to confirm this finding. It has been agreed by the Health and Safety Executive that PFA can be regarded as a nuisance particulate and handled accordingly.

LEGIONELLA PNEUMOPHILA

A recent example of the involvement of the industry with environmental and community health occurred during the epidemic of legionnaires disease in Staffordshire. During the search for the source of the infection suggestions were made that one of a number of power stations operating large industrial cooling towers and sited in the vicinity may have played some part in spreading the infective agent.

A country wide investigation of the cooling water systems of all power stations on inland waterways and using cooling towers was instituted. This confirmed that the Staffordshire power stations were not a source of infection (Bonnell & Rippon, 1985).

It is of interest to note that since the first identified outbreak of legionnaires disease in Philadelphia in 1976 and the identification of the causative organism (*Legionella pneumophila*) clusters of cases have been associated with hot water systems in hospitals, hotels and office buildings or with air conditioning plant and small cooling towers. These small cooling towers have a throughput of about 3000–4000 gallons of water per hour compared with up to 9 million gallons per hour in the 370 feet high natural draught cooling towers of a large modern power station. These towers are used on inland waterways to conserve water. Evidence obtained as the result of serological studies on power station operatives between 1978 and 1980 had confirmed that antibody titres to *L. pneumophila* were the same as in the general population.

As the result of a proposal to change the process for control of biofouling in the condensers with the abandonment of chlorination it was decided to study the incidence and concentration of *L. pneumophila* in cooling water systems of a representative group of power stations. Seventeen were investigated during the summer of 1983. A further larger number of samples were taken from four power stations in 1984, and a detailed study of a newly commissioned power station was undertaken in 1985. The 1985 study referred to in the previous paragraph was an additional specific investigation. *Legionella* was isolated and identified in a random fashion in most power stations at one time or another. The numbers were always low and the distribution within the system was patchy. The number of colony forming units are in general lower than those found in hotels and hospital waters and no concentration of organism have yet been found which could constitute a source of infection of the general public (Bonnell et al, 1986).

ELECTROMAGNETIC FIELDS

Reports from the USSR during the 1960s and 1970s suggested that electric fields emanating from high voltage transmission plants operating at 500–750 kV (Korobkova et al, 1972) had given rise to non-specific symptoms among workers employed in the vicinity of such plants. Extensive studies in Western Europe and North America have failed to substantiate these claims (Knave et al, 1979; Strumza, 1970). It is likely that the symptoms could be associated with perception of the field arising as secondary effects such as induced microshocks giving rise to annoyance or irritation (Bonnell, 1983; WHO, 1984).

More recently reports have appeared in the USA and UK suggesting an association between childhood cancer and electro magnetic fields probably the magnetic component of the field (Wertheimer & Leeper, 1979).

In view of these reports and following a public inquiry in South West England the CEGB in association with the Electricity Council undertook a specific research programme. This involved the following.

a. The study of men exposed to electric fields from high voltage transmission systems and the distribution network in South West Britain (Broadbent et al, 1985).

b. The study of volunteers subjected to the passage of electric currents under controlled laboratory conditions using psychometric tests (Stollery, 1986).
c. The study of childhood cancer in one health authority region of the UK using a case control study (Myers et al, 1985).
d. A study of the response to electric fields of cardiac pacemakers of different design in patients suffering from a variety of cardiac diseases (Butrous et al, 1983).

The findings confirmed that power frequency electric fields at strengths encountered in the occupational and general environment do not have adverse effects on the health of persons exposed at work or in the community.

Various designs of cardiac pacemaker are used for the treatment of heart disease involving different conduction defects. These are multiprogrammeable, on-demand devices which are extremely sophisticated. It was found that some models were sensitive to very modest currents induced by electric fields of 2–3 kV per metre. The problem is readily resolved by ensuring proper design of pacemakers.

However in view of the continued publication of reports suggesting an association between adult leukaemia and electrical occupations (Coleman et al, 1983; McDowall, 1983; Milham, 1982; Wright et al, 1982) further studies are taking place. It should be borne in mind that the electrical occupations referred to include assemblers of electronic equipment, telecommunication engineers and electricians. The nature of their work suggest no exposure to electromagnetic fields but other environmental factors need to be taken into account (Bonnell, 1983). A more detailed analysis of UK OPCS data (McDowall, 1986) failed to establish any significant association in the mortality experience of persons resident in the vicinity of electricity transmission plant.

EPIDEMIOLOGICAL STUDIES

In view of the variety of jobs and the potential health hazards associated with electrical power generation the Board established an Epidemiology Advisory Committee within the medical branch. Its terms of reference was to comment upon and advise the Board on the setting up of long and short term studies on the general health of the work force. The working population during the years 1972–1975 have been identified and in the first instance a mortality study will be carried out at which it will be possible to examine the prevalance and incidence of diseases in the various working groups. The work is fully supported by the industry's trade unions.

DISTRIBUTION OF ELECTRICITY

Whilst the distribution Area Boards do not have the variety of occupational health problems encountered in the generation of electricity, there are a number of problems which are common to both sides of the industry. This is particularly so in the study of possible health related effects of electromagnetic fields. The need for safety rules to ensure safe practice in the handling of electricity also applies, as do problems associated with lifting and carrying, noise and vibration in tree felling and road crushing equipment. Cable jointing in area boards is a chemical process as opposed to mechanical jointing as practiced in the generating board. Cable jointing is frequently carried out on live 11 kV cables, the cables have to be isolated and exposed by road drilling.

The aluminium cable joint is made using fluoride fluxes and silver solders which can give rise to noxious fumes. The junction boxes are sealed using polyurethane resins together with various chemical hardeners. There is a problem associated with the development of skin sensitivity to these various materials.

The greatest single difference between the responsibilities of the generating board and the area boards is the latter's close and direct contact with consumers and the problems associated with working in domestic premises. This has been particularly highlighted of late by the requirements of the Asbestos Licencing Regulations in relation to inspection and maintenance of electric storage radiators. These are extremely heavy pieces of equipment consisting of a core of fire bricks interspersed with electric wiring and enclosed in heat insulating materials. Many of the earlier designs contain asbestos as the thermal insulant thus causing problems during dismantling prior to removal.

OVERSEAS CONSULTANCY SERVICE

The Industry owns and operates an overseas consultancy service (British Electricity International), staff from all electricity boards including Scotland are seconded on a voluntary basis to work for BEI. This may require brief visits abroad to advise in a consultant capacity or it may require protracted employment overseas. Such absences requiring family transfer with all it entails for the provision of adequate family health care in addition to ensuring proper occupational health facilities.

INTERNATIONAL ORGANISATIONS

There are various international organisations which have been set up to exchange views and experiences in electricity generation and transmission. These organisations provide for broad research programmes to be instituted which reduce reduplication of effort, there are also many collaborative research programmes. The main organisations involved are the following:

UNIPEDE — Union Internationale des Producteurs et Distributeurs D'Energie Electrique;
CIGRE — Conference Internationale des Grands Reseauz Electriques;
EPRI — Electrical Power Research Institute.

CONCLUSIONS

The industry will continue to expand as electricity features more and more as the main source of power for industry and the continuing expansion of domestic needs. The basic problem facing the industry is the decision regarding the basic future source of fuel. Nuclear power presents the most likely practical source for the future electrical needs of society; various studies have been undertaken to develop alternative sources of energy, for example, windpower, wave power and solar heat. Only the latter has been shown to be a possible method for production of domestic electricity in countries with long hours of sunshine but it is unlikely ever to be more than a supplementary source of power. Deposits of coal, oil and gas are limited. The development of nuclear power would seem at the present time to be the only practical answer to the profligate waste of valuable resources, the fossil fuels currently used in large quantities contain valuable chemical substances.

In most of the developed countries the hydroelectric capability has been exhausted. Pump storage schemes allow for utilising a higher load factor for existing power plant. Electricity cannot be stored, but water as a source of power can be used in a recycling process. Two lakes at different heights are necessary, water being used to generate electricity when it is required at peak periods, and during the silent hours the water is pumped back to the upper lake. Suitable sites are hard to find but the CEGB has one very large 2000 MW installation in North Wales which has proved most successful. Research efforts will undoubtably continue to be made both to develop alternative sources of power and to improve the efficiency of existing generating plant.

REFERENCES

Adams C A, Bonnell J A 1962 Administration of stable iodide as a means of reducing thyroid irradiation resulting from inhalation of radioactive iodine. Health Physics 7: 127–149

Alkali Etc and Works 1974 IIIth Annual Report of the Chief Alkali Inspector. Appendix vii p 86. HMSO, London

Blank L W 1985 A new type of forest decline in Germany. Nature 314: 311

Bonnell J A 1983 Leukaemia in electrical workers. Lancet i: 1168

Bonnell J A 1982 Effects of electric fields near power-transmission plant. Journal of the Royal Society of Medicine 75: 933–941

Bonnell J A, Bowker J R, Browne R E, Erskine J F, Fernandez R H P, Massey P M O 1975 A review of the control of asbestos processes in the Central Electricity Generating Board. XIII International Congress on Occupational Health, Brighton

Bonnell J A, Reid A R, Rippon J E 1986 In press

Bonnell J A, Rippon J E 1985 Legionella in Power Station Cooling Waters. Lancet ii: 327

Bonnell J A, Schilling C J, Massey P M O 1980 Clinical and experimental studies of the effects of pulverised fuel ash — a review. Annals of Occupational Hygiene 23: 159–164

Broadbent D, Broadbent M, Male J, Jones M R C 1985 Health of workers exposed to electric fields. British Journal of Industrial Medicine 42: 75

Butrous G S, Male J C, Webber R S, Barton D G, Meldrum S N, Bonnell J A, Camm J 1983 The effect of power-frequency high-intensity electric fields on implanted cardiac pacemakers. Pace 6: 1282–1292

Chadwick C, French M J 1976 The impact of a power station on Gwynedd. Planning Department, Gwynedd

Chrisp C E, Fischer G L, Lammert J E 1978 Mutagenicity of filtrates from respirable coal fly ash. Science 199: 73–75

Coleman M, Bell J, Skeet R 1983 Leukaemia incidence in electrical workers. Lancet 8: 982

Fisher G L, Chrisp C E, Raabe O G 1979 Physical factors affecting the mutagenicity of fly ash from a coal fired power plant. Science 204: 879–881

Howells G D, Kallend A S 1983 Acid rain — the CEGB view. Chemistry in Britain 19: 504

Knave B, Gamberale F, Bergstrom S, Birke E, Iregren A, Kolmodin-Hedman, Wennberg A 1979 Long-term exposure to electric fields — a cross sectional epidemiologic investigation of occupationally exposed workers in high-voltage substations. Electra. 65: 41–54

Korobkova V P, Morozov Yu A, Stolarov M S, Yakub Yu A 1972 Influence of the electric field in 500 and 750 kV switchyards on maintenance staff and means for its protection. CIGRE 23–26

Lewis P M 1985 The economic impact of the closure, without replacement of Trawsfynydd power station. Institute of Economic Research University College of North Wales, Bangor

Looms J S T 1971 Research into live working for UK transmission lines. CERL report RD/L/R — 1717

McDowall M E 1983 Leukaemia mortality in electrical workers in England and Wales. Lancet i: 246

McDowall M E 1986 British Journal of Cancer 53: 271–279

Milham S Jr 1982 Mortality from leukaemia in workers exposed to electrical and magnetic fields. New England Journal of Medicine 307: 249

Milham S 1985 Leukaemia mortality in amateur radio operators. Lancet i: 12

Myers A, Cartwright R A, Bonnell J A, Male J C, Cartwright S C 1985 Overhead power lines and childhood cancer. Int. Conf. on Electric and Magnetic fields in Medicine and Biology, IEE, London, 4–5 Dec

Programme of Nuclear Power 1955 UK Govt. White Paper, Cmnd. 9389

Pearce N E, Sheppard R A, Howard J K, Fraser J, Lilley B M 1985 Leukaemia in electrical workers in New Zealand. Lancet i: 811

Reid A R 1984 A review of toxicological testing of pulverised fuel ash. Ash Technology '84

Stollery B 1986 Effects of 50 Hz electric currents on psychological functioning: Part 1 Mood and verbal reasoning skills. British Journal of Industrial Medicine 43: 339–349
Strumza M V 1970 Influence sur las sante humaine de la proximite des conducteurs de l'electricite a haute tension. Archives Maladies Professionnelles 31: 269–276
Wertheimer N, Leeper E 1979 Electrical wiring configurations and childhood cancer. American Journal of Epidemiology 109: 273
WHO 1984 Extremely Low Frequency (ELF) Fields, Enviornmental Health Criteria No. 35, Geneva
Wright W E, Peter J M, Mack T M 1982 Leukaemia in workers exposed to electrical and magnetic fields. Lancet ii: 1160–1161

4. Modern farming

D. M. Smith

INTRODUCTION

Agriculture is the oldest and most basic industry of all and is essential for providing food and clothing worldwide. It is also a way of life in which the occupational aspects are difficult to separate from the effects of living in a rural environment. Members of the family often take part in the farming activities and it differs from other occupations in having active participants who are both younger and older than the normal working population.

It is not, however, the ancient and unchanging activity which may be imagined by town dwellers. In much of the developed world it is a modern, capital-intensive industry, whose labour requirements have decreased markedly over the last 20 years and which changes rapidly to meet changes in demand for different types of product. In developing countries, farming may cover the full spectrum from subsistence farming to intensive production of cash crops by multi-national companies.

The general adoption of intensive systems of farming has led to a number of changes, some desirable and others less so. Intensive cultivation of land demands intensive application of fertilisers, weed killers and pesticides. Intensive animal rearing brings animals into close contact with large numbers of their fellows and this increases the likelihood of infection among the herd. This leads to the use of prophylactic antibiotics and immunisation which may themselves exert effects which require further methods to control them. Such farming methods also lead to the necessity of food additives and for disposal systems for slurry. For example, intensive poultry rearing may lead to infestation of the houses with a variety of mites and other pests which require control whilst pig rearing in confinement buildings leads to a number of problems, including a build up of carbon monoxide (CO) and ammonia (NH_3) as well as dust.

Some advances, however, are likely to prove beneficial. The green revolution which has resulted in some Asian countries now being self-sufficient in food, would have been impossible without research into resistant strains of plants and animals and the application of fertilisers and pesticides. The development of ultra low-volume and controlled droplet methods of application of pesticides may lead to a decrease in environmental pollution since these methods require a smaller quantity of chemical. However, this may be offset by more danger to the operative from the concentrated solution used. Different methods of silage manufacture are currently being investigated. For example, the hazards of adding formic acid and sulphuric acid to vegetable matter to make silage may be avoided by methods which simply contain the silage under plastic or by those which use bacterial and enzyme inoculants. Many fruit growers are now turning away from powerful organophosphorus chemicals towards using less hazardous materials in a controlled fashion at appropriate times in a planned controlled system. Also in fruit growing, a commonly occurring fungus, *Trichomona viride* is being used to prevent other fungus infections of some fruit trees and evaluation before its use has so far shown no human ill-effect.

The fall in numbers engaged in agriculture is accompanied by a change in the composition of the labour force. The 1985 figures for the UK, based on a Ministry of Agriculture, Fisheries and Food (MAFF) census, are shown in Table 4.1.

Table 4.1 Numbers of workers engaged in Agriculture in 1985

Regular hired workers	162 500
Seasonal/casual	98 100
Family workers	55 390
Salaried managers	8 270
Total employed	324 260
Employers and self employed	368 900
Total labour force	693 160

Since less than half of those engaged in agriculture in the UK are employed persons, any figures for prescribed industrial diseases must not be taken as representing an accurate picture of the toll of these diseases on the farming community. Those engaged in agriculture are commonly stoical about health complaints and regard them as part of the job. It may be very inconvenient to take time off when animals need to be fed every day or a crop must be harvested. These factors combine to make an accurate estimate of occupational disease in this sector difficult.

Agricultural practices vary greatly in different countries. The UK's entry into the EEC has meant a change in agricultural policy. Before 1973, food imports accounted for 22% of Britain's import bill. The figure is now about 12%. In 1982 Britain became a net exporter of grain for the first time since the repeal of the Corn Laws in 1854. The farming industry in the UK is recognised to be one of the most efficient in the world.

In some Common Market countries, however, mechanisation is at a lower level than in Britain. In most, the major part of farming is done in small self-sufficient units. North American farming is very different with a labour force augmented by migratory workers who are often poorly housed and educated and who take the greatest risk in farming. In the UK, the reported cases of occupational disease are lower than in other parts of industry, averaging about 200 cases per year in a population of 340 000 employed persons. In some states of the USA, however, agriculture has the highest occupational disease rate of any industry. This may be accounted for in part by difference in reporting, but much of it is probably real. The Bureau of Occupational Health of the State of California (undated) states that most of the occupational diseases and injuries occur in farm labourers and that this group needs the most protection from work hazards, for they are least able to protect themselves. Because of their inadequate education, language problems, migratory status, substandard health and poor hygiene, they are the least likely of any group of workers to be able to work safely with farm chemicals.

In the Third World, modern pesticides have had a considerable impact on farming practice and the same consideration would apply to many agricultural workers there. Chemicals are intensively marketed and standards of personal protection are low for reasons not only of cost, but because it is extremely uncomfortable to wear protective clothing in hot climates and it may cause heat stress (Jeyaratnam, 1985).

Interest in agricultural health has been growing in recent years, as is evidenced by the number of publications in medical journals. The Ninth International Congress of Agricultural Medicine and Rural Health was held in New Zealand in 1984 (Proceedings of the 9th International Congress on Agricultural Medicine and Rural Health; Höglund, 1984) and the proceedings include many papers on pesticides, respiratory hazards and other subjects of interest to occupational health practitioners. The largest number of papers in a literature search referred to respiratory disease.

There were others on pesticides and allergies, including skin problems. Very few mentioned any form of occupational hygiene applied at the workplace; those that did were from the USA, Sweden and Finland. Occupational health services are also lacking because of the nature of the industry, but papers from the Swedish, Finnish and New Zealand representatives to the Symposium describe different ways of providing occupational health services in agriculture. Although accidents are very much more common than occupational disease, few papers have been written on accidents and their prevention, although there are some medical papers descriptive of particular types of accidents.

NEW FARMING METHODS — HAZARDS

The reduction of the labour force and the increase in mechanisation has meant that each individual working on the land has to be able to do a number of jobs, and this requires a degree of adaptability and literacy which has not always been present in the past. Mechanisation has led to an increase in noise-producing machinery on the farm and legislation in Britain has required the provision of quiet cabs on all new tractors since 1982. In these cabs noise levels should not exceed 90 dBA. Rural workers have also been educated to wear hearing protection. Furthermore, large harvesting machines produce a great deal of dust which may lead to respiratory disease of different sorts.

Animal rearing

The mass housing of animals leads to increased risks from dust and toxic gases, from zoonoses and from accidental self-injection and from skin conditions. The need to dispose of slurry in particular leads to the hazards of toxic gases and of infection. Noise levels of up to 102 dBA have been measured by HSE in intensive pig-rearing houses, making hearing protection necessary. Attention to feeding methods may cut down the squealing, but there will always be some some squealing! For example, piglets also squeal when handled for any intervention such as injection.

Pesticide application

There has been an increase in aerial application, both from fixed wing aircraft and helicopters. The operators are at risk, both from the hazards of low flying in dangerous terrain, beset with trees and low wires, and from contamination of the vehicle with chemicals, particularly organophosphorus compounds. Great care is needed in loading and in maintenance of machines if contamination is to be avoided. Considerable worries are expressed in rural communities about the level of air pollution with pesticides by aerial spraying, but few cases of illness resulting from accidental overspraying are confirmed.

Other hazards

Older hazards which persist include tenosynovitis in such jobs as sprout picking and onion tying. Exposure to heat and cold are frequent and work in hot climates is discussed in Chapter 11. The wind-chill factor on some hill shepherds may be considerable and adequate clothing and working methods are essential. An all-terrain powered tricycle vehicle has recently been introduced, and may be used by shepherds, but most still walk over the hills.

Identification of problems in zoonosis

Cohen (1978) reviews the risks of intensive farming. He gives a check list of factors which should be taken into consideration in planning, and recommends action to be taken. This is an excellent practical approach.

PESTICIDES

Large amounts of chemicals are used each year. Pesticides accounted for £330 million in the UK in 1984; insecticides £31.6 million, herbicides £172.4 million, fungicides £90.9 million, seed treatment £15.1 million and others £20.3 million.

The estimation of toxicity must be based not only on LD_{50}, but on ease of absorption, and therefore on the type of formulation and the method of application. Many pesticides are absorbed through intact skin as well as by inhalation. Formulations include powders, wettable powders, emulsions, solvent suspensions, solutions and granules.

New methods of application involve small drops (controlled droplet application, ultra low volume). These are produced from a spinning disc. The object is to get the pesticide on to the target accurately, and to use less pesticide. Although less pesticide is deposited in the environment, the active chemical is much more concentrated than in conventional spraying. Any spray deposited on the skin or inhaled is therefore likely to give rise to significant absorption and possible poisoning (Smith, 1977).

The concentration of chemicals used in aerial spraying is also higher than in tractor spraying, e.g. for aerial spraying of demeton-s-methyl (an organophosphorus insecticide) on cereals the recommended concentration is 1/50 as against 1/350 for low volume tractor spraying. (1/2500 for high volume tractor spraying may be used for other crops, but is not used on cereals.)

Synthetic pyrethroids

Many new formulations appear each year, but few new types of pesticide have been produced. The first synthetic pyrethroid, barthrin, was patented in 1959 and permethrin and decamethrin were described in 1973 and 1974; their use has become more common since then. These are neurotoxic, but of low human toxicity.

The active pyrethroids are esters of cyclopropane carboxylic acids with alkenylmethyl cyclopentenilone alcohols. They fall into two groups; alpha-cyano derivatives and non-cyano pyrethroids. Human toxicity is minimal and they are not teratogenic or mutagenic. Symptoms of burning and tingling of the skin after human skin contact have been described, and this was investigated by Flannigan et al (1985). They found that the paresthesia were much more pronounced with the cyano derivatives than with the non-cyano compounds.

It is thought that the biological action of the pyrethroids is a rapid transient change in the permeability of the nerve membrane to sodium and potassium ions, and that the cyano compounds are more active in this respect. The use of vitamin E ointment as a prophylactic application to exposed skin is advocated by Flannigan et al (1985). They argue that the pyrethroids have been shown to be of low toxicity and if the cutaneous sensation was unacceptable, more toxic products might be used. However, the sensation is a pharmacological effect of the chemical and therefore a warning of exposure and absorption. It is not acceptable merely to treat such an effect; attention should rather be directed to reducing exposure and absorption.

Chlorinated phenoxy acids

This subject was fully reviewed by Axelson (1984). Since then Balarajan & Acheson (1984) used the National Cancer Register to investigate the relative risk of soft tissue sarcomas amongst farmers, agricultural workers and related occupational groups. They found that the relative risk for the group as a whole was 1.5 which is not statistically significant, but one of the 4 sub-groups, farmers, farm managers and market gardeners, experienced a relative risk of 1.7 which just reaches significance at the 5% level. There was no excess risk in the other three sub-groups, agricultural workers, gardeners and groundsmen, and foresters and woodmen. This would not support an association with phenoxy herbicides, although no attempt was made in the study to determine such exposure in cases or controls.

Studies which are able to quantify the exposure of a population may give information which is so far lacking.

Health surveillance

Biological monitoring is effective in relatively few groups of pesticides.

Cholinesterase estimation

Measurement of cholinesterase estimation (cHE) is still the main biological method used for organophosphorus exposure. In the investigation of poisoning, it is preferable to measure plasma and red blood cell cholinesterase, but for biological monitoring, plasma is sufficient. In the UK, the Health and Safety Executive (HSE) recommends biological monitoring using plasma cholinesterase before exposure and at intervals depending on the nature of the exposure during working. In the case of accidental over-exposure, both red cell and plasma cholinesterase should be measured (HSE 1985d). Field methods of monitoring cHE in whole blood by a colorimetric method have been evaluated and found to be suitable for routine monitoring in the field (Miller & Shah, 1982). This will be particularly useful in developing countries where laboratory facilities may be absent or distant.

Electromyographic measurements

Electromyographic (EMG) measurements have been used (Roberts, 1977) in pesticide production, but are not suitable for general biological monitoring of field workers.

Kelman (1984) reviewed their use and concluded that while EMG might be of some value for group studies, it is of little use for individual surveillance.

Laboratory methods for the detection of various pesticides in blood and urine have been developed for use in cases of suspected poisoning, but are not, in general, used for biological monitoring.

Guidance for medical practitioners in the UK is available in a booklet from the Department of Health and Social Security (1983) entitled Pesticide Poisoning. This classifies pesticides according to their medical effects, listing trade names as well as active ingredients. This is essential for doctors faced with suspected poisoning cases. The only weakness of this is that the trade names are apt to change every year and new names are added. It will, therefore, require regular updating.

Packaging

Many chemical containers are difficult to open and pour without contaminating the operative, especially when wearing protective gloves. A report from the Swedish Packaging Research Institute (1983) details some of the problems and solutions.

A new scheme is in operation between UK manufacturers and a working party of the Agriculture Industry Advisory Committee (AIAC) by which complaints are referred and improvements made. (AIAC is an advisory committee of the Health and Safety Commission, chaired and staffed by the Health and Safety Executive and having members from both sides of industry.)

Another common problem is leaking knapsack spray containers. This is also under consideration by the working party. HM Agricultural Inspectorate has carried out an investigation of manufacturers and importers (results are not yet available).

Personal protective equipment

Advances have been made in providing air filtration for tractor cabs. This reduces the necessity for wearing protective clothing or respiratory protective equipment in some dusty atmospheres and in some cases exemptions from such provisions of the regulations may be granted.

A study carried out by the British Agrochemical Association in 1983 in conjunction with the Robens Institute of Industrial Hygiene Services, University of Surrey, showed an acceptable degree of protection by protective clothing recommended in the UK. The World Health Organisation in 1982 produced a protocol for assessment of exposure to pesticides, which was used in this study.

INFECTIONS

Brucellosis

A brucellosis testing and slaughter programme in Britain has led to the virtual eradication of this infection in cattle. The number of human infections contracted annually in Britain has fallen from 186 in 1976 to 8 in 1983. There were 9 cases in 1984 and 4 in 1985. (Public Health Laboratory Services, Communicable Disease Centre—Personal Communication.)

Leptospirosis

There has been a change in the type of leptospirosis infection reported in Britain and in other countries. Until the 1970s, most human cases were of *Leptospirosis icterohaemorrhagiae* acquired from rats or from water contaminated by rat urine. Most cases were in sewer workers and miners. Rodent control and the increased amount of detergent in sewage have lessened the hazards to miners and sewer workers. Farmers and farm workers are now the largest occupational group affected by *L. icterohaemorrhagiae*. In 1969 Sakula & Moore reported a case of human *Leptospirosis hebdomadis* and since then further reports have appeared (Crawford & Miles, 1980; Wilson & Wetson, 1981; English et al, 1983). More cases of *L. hebdomadis* in cattle have been diagnosed. Most of these are of the sero group hardjo and most of the cases of human *L. hebdomadis* are of this group; nearly all of these had been associated with cattle, either as farm workers or other cattle workers. In the 5 years from 1978 to 1983, there were 50 cases of *L. icterohaemorragiae*, 1 case of *Leptospirosis canicola* and 119 cases of *L. hardjo* in farmers. The latter illness is, in general, milder than that caused by *L. icterohaemorrhagiae*, but some of the cases had meningitis, many had meningism and severe headache and some had a mild influenza-like illness. Antibody titres may be detected in the blood. The increased frequency of diagnosis of *L. hardjo* infection in cattle may be because more attention is being directed to this as a cause of abortion and ill health in cattle, now that brucellosis is eradicated. Reasons for its spread to humans may lie in methods of dairy farming. Many modern milking parlours are designed so that the milker is in a well and is thus exposed to massive contamination with cow urine (Fig. 4.1). However, in an unpublished HSE survey in one county of England (Hodnett, 1985) most cases of human infection occurred in farms with old-fashioned milking sheds. This may be because these are small units where husbandry is less advanced and there is more likelihood of cattle infection and of passing the infection to the milkers.

In New Zealand, leptospirosis is the main dairying zoonosis. *L. hardjo* accounted for two-thirds of human cases; *L. pomona* is also present (Bettelheim, 1984). In New Zealand, human incidence of leptospirosis is related to high early spring rainfall and this may mean that the organism is easily passed in wet conditions, or may be related to its prevalence in small mammals. In Israel also, there has been a change in the pattern of infection. There were no reported cases of *L. hardjo* infection until 1973 when five cases were reported. Now most cases of leptospirosis are of this sero type (Sehnberg et al, 1982). All the patients with *L. hardjo* were dairy workers and cattle are the principal source of *L. hardjo* infection for man.

Vaccines for cattle immunisation are being produced and farmers are encouraged to free their herds from this infection.

Streptococcus suis

Streptococcus suis type II causes meningitis in piglets, mainly young piglets. Human cases have occurred mainly in butchers and abattoir workers and it is now a prescribed industrial disease in Britain. The pyramidal way the British pig industry is structured makes its spread among piglets likely, as survivors may be sold to other farms. In humans it causes meningitis, often resulting in lesions of the eighth cranial nerve causing permanent deafness and affecting the sense of balance. Such a case has been

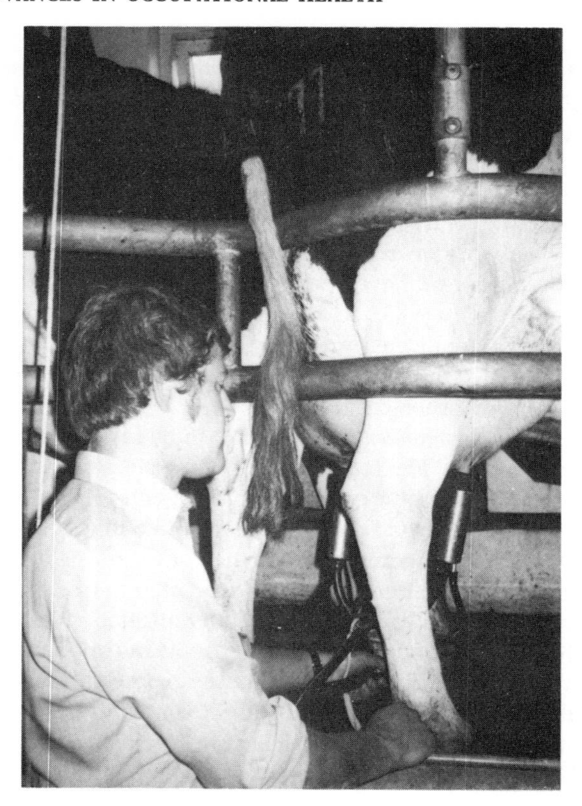

Fig. 4.1 Herring bone milking parlour.

investigated by the author. In 1981 10 cases had been reported in the UK and 40 world wide (Twort, 1981).

Chlamydia psittaci
Commercially reared poultry caused outbreaks of human psittacosis in 1979 and 1980 in the UK (Palmer et al, 1981; Andrews et al, 1981). *Chlamydia psittaci* also causes enzootic abortion in ewes, and cases of human illness from this source have been reported (Beer et al, 1982). Non-avian sources were reviewed in a leading article in the Lancet (Editorial 1984). Women often help with lambing because of their small hands; the placenta is heavily infected and this may be a source of infection from the ewe; the selection of this site caused severe illness in pregnancy in the two cases (one fatal) described by Beer et al (1982). In France and Czechoslovakia, human illness associated with bovine chlamydions has been described.

RESPIRATORY DISEASE

There has been a considerable interest in recent years in the allergic respiratory diseases and their investigation. Most of the advances are in immunology. Enzyme-linked immunosorbent assay (ELISA) has been found to be a useful method of investigation in farmer's lung disease (Bamdad, 1980). A comprehensive review of respiratory health risks in farmers was published by Cockcroft & Dosman (1981). The inhalable substances to which farmers may be exposed include organic antigens such as pollen, fungal spore, animal danders, grain dust and mites; synthetic chemicals including insecticides, herbicides and fertilizers, and toxic gases such as nitrous oxide, hydrogen sulphide, methane, carbon monoxide and ammonia. Respiratory manifestations are varied. Modern intensive methods may expose farmers to respiratory allergens in, for example, swine confinement buildings and poultry houses (Donham et al, 1977, 1984a, 1984b; Jones et al, 1984).

Various respiratory symptoms are associated with silos and although silo filler's disease, due to acute intoxication with oxides of nitrogen, is well known, silo emptiers' disease, due to overwhelming concentration of fungal spores, has only recently been described. Pladson (1984) gives a good description of the three separate types of disease which may be encountered in silo workers:

silo fillers' disease;
farmer's lung;
pulmonary mycotoxicosis (silo emptiers' disease).

Silo fillers' disease

Silo fillers' disease results from entering a silo which has been recently filled; oxides of nitrogen are formed within hours. They cause acute irritation and may cause pulmonary oedema in the acute phase; illness may recur after 2 to 5 weeks even after mild initial symptoms. These symptoms include dyspnoea, cough with sputum, fatigue and even bronchilitis obliterans.

Farmer's lung disease

Farmer's lung disease and allergic alveolitis may occur after exposure to mouldy grain in silos as well as in barns and stacks.

Pulmonary mycotoxicosis

Pulmonary mycotoxicosis results from disturbing the top, mouldy layer in the silo. Organic dust is inhaled in massive amounts and the illness may last several days to weeks. Symptoms include upper respiratory irritation, dyspnoea, malaise and fever. In both this and in farmer's lung, immune complexes are involved. Most subjects are not atopic. Parkes (1982) emphasises the dangers of progressive lung disease in cases where there is no acute episode; repeated exposure may then continue. He considers that it is not proved that massive exposure to mycotoxins gives rise to a separate pathological entity.

Differential diagnosis is aided by chest x-ray and investigation for specific antibodies. This is important, since sufferers from pulmonary mycotoxicosis are not sensitised as are farmer's lung victims. Indeed, both categories should receive appropriate advice so that they may avoid further harmful exposure. If entry into silos cannot be avoided by a reconsideration of the work, appropriate respiratory protection must be provided.

Other causes of agriculturally induced lung disease may be cited: Cuthbert (1981) and Cuthbert et al (1980, 1984) investigated barn allergy in Scottish farmers and found it to be related to the mite content of their hay. The mites *Glycophagus destructor* and *Tyrophagus longior* gave strongly positive skin prick results in farmers. Another fungal cause of allergic alveolitis, *Botrytis cinerea*, a mould of grapes, is described as spaetese lung by Kummer et al (1984).

In 1982, the European Economic Community (EEC) set up a working party into the epidemiology of respiratory disease among agricultural workers. A comparison was made of mortality data from each of the eight member countries. There were differences in reporting and classifying respiratory disease, but comparisons were possible. The figures from France, and for England and Wales, suggest an excess of deaths from respiratory disease, mainly pneumonia and influenza, among agricultural workers (Heller & Kelson, 1982).

A study from Poland (Dutkiewicz, 1978a) found large quantities of air-borne bacteria in grain storing and processing plants and in animal houses. On the basis of this, he concluded that high exposure to dust-borne bacteria constitutes a hazard to agricultural workers. However, he does not analyse working practices or individual exposure and does not show any evidence of increased ill-health in these workers. In a second paper (Dutkiewicz, 1978b) he showed that many grain workers had positive reactions to *Erwinia herbicola* and that those with respiratory symptoms reacted more frequently than those without. This may prove to be an important antigen, but is not often found in association with British or Canadian grain.

A comprehensive account of the microflora of grain dusts, together with sampling methods is given by Lacey (1980).

SKIN DISEASE

A number of papers have been published on various plant allergies including phototoxic dermatitis, the occurrence of which is already well known (Austad & Kavli, 1983; Smith, 1985). Cotterill (1981) described a contact dermatitis following exposure to tetrachloronitrobenzene. This is used in the form of granules to prevent sprouting in stored potatoes and is in common use in the UK. Pasricha & Gupta (1983) reported the unusual occurrence of contact dermatitis to a fertilizer, calcium ammonium nitrate.

Farmers now obtain and administer many veterinary medications themselves, and cases of contact dermatitis from antibiotics and other medications are likely to increase. Peachey (1981) reviewed skin hazards in farming. The plants and pesticides likely to cause dermatitis are also listed by Cronin (1980) whilst Rycroft (1977) reviewed cases of dermatitis due to organophosphorus chemicals.

ACCIDENTS

Accidents are relatively common in farming. Accident figures for 1984 in Britain, published by the Health and Safety Executive (1985a) show that there were 67 fatal accidents in agriculture. Twenty-six of those killed were employees; the other 41 included farmers, family members and children. The causes of the fatal accidents included: self-propelled machinery 16; falls and falling objects 8; other machines 4; asphyxiation 1, gases 4; and drownings 5.

In the non-fatal category figures are not available for 1984, but in 1980 machinery accidents totalled 1057, of which tractors were involved in 268. There were 856 falls and 30 cases of poisoning (HSE, 1982).

In addition, farmers and farmworkers are frequently victims of road traffic accidents associated with living in a rural area. The sharing of the work environment with the family and with the public means, unfortunately, that children are also victims of accidents. Recent figures show a decline in this, but there are still appreciable numbers. More accidents (268 in 1980) occur with tractors than with any other machine. The introduction of safety cabs has led to a considerable decrease in fatal and serious injuries from this cause. Accidents with machines such as hay balers can cause severe injury, including avulsion of limbs. In many cases it is impossible to guard the intake adequately because of the nature of the work. Power take-off shafts also cause revolving machinery accidents; although guarding is mandatory, guards are often broken or inefficient.

Animals

Accidents with animals have shown in recent research to be more frequent with older people. This may reflect a decline in speed of reaction or muscular strength. People who have been used to dealing with animals all their life may not realise that they themselves are not as strong and as fast as they used to be (HSE, 1985b).

Enclosed spaces

In the last 5 years in the UK there have been 19 deaths in tower silos caused by asphyxiation in oxygen deficient atmospheres, or by gassing. Recent advice from HSE strongly advises farmers not to enter such silos unless wearing breathing apparatus and in the presence of an adequate rescue team (HSE, 1985c). It is extremely difficult to lift an inert body out of such a silo, particularly through the small access hole commonly provided, even if a safety harness is worn.

Accidents have also occurred in underground slurry pits where complex mixtures of gases may be present. These may include hydrogen sulphide (H_2S), carbon dioxide (CO_2), carbon monoxide (CO) and oxides of nitrogen. Unconsciousness can be very rapid. There is a USA report of a fatal illness following rescue from such a pit (Hagley & South, 1983). A 16-year-old boy was found unconscious at the bottom of a mobile tank used for spreading liquid manure. He was apnoeic, but after mouth to mouth ventilation he began breathing spontaneously and was transferred to hospital. High concentrations of hydrogen sulphide were subsequently found in the tank. The boy died from cerebral damage 5 days after the accident. Between 1975 and 1982 14 deaths in the USA were attributed to the inhalation of liquid manure gas (Donham et al, 1982). In one of the cases reported, a farmer's wife ran to the road and stopped two passing motorists who then entered the tank in a rescue attempt and died. It is common knowledge in chemical factories that H_2S rapidly causes unconsciousness, but this is not so in farming circles; education about this hazard is urgently required. Prevention requires planning to prevent the need for entry as far as possible, provision of self-contained breathing apparatus if entry is unavoidable, and provision of adequate rescue facilities.

Aerial spraying

Accidents are frequent in aerial spraying because of the inherent dangers in flying low, often in terrain with obstructions such as trees, hedges and electric and telephone lines. The difficulties are added to by contamination of the aircraft by pesticides. Strategies for identifying and eliminating the various hazards are described in an article by Richter et al (1981). They point to the dangers of fixed obstacles, low altitude runs and heavy work schedules and the exposure to noise and vibration,

G Forces, heat stress and dehydration as well as pesticides. The addition of these exposures may well produce a decrease in pilot performance, alertness and skill. Prevention strategies described include cooling and filtering of cockpit air to prevent pesticide exposure. Design changes to reduce injuries in the case of accident are also described. Preventive measures also include attention to dangerous ground obstacles and consideration of the action to be taken in emergencies.

Chain-saws

A survey conducted by Her Majesty's Agricultural Inspectorate showed that most chain-saw injuries are to the left hand and left leg and this finding was followed by a campaign for training and protective clothing. Though the results might seem obvious, the graphic demonstration of the figures has increased awareness of the hazard, and use of preventive strategies. Much better protective gloves, overalls, trousers and leggings are now available, incorporating plastics materials such as Kevlar.

Animal injections

Poultry, pigs, cattle and sheep reared intensively are given injections of various vaccines routinely during rearing and may require injections of antibiotics. A multiple dose system is in common use, consisting of a reservoir, tubing and a gun activated by a trigger mechanism. The risk of self-injection varies with the method used; in poultry breeding for egg production injections are safely given into the breast; in those for consumption the usual method is for a catcher to hold four/five birds by the legs in each hand and the injector to feel among the feathers with his left hand to locate the thigh to be injected. Huge numbers are injected by this means in a short time. (Fig. 4.2). Some vaccines such as those used for Newcastle disease (fowl-pest) are

Fig. 4.2 Multidose chicken injection.

oily. If this is injected under pressure into the terminal phalanx of a finger, it may cause tissue necrosis; prompt preventive treatment is necessary and consists of incision and removal of the oily material by washing with detergent (Hardy, 1981 and personal communication). Several cases have been seen by the author; one woman has a permanent deformity of a terminal phalanx following extrusion of a bony sequestrum; treatment had been delayed for 5 days in spite of her complaints of increasing pain. One man had had extensive incisions extending up the tendon sheaths, such as are used for grease gun injuries. The much lower pressure of the vaccine gun renders this unnecessary. Others promptly treated have had no sequelae.

In a pig rearing unit with 20 sows, over 300 injections a year will be given. Adequate provision for restraint of animals will minimise injuries. A new self-cleaning multiuse needle (Sterimatic) has a guard which may decrease the likelihood of self-injection injuries. The author knows of no cases of systematic disease resulting from accidental self-injection, but anaphylaxis remains a possibility and calls into question the administration of veterinary products by people without professional training. HMAI is to conduct a survey in 1986 to determine the extent of the problem, as reporting has not been good so far.

In farm work there is always pressure to get a particular job completed because of the vagaries of the weather or the market, whilst long hours and fast work may lead to accidents. Procedures for dealing with emergencies are often not well thought out. In many cases farmers are unaware of where the nearest accident and emergency department is and have little provision for first aid on the premises (Bamford, 1985).

OCCUPATIONAL HEALTH AND SAFETY SERVICES IN AGRICULTURE

Because most farming units employ few workers, it has been very difficult to organise occupational health and safety services in agriculture and few countries have a comprehensive service. In the UK there is little provision of organised occupational health services to farms and farmers. Primary and treatment care is provided by rural general practitioners (GPs) and although advice on matters affecting health may be sought from the Employment Medical Advisory Service (EMAS), few farmers or GPs yet avail themselves of this. Companies manufacturing agricultural chemicals commonly have occupational medical advisers and these doctors may be prepared to give advice to users of chemicals, but do not, in general, conduct formal occupational health surveillance of pesticide users. Larger spray contractors do make arrangements for medical surveillance. Little use is made in UK agriculture of occupational hygiene services in measuring levels of dust or toxic materials at the workplace. Some agricultural suppliers are now selling a portable meter for toxic gases for use in silos and slurry pits, but this is not backed up by assessment by an occupational hygienist.

In other developed countries, different types of occupational health services have developed. In the Netherlands there is little organised occupational health attention to agriculture, but some local groups are being formed with a view to such provision. In Sweden, branch orientated occupational health is organised in some branches of industry such as the construction industry, forestry and transportation; agriculture has recently joined this group. Farm health units, each serving about 1000 farmers, have been set up, each of which employs a full-time occupational nurse, half-time physiotherapist, secretary and safety engineer and about a quarter of the time of an occupational physician. Regular health checks are carried out and the work environment is studied by farm visits and special projects (Höglund, 1984). Similar programmes are in progress in Norway and Finland. In Finland, an experimental survey was made of farm workers to determine the main medical problems: 3500 health examinations were done as well as 1300 farm visits. The main problems were found to be in heavy lifting and bad posture: 29% of those examined had low back pain; this percentage rose to 45% over the age of 50. Similar figures were found for pain in the arm and twice as many cases of musculoskeletal disorder were found in the farming population as in the reference population (Nuutinen, 1984). In France the medical service to farmers is organised mainly on an insurance basis with many individual health checks, but few work place visits; however, the French medical services recently gathered together some of its scattered statistics and has begun to be engaged in epidemiological studies (Dubrisay, 1984).

In the UK, a working party of the Agricultural Industry Advisory Committee is studying the problems of setting up occupational health and services. The Employment Medical Advisory Service has been involved in a number of studies, some of which have been published, and the Health and Safety Executive sponsors research. A recent Employment Medical Advisory Service study (Wells & Swift, 1985 unpublished) bears out the Finnish findings on the high incidence of musculoskeletal and respiratory disorders. One district health authority has studied agricultural accidents and ill health in its district (Rowe et al, 1980). It is likely that in the future much more use will be made of occupational health and hygiene services.

In the USA, practices vary from state to state: in California it is mandatory for all organophosphorous workers to be medically supervised, whilst in the Mid West systematic studies have been made on swine confinement buildings. The US National Centres for Disease Control have taken an interest in the use of agricultural chemicals and chemicals are registered with the Environment Protection Agency.

In the developing countries, some multi-national companies raising cash crops such as sugar, tea, coffee and pineapples have occupational health services and these may be extended to the rural population. In addition, some African countries are forming their own occupational health groups, sometimes doctor-based and sometimes nurse-based (Douglas & Selema, 1985 personal communication). There are considerable problems to be overcome in educating such a population to the hazards of pesticides, for example, when these have brought great benefits in the form of increased yield and in elimination of vector borne disease. The problems faced by such physicians include the shortage of library facilities so that they have difficulty in keeping up to date with modern developments. Deficiencies in hygiene and water supply, climate extremes and lack of education add to the difficulties in health education.

In the Eastern block, farms tend to be large collectives and the occupational health care follows the industrial model.

Australia faces considerable difficulty in collecting information in a large country in which many farms are remote from towns. So far, limited agricultural preventive medicine is practised, but planning for services in remote areas is being undertaken by several states and Federal Government initiatives have supported this (Ferguson, 1984).

The International Labour Organisation (ILO) in June 1985 adopted the Occupational Health Services Convention 1985. This defines the role and function of occupational health services and calls on all members of the ILO to formulate, implement and periodically review a coherent national policy on occupational health services and progressively develop occupational health services for all workers in consultation with organisations representing employers and workers. This may encourage governmental agencies to draw up plans for occupational health services.

LEGISLATION — UK

The earliest legislation in the UK to protect the health of agricultural workers dates only from 1952 when the Agriculture (Poisonous Substances) Act was passed. This was followed by the Safety, Health and Welfare Provisions Act in 1956 whose regulations included machinery and first aid regulations. These were administered by the Safety Inspectors of the Ministry of Agriculture, Fisheries and Food (MAFF). Regulations under these acts was passed to the Health and Safety Executive (HSE). The safety inspectors of MAFF acted as agents of HSE for this purpose until 1977 when the MAFF safety inspectors were transferred to HSE. The general provisions of the Health and Safety at Work etc Act 1974 also apply to those engaged in agriculture.

Poisonous substances in agriculture regulations

The Regulations now in force are the Poisonous Substances in Agriculture Regulations 1984, of the Health and Safety at Work etc Act. When using any chemical listed in the four parts of schedule 2 of the Regulations, specified precautions are required. These include provision of personal protective equipment, keeping records of exposure, notification of sickness and absence, training and supervision, provision of washing facilities and they also forbid the employment of persons under 18 with these chemicals.

Food and Environmental Protection Act 1985

The Pesticide Safety Precautions Scheme (PSPS), administered jointly by the MAFF and Department of Health and Social Security (DHSS) with the advice of the Independent Advisory Committee on Pesticides and other Toxic Substances, was a voluntary scheme in which each pesticide was submitted for appraisal before introduction to the market. The 1985 Food and Environmental Protection Act has made the PSPS a statutory requirement and in addition has reinforced statutory control over the use of pesticides. It is intended that regulations and codes of practice will be made. These are likely to reiterate the provisions of Section 2 of the HASAW Act in requiring adequate training and supervision, protective clothing and record keeping. A clause is included empowering ministers to make information available to the general public, including safety data submitted in support of an application for approval of a pesticide. The mechanism is under discussion but one proposal is to provide on demand an evaluation based on the documentation considered by the Advisory Committee on Pesticides and Other Chemicals.

A certificate of training in pesticide application will be required under this Act.

There is at present no statutory requirement for health surveillance other than the maintenance of exposure records. However, some Guidance Notes from HSE, e.g. MS 17 (1980) on organophosphorus chemicals, advise medical surveillance and biological monitoring for users of certain chemicals.

Proposed legislation

The subject of a consultative document for proposed regulations on the Control of Substances Hazardous to Health (COSHH) would require the keeping of occupational health records for anyone exposed to any chemical which is in one of the categories specified, that is: very toxic, toxic, harmful, corrosive or irritant. These include carcinogenic, teratogenic and mutagenic chemicals. The minimum requirement is for the maintenance of a record of identification and exposure on each worker and in some cases full medical and biological surveillances would be required. New provisions in these regulations would mean that regular users of organophosphorus chemicals and users of mercury would require medical surveillance by EMAS or an appointed doctor.

Civil Aviation Authority

The Civil Aviation Authority (CAA) licenses aerial spray operatives. To obtain a licence the operative must provide a manual detailing his precautions and arrangements for such mattters as personal protective equipment, health surveillance and provision for emergencies, as well as training and supervision. There are guidelines about spray drifts and a prohibition on flying directly over dwelling houses and their gardens which are occupied, at a height below 200 feet. They are required to include detailed instructions for restriction of the chemicals to the target areas. Enforcement and supervision are, however, difficult, not least because there are only six CAA inspectors for the UK.

EEC Directive (80/1107 EEC)

EEC Directive 80/1107 (1980) calls on governments to adopt measures for the protection of workers against risks at work from exposure to chemical, physical and biological agents considered harmful. These include preventive strategies and health surveillance.

CONCLUSIONS

Modern farming exposes the farm worker to physical, chemical and biological hazards. At one extreme, agribusiness has absentee landlords and factory methods with many employees, while at the other end the small farmer has no employees and has to do all sorts of tasks himself.

Standardised mortality ratios (SMRs) for all diseases are low, but certain categories show elevations deserving further study. No satisfactory explanation has been given for the high SMR (135) from accident, suicide and violence (Office of Population Censuses and Surveys, 1978). In the UK, deaths and illness from occupational chemical usage remain low, but world wide there is an immense problem (Jeyaratnam, 1985).

Accidental mortality and morbidity have been greatly reduced in the UK by legislation and cooperation from the industry in the introduction of safer tractors, power take off guards, chain-saw guards and personal protective equipment.

The virtual elimination of brucellosis is a major step forward. This disease has caused a great deal of misery, from joint pains, and from depression and mood changes. The debility can go on for years, affecting every aspect of working life.

The newer zoonoses, *Leptospirosis hebdomadis* and chlamydiosis, deserve further study and attention to methods of prevention.

Respiratory disease is a major cause of disability and we do not yet know how many are affected by one or other of its manifestations.

The most frequently reported prescribed industrial diseases, dermatitis and tenosynovitis, continue to cause ill-health and although preventive methods are well known, their implementation is patchy. Back strains and hip disorders are common causes of disability (Axmacher et al, 1984).

The occupational health and safety services are in their infancy, but many bodies are prepared to provide them. The tasks for the future will include provision of much better information for the farming community on health risks, and on when to call in occupational health and hygiene advisers. It will be necessary to devise methods of delivery of occupational health and safety services at the work place, and group schemes will probably play a large part. The Swedish and Finnish models show that this approach can bring benefits in increased knowledge and in preventive strategies. The formation of the Agriculture Industry Advisory Committee has brought together health and safety professionals, farmers and unions. The appointment of health and safety advisers to national unions has greatly stimulated members' interest, but in agriculture union pressure is generally low. The UK farming community has traditionally been reluctant to pay for advice.

The occupational health community in developed countries has a duty to communicate knowledge on medical aspects and on safer working to the third world, where the workers face the highest risks.

NOTE

Any opinions expressed in this chapter are those of the author and not necessarily those of the Health and Safety Executive.

REFERENCES

Andrews B, Major R, Palmer S R 1981 Ornithosis in poultry workers. The Lancet: 632–634

Austad J, Kavli G 1983 Phototoxic dermatitis caused by celery infected by sclerotinia sclerotiorum. Contact Dermatitis 9: 448–451

Axelson O 1984 The health effects of phenoxy acid herbicides. In: Harrington J M (ed) Recent advances in occupational health 2. Churchill Livingstone, Edinburgh

Axmacher B, Dahlsjo L, Akesson B 1984 Coxarthrosis among farmers and farm workers, a case referent study. Proceedings of the 9th International Congress of Agricultural Medicine and Rural Health

Balarajan R, Acheson E D 1984 Soft tissue sarcomas in agriculture and forestry workers. Journal of Epidemiology and Community Health 38: 113–116

Bamdad S 1980 Enzyme-linked immunosorbent assay (ELISA) for IgG antibodies in farmer's lung disease. Clinical Allergy 10: 161–171

Bamford M 1985 First aid in fruit farming. Occupational Health 37: 162–167

Beer B J S, Bradford W P, Hart R J 1982 Pregnancy complicated by psittacosis acquired from sheep. British Medical Journal 284: 1156–1157

Bettelheim K A 1984 Antibody levels of leptospirae among milkers in the South island of New Zealand. Proceedings of the 9th International Congress of Agricultural Medicine and Rural Health

British Agrochemicals Association, Robens Institute of Occupational Health and Safety University of Surrey 1983 Spray operator safety study. British Agrochemicals Association Ltd, London

California Department of Public Health Bureau of Occupational Health undated questions and answers about medical supervision of workers using organic phosphate pesticides

Cockcroft D W, Dosman J A 1981 Respiratory health risks in farmers. Annals of Internal Medicine 95: 380–383

Cohen D 1978 Identification of the problems. Ann 1st Super Sanita 14: 197–206

Cotterill J A 1981 Contact dermatitis following exposure to tetrachloronitrobenzene. Contact Dermatitis 7: 353

Crawford S M, Miles D W 1980 Leptospira hebdomadis associated with an outbreak of illness in workers on a farm in North Yorkshire. British Journal of Industrial Medicine 37: 397–398

Cronin E 1980 Contact dermatitis. Churchill Livingstone, Edinburgh

Cuthbert O D 1981 The incidence and causative factors of atopic asthma and rhinitis in an Orkney farming community. Clinical Allergy 11: 217–225
Cuthbert O D, Brighton W D, Jeffrey I G, McNeil H B 1980 Serial IgE levels in allergic farmers related to the mite content of their hay. Clinical Allergy 10: 601–607
Cuthbert O D, Jeffrey I G, McNeill H B, Wood J, Topping M D 1984 Barn allergy among Scottish farmers. Clinical Allergy 14: 197–206
Department of Health and Social Security 1983 Pesticide poisoning: notes for the guidance of medical practitions. HMSO, London
Donham K J, Rubino M, Thedell T D, Kammermeyer J 1977 Potential health hazards to agricultural workers in swine confinement buildings. Journal of Occupational Medicine 19: 383–387
Donham K J, Knapp L W, Monson R, Gutafson K 1982 Acute toxic exposure to gases from liquid manure. Journal of Occupational Medicine 24: 142–145
Donham K J, Zavala D C, Merchant J A 1984a Respiratory symptoms and lung function among workers in swine confinement buildings: A cross sectional epidemiological study. Archives of Environmental Health 39: 96–101
Donham K J, Zavala D C, Merchant J 1984b Acute effects of the work environment on pulmonary functions of swine confinement workers. American Journal of Industrial Medicine 5: 367–375
Douglas G A, Selema R 1985 Occupational health services in developing countries. Personal communications
Dubrisay J 1984 Health and work in agriculture — a French survey. Proceedings of the 9th International Congress of Agricultural Medicine and Rural Health
Dutkiewicz 1978a Exposure to dust-borne bacteria in agriculture I environmental studies. Archives of Environmental Health 33: 250–259
Dutkiewicz J 1978b Exposure to dust-borne bacteria in agriculture II immunological survey. Archives of Environmental Health 33: 260–270
Editorial 1984 Psittacosis of non-avian origin. The Lancet: 442–443
EEC Directive 80/1107 1980 on the protection of workers from the risks related to exposure to chemical, physical and biological agents at work. Official Journal of the European Communities 327: 8–13
English J, Molloy W, Murnaghan D, Whelton M 1983 Leptospirosis in South–West Ireland. Irish Journal of Medical Science 152: 145–148
Ferguson D 1984 Agricultural medicine in outback Australia. Proceedings of the 9th International Congress on Agricultural Medicine and Rural Health
Flannigan S A et al 1985 Synthetic pyrethroid insecticides: a dermatological evaluation. British Journal of Industrial Medicine 42: 363–372
Hagley S R, South D L 1983 Fatal inhalation of liquid manure gas. Medical Journal of Australia 2: 459–460
Hardy R H 1981 Accidents and emergencies. A practical handbook for personal use 3rd edition. Oxford University Press, Oxford p 88
Hardy R H 1981 Accidental self-injection. Personal communication
Health and Safety Executive 1982 Health and safety statistics 1980. HMSO, London
Health and Safety Executive 1985a HSE News Release, 2 July 1985
Health and Safety Executive 1985b Guidance note GS35: Safe custody and handling of bulls on farms and similar premises. HMSO, London
Health and Safety Executive 1985c Agricultural health and safety topics: HSE warns on invisible dangers in sealed forage and grain silos. AST2: 85
Health and Safety Executive 1985d Guidance note MS 17 Health surveillance and biological monitoring of workers exposed to organo-phosphorus pesticides. HMSO, London
Heller R F, Kelson M C 1982 Respiratory disease mortality in agricultural workers in eight member countries of the European community. International Journal of Epidemiology II: 170–174
Hodnett T 1985 Leptospirosis hebdomadis hardjo; A study amongst the dairy farm workers in Cheshire. HSE, unpublished
Höglund S 1984 Health Services through the world — Sweden. Proceedings of the 9th International Congress of Agricultural Medicine and Rural Health.
Jeyaratnam J 1985 Health problems of pesticide usage in the third world. British Journal of Industrial Medicine 42: 505–506
Jones W, Morring K, Olenchock S A, Williams E, Hickey J 1984 Environmental study of poultry confinement buildings. American Industrial Hygiene Association Journal 45: 760–766
Kelman G R 1984 Use of electromyography to assess exposure to organophosphorus insecticides. Proceedings of the 9th International Congress of Agricultural Medicine and Rural Health
Kummer F, Schnetz E, Ebner H, Poitschek Ch, Scheiner O, Kraft D 1984 Spaetlese lung. New England Journal of Medicine 311: 1190
Lacey J 1980 The microflora of grain dust. In: Dosman J A, Cotton D J (eds) Occupational pulmonary

disease: focus on grain dust and health. Academic Press, New York, Ch 41, 417–440
Miller St, Shah M A 1982 Cholinesterase activities of workers exposed to organophosphorus insecticides in Pakistan and Haiti and an evaluation of the tintometric method. Journal of Environmental Science and Health 17: 125–142
Nuutinen J J 1984 Musculoskeletal diseases and symptoms of Finnish farmers. Proceedings of the 9th International Congress of Agricultural Medicine and Rural Health
Office of Population Censuses and Surveys 1978 Occupational mortality decennial supplement 1970–1972. HMSO, London
Palmer S R, Andrews B E, Major R 1981 A common source outbreak of ornithosis in veterinary surgeons. The Lancet: 798–799
Parkes W R 1982 Occupational lung disorders 2nd edition. Butterworth, London
Pasricha J S, Gupta R 1983 Contact dermatitis due to calcium ammonium nitrate. Contact Dermatitis 9: 149
Peachey R D G 1981 Skin hazards in farming. British Journal of Dermatology 105: Sp 21 45–49
Pladson T R 1984 Silo emptiers' disease. Minnesota Medicine 67: 265–269
Public Health Laboratory Service Communicable Diseases Surveillance Centre 1985 Brucellosis figures. Personal communication
Richter E D, Gordon M, Halamish M, Gribetz B 1981 Death and injury in aerial spraying: pre-crash, crash and post-crash prevention strategies. Aviation, Space and Environmental Medicine 52: 53–56
Roberts D V 1977 A longitudinal electromyographic study of six men occupationally exposed to organophosphorus compounds. International Archives of Occupational and Environmental Health 38: 221–229
Rowe R G, Cliff K S, Gill J 1980 A review of the occupational hazards in the agricultural industry. Dorset Area Health Authority and Department of Community Medicine Southampton University
Rycroft R J G 1977 Contact dermatitis from organophosphorus pesticides. British Journal of Dermatology 97: 693
Sakula A, Moore W 1969 Benign leptospirosis: first reported outbreak in British Isles due to strains belonging to the hebdomadis serogroup of Leptospira interrogans. British Medical Journal I: 226–228
Shenberg E, Gerichter Ch B, Lindenbaum I 1982 Leptospirosis in man, Israel 1970–79. American Journal of Epidemiology 115: 342–358
Smith D M 1977 Organophosphorus poisoning from emergency use of a handsprayer. The Practitioner 218: 877–883
Smith D M 1985 Occupational photodermatitis from parsley. The Practitioner 229: 673–675
Swedish Packaging Research Institute 1983 Packages for chemical pesticides. Publication No. 86 ISSN 0280–5545
Twort C H G 1981 Group R streptococcus meningitis (Streptococcus suis type II): a new industrial disease. British Medical Journal 282: 523–524
Wells G C, Swift G 1985 Health problems of farm machinery drivers. HSE, unpublished
Wilson D, Wetson R 1981 Leptospirosis — a diagnostic problem and an industrial hazard. Journal of the Royal College of Practitioners 31: 165–167
World Health Organisation 1982 Field surveys of exposure to pesticide standard protocol. VBC/82.1

5. Repetition strain injuries

D. Thompson   A. J. Rawlings   J. M. Harrington

INTRODUCTION

Occupational disorders of the musculo-skeletal system are not new. Ramazzini (1713) recognised such conditions whilst Velpeau (1841) gave an actual description of tenosynovitis. Today, the incidence of such conditions appears to be increasing particularly it seems in Australia where Ferguson (1984) reports repetition strain injury (RSI) to be reaching epidemic proportions. The New South Wales Workers Compensation Commission statistics (Fig. 5.1, Table 5.1) show an increased number of successful claims from 526 in 1978–1979 to 1499 in 1981–1982.

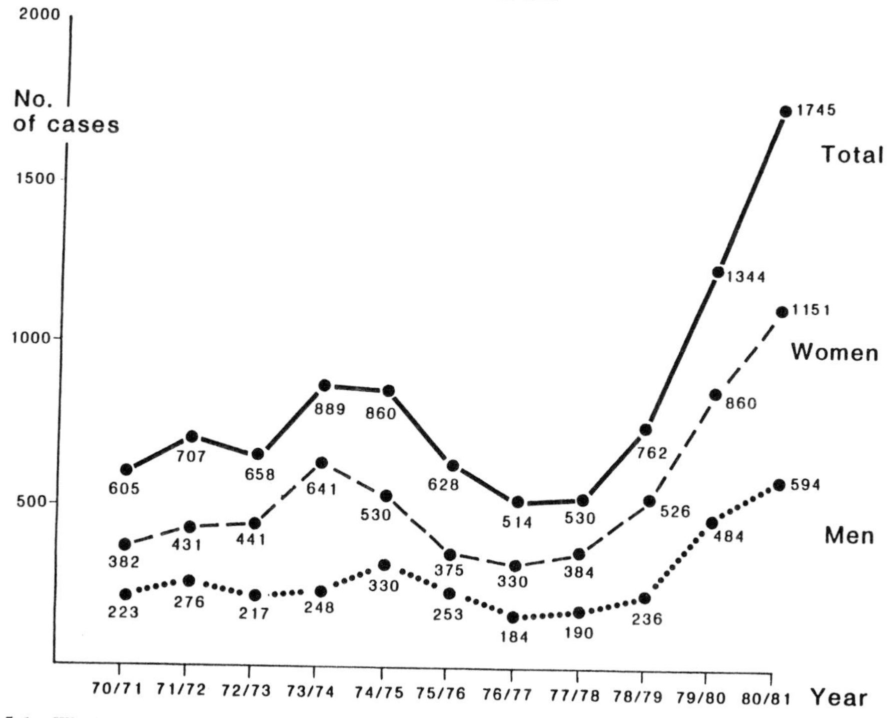

Fig. 5.1 Workers compensation statistics (New South Wales, Australia) for new compensation cases, Synovitis, Bursitis, Tenosynovitis (1970–1981; reproduced with permission from McPhee 1983).

The extent of the problem and the very large sums being paid in compensation prompted the Australian National Occupational Health and Safety Commission to establish the RSI Committee to examine the issue: an Interim Report was published in 1985.

Table 5.1 Number and percentage of new workers' compensation cases in New South Wales as ICD 9 Code 727, classified by sex, financial year (1978–1979 to 1982–1983) and occupational group professional includes clerical and administrative

	Manufacturing		Professional		Other		Total	
	Number	Percentage of all disease cases	Number	Percentage of all disease cases	Number	Percentage of all disease cases	Number	Percentage of all disease cases
Males								
1978–1979	169	2.9	2	1.4	65	2.3	236	2.7
1979–1980	338	4.9	8	3.6	138	3.0	484	4.1
1980–1981	400	5.5	20	6.9	174	3.7	594	4.8
1981–1982	507	7.1	22	7.0	235	5.0	764	6.3
1982–1983	335	5.2	30	7.7	190	3.7	555	4.6
Females								
1978–1979	412	50.9	40	28.6	74	17.7	526	38.5
1979–1980	645	55.5	78	38.5	137	23.6	860	43.8
1980–1981	814	61.7	163	48.2	174	24.0	1151	48.3
1981–1982	973	68.1	273	55.0	253	31.2	1499	54.8
1982–1983	787	54.3	317	53.4	287	32.9	1391	47.8
All persons								
1978–1979	581	8.7	42	15.1	139	4.3	762	7.5
1979–1980	983	12.2	86	19.4	275	5.3	1344	9.8
1980–1981	1214	14.0	183	9.8	348	6.3	1745	11.8
1981–1982	1480	17.0	295	29.2	488	8.9	2263	15.2
1982–1983	1122	14.3	347	36.4	477	3.4	1946	13.7

Source: Published statistics of NSW Workers' Compensation Commission 1978–1979 and 1982–1983.

Kilbom (1983), citing statistics on the Swedish population from their Occupational Injury Information System (Fig. 5.2) states:

> 'more than 50% of reported cases of occupational diseases are caused by ergonomic factors such as physically heavy work, manual materials handling, repetitive work and unsuitable work postures and that employees injured in these categories have 20 to 26 days of sick leave per year more, depending on a number of harmful factors, than others'.

The causes suggested for this marked increase in occurrence of RSI in recent years, its recognition and possible strategies for its prevention are reviewed in this paper.

TERMINOLOGY AND DIAGNOSIS

Generic terms in common use are repetition strain injury (RSI) and tenosynovitis, although such syndromes are not necessarily caused by repetition, nor are they restricted to tendons or synovial membranes. What is being described is, in fact, a whole range of symptom/sign complexes involving part or all of the upper limbs including the cervico-brachial region (Thompson et al, 1951; Kurppa et al, 1979; Armstrong & Chaffin, 1971). Elenor (1981) suggests that in reporting the occurrence of cases, there is a general lack of knowledge about the anatomy and physiology

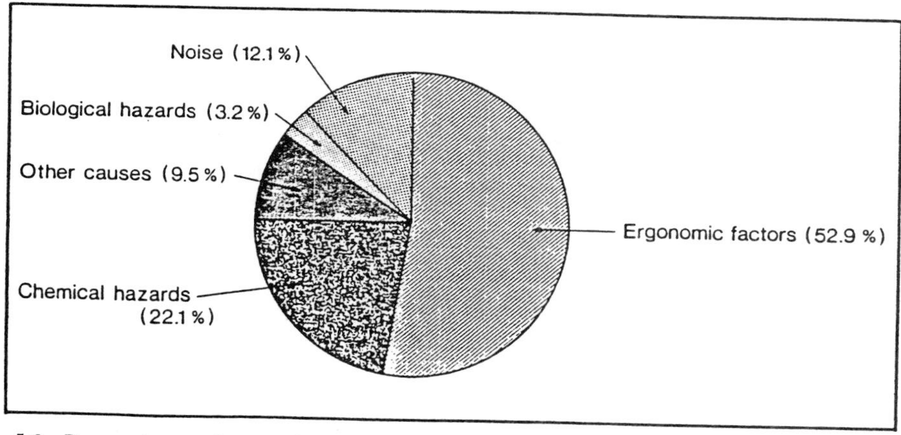

Fig. 5.2 Reported cases of occupational diseases in 1980. Source: The Official Statistics of Sweden: Occupational Injuries 1980.

of the conditions. Furthermore the terminology used to define them is confused. Ferguson (1984) claims that such terminological confusion hampers the compiling of statistics relating to the occurrence of RSI.

Browne et al (1984) have proposed new guidelines for diagnosis, at the same time stating that there are no specific pathological or radiological features to support a diagnosis of repetition injury. This underlines the difficulty of providing injury compensation which may be based on semi-objective findings such as local tenderness or pain. Until these matters are resolved, good quality epidemiological studies are unlikely to ensue, thus preventing the clarification of the biomechanical factors involved in the various aetiologies. This will, in turn, limit the establishment of well founded preventive measures.

In essence, this group of conditions arbitrarily subsumed under the rubric RSI are soft tissue disorders. They include a variety of musculo-tendinous conditions, which, depending on symptoms and signs may be classified as tenosynovitis, tendinitis, peritendinitis and epicondylitis. Other classifications concentrate on anatomical sites in the upper limb such as rotator cuff syndrome and bicipital tendinitis. It is, however, possible to envisage single muscle or tendon disorders and the authors have seen such manifestations. A common example concerns the tendon of extensor pollicis longus.

In addition to these fairly well defined conditions, some medical workers believe that repetitive work can result in vague disorders of the thoracic outlet or even diffuse sympathetic dystrophy. Psycho social factors have been invoked in all the disorder groups, but particularly in the latter more diffuse types. Such invokations should always be treated with great circumspection in medicine but there is no doubt that psycho social factors do play a part in the progression (and regression) of even the most clear cut disorders.

A further complication is that humans have the infuriating habit of responding differently to the same injurious agent. We have seen groups of workers exposed to identical repetitive work which has resulted in all manner of health effects ranging from continued good health to semi-invalidity with multiple system disorders and an almost total upper torso paralysis. Clearly, previous health status will play an important part in the pathogenesis of the resulting syndrome thus a detailed past medical history is an essential prerequisite to the usual occupational history and physical examination.

Differential diagnoses must include inflammatory and degenerative arthoses as well as spinal disorders and connective tissue disorders. Laboratory tests may well be essential before an occupational aetiology is accorded the prime suspect cause.

OCCUPATIONAL CAUSES

Task related

RSI has probably been around since Man undertook repetitive tasks. Early examples of the injury include the various craft palsies or cramps, such as threader's wrist and brewer's arm. More recently, Ferguson (1971) reported telegraphist's cramps in 14% of 516 Australian telegraphists.

Nevertheless, despite increased mechanisation, RSI has increased dramatically in recent years. The increase can indeed be linked to the changes in technology which have taken place. However, Kurppa et al (1979) emphasise that the cause-effect relationship between work and injury has not been substantiated by analysis of the task with regard to its biomechanical components, the disorder cannot be attributed to repetitive work unless symptoms and signs can be clearly linked anatomically with particular activities.

In the past, less sophisticated manufacturing processes meant that a heavy to moderate physical stress was exerted onto the operator's body as a whole, whereas more recently the workload, although much reduced, is often concentrated on small portions of the worker's anatomy. Thus in addition to traumatic injury, it is necessary to be aware that injury can be caused insidiously over a long period of time due to cumulative overstrain of small areas of the body (Tichauer, 1966).

In industry

RSI has been reported as occurring in a wide range of industries (Elenor, 1981; Walker, 1979) and Elenor names six industries as having particularly high risks:

(i) the electronics industry, e.g. telecommunications assembly;
(ii) domestic appliance manufacture involving mechanical assembly;
(iii) poultry processing and packing;
(iv) clothing, carpet and bag manufacture requiring manual sewing;
(v) manufacture and packaging of small consumer products e.g. biscuits, cigarettes and sports goods;
(vi) cleaning operations requiring heavy scrubbing and polishing machines;

Walker considered those most at risk to be process workers, cleaners or those employed in clerical jobs and specifically speed typists. Today, of course, with the increased application and use of computer technology, keyboard operators are similarly at risk. Studies in this sector are reported below. According to the Australian Council of Trades Unions (ACTU-VTHC) Guidelines for the Prevention of Repetition Strain

Injury (1982) reasons for the RSI epidemic in the industries quoted above are the increasing tendency with new technology, for work to become segmented and thus repetitious. Ferguson (1984) suggests that RSI may be endemic particularly where it arises from 'the fragmentation and specialisation of tasks in short repeated cycles.' In other words, there has been a 'de-skilling' of the work, which has led to a short cycle time and often high speed repetitive movements which put a large repeated load on a specific anatomical group of muscles or tendons. In other words, the individual tasks may not of themselves be strenuous, but their frequent repetition greatly augments the small musculo-skeletal work load involved in each task.

It is generally agreed that there is no single cause of RSI, but rather a combination of factors (Eckles, 1986; Luopajarvi et al, 1982; Browne et al, 1982). Peres (1961) said:

> 'the continuous use of the same body movement and sets of muscles responsible for that movement during the normal working shift can lead to the onset initially of fatigue and ultimately of immediate or cumulative muscular strain in the local body area concerned'.

He maintained that these movements are frequently associated with sustained tension in the muscles involved

> 'such as gripping objects tightly between thumb and finger for long periods, or where particular movements like hand and forearm rotation (pronation or supination) are combined with the lifting of weights or the application of grip pressure or torque many times during a shift beyond the natural limit of recovery rate of the body'.

Luopajarvi et al (1982) add to the list those repetitive motions carried out at high speed, static muscle work combined with extreme work positions of the hands. Silverstein et al (1985) suggest that one cause is a combination of high force and high repetition, whilst Stevenson (1985) suggests jerky work movements should be considered, and inefficient posture is suggested by Browne et al (1984).

The problem has been made worse by the economic climate in which there is high unemployment, and the need to be competitive leads to small profit margins and an emphasis on worker productivity. In a buoyant labour market, employees tend to leave their jobs if they experience pain due to work, and indeed many of the industries cited above traditionally have a high labour turnover. In times of high unemployment, however, employees stay longer in a job and are generally less critical of working conditions (Eckles, 1986). In addition, the human hand is still a more adaptable tool than any machine, and thus, especially with short production runs, it is often more economical to let a worker adapt rather than adapt a machine. Thompson et al (1951) said that the incidence of repetition injuries increased during periods of economic stress, such as during the depression of the 1930s when it rose to 0.99%, and then fell to 0.15% when there was full employment. A further factor is that many of the repetitive tasks in industry are very often done by women, and in times of high unemployment the woman's income may be critical to the family's standard of living. Therefore, not only will there be a reluctance to complain of injury for fear of recommended medical dismissal, but the condition may well be exacerbated by the manual domestic tasks she additionally carries out in the home. A further factor is the relatively unstable employment position held by many of the female workers. They are often part-timers who would be even easier to replace in present times than the full-time employee.

In offices

One of the main problems caused by new technologies is that in the office as well as the industrial setting, faster work rates and greater job specialisation is needed. In recent years there has been a large rise in the incidence of RSI amongst keyboard operators, particularly with the introduction of the electronic keyboard and visual display units (VDUs). Ross (1985) notes that New South Wales statistics 1982–1983 show that cases in the occupational category administrative, executive, managerial and clerical have increased markedly in recent years. He suggests this may be due to an increased awareness amongst those at risk, and thus an increase in reporting of injury, and to the rapid introduction of new technology, in particular, keyboards. A survey of keyboard operators in the Australian civil services found 17% suffered from RSI, whilst a similar survey by the American post office found an incidence of 54% amongst post office coders. Also a Civil and Public Services Association (CPSA) study at the Government's Bristol Computer Centre found up to 48% of data processors suffered from RSI (Eckles, 1986).

Work organisation related

Apart from the activities involved in the job, organisational factors have been found to increase the incidence of RSI. These include duration of work without rest, bonus and overtime incentives, lack of training and failure of supervision (Browne et al, 1984). To these can be added unsuitable tools, faulty, tight-fitting component elements in assembly jobs, return to repetitive work after absence and full-time employment of inexperienced operators on repetitive work (Welch, 1971).

INCIDENCE

Reliable statistics are necessary if the incidence and prevalence of cases of RSI by occupation or industry are to be effectively monitored. Such statistics are also required for effective strategies to be promulgated for reducing or preventing injury. However, there is a general lack of good quality statistics on the incidence of RSI.

In Britain, statistics are collected by the Department of Health and Social Security of the number of claims for industrial injury benefit under the heading of Inflamation of the tendons of the hand or forearm or the associated tendon sheaths: the number in 1980 was 2957 (Eckles, 1986). In Finland, the National Board of Labour Protection collects statistics provided by doctors who report centrally all cases of occupational disorders. In the USA, The National Institute of Occupational Safety and Health (NIOSH) states that the national surveillance and reporting system for occupational injuries and illnesses is not sufficiently refined to provide figures for repetition strain injuries in specific occupational groups. In Sweden the Occupational Injury Information System gives statistics quoted by Kilbom (1983) shown in Figure 5.2.

In Australia, Ferguson (1984) states that 'the real statistics are unknown' and the RSI Committee (1985) says that data on the Australian workforce is incomplete and fragmented. The most comprehensive source of statistics is produced by the Australian Bureau of Statistics, and compiled from claims submitted to the Worker's Compensation Board. However, inaccuracies may result from the attempt to relate the true incidence of RSI to compensation claim statistics. Possible reasons are:

a. inaccurate reporting due to lack of knowledge of the conditions and confusion in the terminology used to define them (Elenor, 1981; Ferguson, 1984; RSI Committee, 1985);
b. omission of categories such as cases of fewer than three days incapacity, cases where no claim is made, the self-employed, and casual workers;
c. conditions not involving time off are not recorded (McPhee, 1983);
d. the exclusion of many cases of repetition injury while including some that are not (Browne et al, 1984).

There is considerable anecdotal evidence of physicians over-diagnosing the condition when they see a patient with a painful upper limb condition who works in an industry where other cases of RSI have been reported. Similarly, evidence exists of some physicians who virtually deny the existence of the condition. Patients referred to this latter group of doctors never come away with a diagnosis of RSI!

Trends in the incidence of RSI can however be postulated from the increase in compensation claim statistics. Ross (1985) quoting the number and percentage of new workers' compensation cases reported in New South Wales (Table 5.1) suggests that although there was an apparent decrease in the number of RSI injury cases in the year 1982–1983, the number of people employed in high-risk occupations also declined. Furthermore as the classification of injuries has become more accurate, some cases may be categorised under alternative headings. The same table also shows a rise in the number of cases in the clerical sector, and that there are more reported cases amongst women than amongst men. This does not imply that women are more susceptible to RSI than men, but that significantly more women than men are employed in high-risk occupations. Such numerator data are, however, of limited value without an accurate denominator base.

Peres (1961) suggests the greater incidence of injury to women is due to their weaker musculature compared with men. This can be a critical factor where the machinery in production processes and in particular hand tools have been designed for the relatively stronger male work force, and are therefore often not suitable for women. It does not, however, explain away the upsurge in cases reported.

Potential sources of incidence data other than compensation statistics are company records of work-related accidents and diseases; industrial surveys; reports in the medical literature; and hospital and private practitioners' medical records. The difficulties of relating small samples to the total eligible population, of the confidentiality of medical records, and the inadequacy of such records in relating injury to occupation, militate against the usefulness of these sources.

In short, there is evidence of a great increase in claims for RSI in certain industries and in certain countries. The absence of comparable rate calculations and the potential confounding factor of the reasons for claims, does not enable an accurate picture to be assembled.

PREVENTION

The effective implementation and evaluation of measures to prevent RSI requires a multidisciplinary approach, involving ergonomics, occupational health, epidemiology, engineering, administration and management (McPhee, 1985). To this list can be added industrial relations, biomechanics and bioengineering (Browne et al, 1982). The roles of occupational health, ergonomics, and pre-employment screening are considered below.

Occupational health

An important factor in the prevention of injury is the keeping of records on incidence by the employer. Such records tend to be incomplete because they are generally kept only by larger companies with an occupational health department, and even then injuries may not necessarily be reported. However, the maintenance of good records enables faults in the workplace design and activity to be identified, and if these are corrected, the incidence and severity of injury has been shown to decrease (Flynn, 1982). Such records however should accurately state the site and type of injury, its mode of onset and the job history of the employee. Few company recording systems, in our experience, even approach these desiderata.

Successful treatment depends on early detection and accurate diagnosis. Thus workers at risk should be encouraged to make immediate report of repetition injuries. Similarly, regular medical monitoring for neck and upper limb repetition injuries should be an integral part of any prevention programme. This should be carried out on workers engaged continuously on tasks designated as particularly hazardous for repetition injuries. The frequency of examination would vary according to the risk involved, and two-yearly intervals have been suggested as the minimum requirement by the Australian Council of Trades Unions. The RSI Committee (1985) also recommends early reporting of symptoms, assessment by an occupational health professional, and referral to a specialist occupational physician.

Ergonomics

The multidisciplinary approach of ergonomics is essential in the prevention of RSI. However it should be noted that the proven value of ergonomics is in its involvement at the design stage of the operation so that the relevant human factors information can be incorporated. The areas relevant to ergonomists are: task design, workplace design and work organisation as these are the essential factors involved in work analysis and synthesis.

Workplace design

Workplace design and layout should conform to ergonomic principles, whilst both physical and social aspects of the working environment should be considered. The use of ergonomics is essential in such areas as seating, posture, bench and tool design, together with environmental factors such as temperature, noise and lighting.

Task design

This involves the job itself and the design of equipment, machinery and tools in accordance with ergonomic principles. This analysis will attempt to ensure that the jobs are within the physical and mental capacities of the workforce. Where possible, the job should include a mixture of repetitive and non-repetitive work in which recovery from the effects of the repetitive work is possible. The task should not require the muscles to be used repetitively in a forceful manner, at a mechanical disadvantage or at the limit of their range of movement. Abnormal work postures, caused by poor task or workplace design, should not be required of the worker, as this may well increase the static load factor on a specific muscle group.

Work organisation

Job rotation and job enlargement spread the strain over a larger number of musculo-skeletal groups. However, job rotation can only be used where a survey of all jobs undertaken has been carried out to ensure that the alternative tasks do not make the same physical demands on the same area of the musculo-skeletal system. For example, one case recently referred to us was transferred from electronic assembly with many hand/wrist movements to the lighter job of assembling large cardboard boxes from flat parts. Not surprisingly, this change of job exacerbated the man's symptoms and signs!

Job enlargement will also require careful consideration of the availability in the work force of the skill levels required. More immediate measures include reducing the work rate in paced operations, reducing shift length and the provision of extra rest breaks. There should also be periods of adjustment when new operators, or those returning after a period of absence, are allowed to settle in with a lighter work load. A formal structured training programme for new entrants, and for existing workers before new production methods are introduced is also essential as is subsequent supervision to ensure that incorrect practices do not develop. The common practice of learning the job from Nellie can, and often does, lead to the spread of bad practice as much as good. The fact that Nellie was not affected by such malpractice is not biologically relevant.

The RSI Committee (1985) suggests that a strategy for the prevention of RSI should have an in-built review process. Employees should participate in discussions regarding work organisation, job design consultation and changing work practices. Kemp (1984) also recommends the establishment of review teams, which would also help reintegrate injured workers into the workplace and avoid further injury. Both these are activities that could be undertaken by the Health and Safety Committees in British industry.

Referring to prevention of injury in keyboard operators, Oxenburgh et al (1985) state that up to the early 1980s the emphasis had been on work station factors, such as design, anthropometry and lighting, and less attention was paid to other factors such as workload, work organisation, inter-personal factors and other psycho-social effects. However, the need to attend to these latter factors has become progressively more evident.

The importance of work organisation as well as workplace and task design is exemplified in the risk of development of RSI from too fast a work-rate. Ohara et al (1976) in a study of cash register operators found no change in the reported level of RSI even after ergonomic changes in the work activity. They found that the improvements led to an increase in work rate as less force was needed per entry stroke, and the incidence of injury remained the same.

Another work organisation factor is the organisational climate (Bennett, 1985). He draws a comparison between a basically trusting environment in the company, and a distrustful one and suggests the latter is a factor in the incidence of RSI. This may well be one of the important factors in the early identification of causes at work, where because of an unsympathetic attitude to complaints, the worker does not approach the immediate supervisor. The existence therefore of an understanding personnel department may also be a key to prevention.

Pre-employment screening

It is suggested that individuals react differently to stressful and/or repetitive movements and prolonged static loading, and some workers may be more liable to develop RSI than others even though they are carrying out identical tasks. The higher the levels of physical stress, the more individuals will be injured; therefore susceptibility is probably a continuum with the highly susceptible at one end and the highly resistant at the other (McPhee, 1980). Susceptibility in these terms may well be influenced by personal factors such as physical strength and/or psychologically induced stress. It can be clearly demonstrated that strength varies between individuals and with sex and age (Thompson, 1975), yet one can question whether or not these factors are duly noted in the initial design of the work activity and therefore the physically weaker may prove to be those most liable to physical injury. The psychological stress, though more difficult to quantify, if present will add to the overall stress, both of these factors will require due consideration in establishing a category of susceptible.

Test for susceptibility

Hettinger (1957) extended his earlier work (Hettinger & Beck, 1956) on the susceptibility of miners to mechanical stress to develop a test for predisposition for stenographers and punch card operators. The test measures the change in temperature on the back of the hand in response to vibration. From this they developed an index and suggested that its use was 80% successful in differentiating between the susceptible and non-susceptible. The work was later repeated by other scientists. The findings of Welch (1973) and Sakurai (1977) are consistent in agreement, and to a lesser extent is Schroter (1967). However, completely contrary results were obtained by Kuorinka et al (1981). On-going work, the results of Brown et al (1985) using a more rigorous test methodology show promise, but they also stress improved work method as the main method of prevention of stress injury. Later work using a yet more rigorous test methodology by Gillen (1985) showed poor reliability of the Hettinger test on a test/re-test basis, subjects scoring both above and below the index score which purports to indicate predisposition. On the basis of present evidence it may be that the test will be of use to differentiate between injured and non-injured, but this also has yet to be proven. Thus the present state of scientific uncertainty regarding the Hettinger test validity, prevents its universal adoption as a reliable index of susceptibility.

ALTERNATIVE VIEWS

Bennett (1985) presents a view of RSI from a different perspective to that most commonly found in the literature: he entitles his paper 'the politics of repetition strain injury'. He argues that articles on RSI are being presented as fast as patients suffering from it, and looking at 12 studies published in the 1980s, states that the general themes are of exploitation (particularly of migrant women) and of quality of working life in the white collar sector in particular. There has been a move towards greater worker involvement in decision-making and the running of the organisation, and he sees RSI as a vehicle for seeking a broad range of changes; as it is a health issue it is an emotional area and is not subject to normal industrial relations collective bargaining. In addition, he suggests that specialists and academics, by putting the problem into a medical model, are creating a highly technical mystique around RSI so as to create a dependency by management on them, in order that long term consulting can take place. He thus asks specialists to look at their role in the management of RSI from a wider political standpoint and not to assume that they are ideologically neutral.

Lucire (1985) sees the recent epidemic of RSI as in fact an epidemic of writer's cramp and a parallel epidemic of myalgia nervosa, combined with other physical and mental disorders to give a false category, i.e. RSI. Cases of writer's cramp, incorrectly diagnosed as RSI are managed according to a medical model and surrounded by fear and rumour of permanent injury. There is a belief that technology maims and a fear that the myalgia, if not properly treated, will proceed to dystonic cramp. This, she says, causes a conflict of either continuing work and risking permanent injury, or abandoning work, and thus the development of cramp would thus provide an acceptable escape from the conflict.

Both these authors feel that RSI is being wrongly treated under a medical model at present; one believing the true cause is psychological, the other that it is political. There is no doubt that political motives have been involved during the current debate over the Australian epidemic, but such motives are not only on the side of organised labour. Equally loud voices may be heard from the management camp. This is unfortunate as the political debate has obfuscated the investigation of the medical and ergonomic factors involved in RSI, and thus hampered the establishment of a scientifically defensible stance on the pathogenesis of the condition.

Treatment

There is general agreement that there is a need for early detection of RSI for successful treatment, and that early treatment is an effective way of minimising overall morbidity and economic cost both to the employer and the employee.

RSI has been classified according to the reversibility of the injury and a scheme for such a classification with three degrees of severity is shown in Table 5.2. Due to lack of early reporting, diagnostic awareness or effective management, workers often present themselves at stage II or III when therapeutic intervention is ineffective and rehabilitation virtually impossible (Browne et al, 1982).

There is a range of available treatments from rest to surgery, depending on the type and severity of the injury. The RSI Committee (1985) noted a wide variety of approaches to treatment.

1. Medication, including low-dose tricyclics for their endorphin effect in the control of night pain; non-steroidal anti-inflammatory agents; muscle relaxants; anxiolytics; antidepressants; and local steroidal infiltration.
2. Provision of ortheses.
3. Relaxation training to decrease muscle tone, including biofeedback.
4. Attention to posture, manipulation, mobilization and electrotherapy.
5. Acupuncture.

Table 5.2 Degrees of severity of repetition strain injury

Severity	Symptoms	Duration	Prognosis
Stage I Pain noticeable more as a dull ache which disappears with rest	Aching and fatigue in the affected limb during working hours, which settles at nights and at weekends	Several weeks during performance of repetitive task	If detected early there is usually no reduction in work capacity and condition is completely reversible
Stage II Reduced capacity for work and to perform repetitive daily tasks such as using scissors without pain. Fatigue is often present due to intermittent night pain	Recurrent aching and fatigue occurring shortly after work commences and continuing after work ceases	Up to several weeks after cessation of work	Reasonable if treated early — can easily develop into a Stage III if not reported. Presence of pain can cause depression
Stage III Inability to perform less arduous tasks or light duties. Person's lifestyle is affected by inability to perform daily tasks. Intractable pain may cause depression, anxiety and sleep disturbances which increase muscle tension and further aggravate the injury	Persistent aching and fatigue and weakness while at rest, plus pain with even non-repetitive movements	Months to years, even after retirement from work	Poor prognosis — employee usually suffers anxiety and depression due to continual pain — addiction to sedatives and alcohol is quite common. Family disturbances are a possibility

Reproduced with permission from Kemp 1984.

6. Psychological support.
7. Chiropractice and naturopathy.
8. Surgery.

The Committee considered that surgery and psychotherapy, in the absence of specific clinical indicators to be of little value. The value of the other measures is also open to debate. Taylor et al (1982) used mild active exercise, and Browne et al (1982) ice and pulse ultrasound in stages I or II, as well as occupational health aids and counselling for housework.

Surgery has resulted in varying degrees of success and has often been inappropriately advised for such injuries as carpal tunnel syndrome and thumb extensor tenosynovitis (Brown et al, 1984). Walker (1979) advises that carpal tunnel release operations should only be carried out on patients suffering from carpal tunnel syndrome with severe parasthesiae or muscle wasting with clear evidence of nerve compression from nerve conduction studies. He says the operation can be 'disastrous' for patients with tenosynovitis.

There is general agreement that rest and removal from causative tasks is the best treatment for RSI. Walker (1979) states that patients asked about the best treatment preferred rest and no treatment to a variety of other measures including plaster immobilisation, ultrasound, cortisone injection, physiotherapy and surgery. Patients thought anti-inflammatory and pain-killing drugs were the least helpful. Taylor et al (1982) found a poor response to medical and surgical measures, and suggested avoiding movement causing pain, a cessation of the causative activity and a possible immobilisation of the affected part. However, it should be noted that cessation of the causative task may cause difficulty where alternative employment is unavailable and the income from work a necessity.

There have been few attempts to evaluate the effectiveness of treatment of RSI but given the poor definition of cases and the almost total absence of good evaluative studies the present state of confusion is hardly surprising. Browne et al (1982) say that the proportion of sufferers who recover and return to the workforce is unknown, but thought to be low; thus long term prospective studies are needed to establish such data. Such a general lack of information on the effectiveness of therapeutic measures emphasises the great importance of efficient preventive strategies.

COMPENSATION

In Britain, at present, only tendinitis is a prescribed disease under the Industrial Injuries Scheme. Carpal tunnel syndrome and rotator cuff syndrome were recently turned down by the Industrial Injuries Advisory Council due to a lack of clear cut evidence linking the syndromes *preferentially* with occupation, as opposed to their existence as common musculo-skeletal disorders of the population as a whole. Apparently carpal tunnel syndrome is compensatable if it is considered to follow on from tenosynovitis (Gee, 1985).

Alternatively, or in addition, workers can claim compensation in Britain from their employers if they can prove negligence in common law. Their case is strengthened if they can show that no improvements in working conditions were made after the start of complaints (General Municipal and Boiler Makers Union, 1985). However, proving negligence is always difficult and to date, no worker has been successful in sueing their employer (Times, 1985).

One suggested alternative to this unsatisfactory legal state of affairs is to introduce a no fault system as exists in New Zealand and Australia. A Royal Commission rejected this idea some years ago, thus the likelihood of an early renewal of the debate is remote (see Chapter 14).

The large number of successful claims in Australia (Fig. 5.1 and Table 5.1) with high compensation awards may be in part due to the different compensation schemes. Indeed the Cooney Report (1983–1984) states that the Australian system including the delays involved, encourages the injured worker to delay rehabilitation so as to maximise the physical and financial loss and thus build up a better case in court.

CONCLUSION

This account has shown that repetition strain injuries are not solely a result of work in a modern industrial society. In the past, such conditions were often eponymised by the craft most frequently associated with the injury. Modern technology has largely eliminated these sources of musculo-skeletal ill health but, in some instances, has replaced them with new repetition strain syndromes.

Nevertheless, there is considerable confusion regarding the terminology, the diagnostic labels, and the ergonomic factors responsible for the present spate of reported injuries, which in addition vary greatly in various countries. Not surprisingly, treatment regimes are legion and none is of established efficacy. Nevertheless, the most effective measure would be prevention and this involves a multi-disciplinary approach to the workplace task and its health sequelae. Scientifically robust investigation of repetition strain injuries are almost universally absent. There is, therefore, an urgent need for high quality studies of the pathogenesis and treatment of this potentially disabling group of ailments.

REFERENCES

Armstrong T J, Chaffin D B 1971 Carpal tunnel syndrome and selected personal attributes. Journal of Occupational Medicine 21: 481–486

Australian Council of Trade Unions 1982 Victorian Trade Hall Council Occupational Health and Safety Unit. Guidelines for the prevention of repetitive strain injury (RSI). Mathews J, Calabrese N (eds) Health and Safety Bulletin 18 August

The Australian National Occupational Health and Safety Commission 1985 Interim report of the RSI committee

Bennett M J 1985 The politics of repetitive strain injury. The Journal of Occupational Health and Safety Australia and New Zealand 1: 102–105

Brown D A, Coyle A R, Beaumont P E 1985 The Automated Hettinger Test in the diagnosis and prevention of repetition strain injuries. Applied Ergonomics 16: 113–118

Browne C D, Faithful D K, Nolan B M 1982 Occupational overuse injuries. Report for Medical sub-group, Occupational repetition Injuries advisory committee NSW

Browne C D, Nolan B M, Faithful D K 1984 Occupational repetition strain injuries. Guidelines for diagnosis and management. Medical Journal of Australia Mar 17: 329–32

Cooney K 1983–1984 Report. Committee of enquiry into the victorian workers' compensation system. Victoria

Eckles S 1986 Repetition strain injury–keyboard operator's disease? Environmental Health 94: 7–10

Elenor R C 1981 Tenosynovitis and other repetition injuries of the upper limb. Report. Central Planning and Research Unit NSW Government Department of Industrial Relations

Ferguson D 1971 An Australian study of telegraphist's cramp. British Journal of Industrial Medicine 28: 280–285

Ferguson D 1984 The new industrial epidemic. Medical Journal of Australia 140: 318–319

Flynn R 1982 Personal communication

Gee D 1985 Industrial disease. The Times Sept 13

General, Municipal, Boilermakers and Allied Trades Union 1985 Tackling Teno. A GMB guide to tenosynovitis and repetitive strain injury

Gillen J 1985 Hettinger test/retest reliability. MSc Thesis. Department of Engineering Production, Birmingham University

Hettinger T H, Beck W 1956, Der Einfluss sinusforminger sehwingungen auf die Skelelmuskolatur. International Zeitschrift fur Physiologie und Arbeitsphysiologie 16: 250–264

Hettinger T H 1957 Ein Test zur Erkennung der Disposition zu Schnenscheidenentzundungen. International Zeitschrift fur angewandt Physiologie ensclessende Arbeitsphysiologie 16: 472–479

Kemp M 1984 Work and People. Working environment branch, Department of Employment and Industrial Relations (Australia) 10: 1 25–38

Kilbom A 1983 Occupational disorders of the musculo-skeletal system. Arbetarskyddsstyrelsen Newsletter 4/82 1/83 National Board of Occupational Safety and Health. Sweden

Kuorinka I, Videman T, Lepisto M 1981 Reliability of a vibration test in screening for predisposition to tenosynovitis. European Journal of Applied Physiology 47: 356–376

Kurppa K, Waris P, Rokkanen P, 1979 Peritendinis and tenosynovitis. A review. Scandinavian Journal of Work Environment and Health 3 (Supp): 19–24

Lucire Y 1985 Neurosis in an occupational setting. Submission to the Task Force on RSI in the Australian Public Service

Luopajarvi T, Kourinka I, Kukkonen R 1982 The effects of ergonomic measures on the health of the neck and upper extremities of assembly line packers — a 4 year follow-up study. The 8th Congress of the International Ergonomics Association

McPhee B 1983 The prevention and management of repetition injuries — cause for concern. Paper presented at the Australian Physiotherapy Association Conference, Adelaide

McPhee B 1985 Prevention of occupational upper limb and neck disorders — a review. Paper presented at the International Rheumatology Conference

Ohara H, Aoyama H, Itani T 1976 Health hazards among cash register operators and the effects of improved working conditions. Journal of Human Ergology 5: 31–40

Oxenburgh M S, Row S A, Douglas D B 1985 Repetition strain injury in keyboard operators. The Journal of Occupational Health and Safety Australia and New Zealand 1(2): 106–112
Peres N C 1961 Process work without strain. Australian Factory July 1: 38–49
Ramazzini B 1713 De Morbis Artificium. Translated by W C Wright, University of Chicago Press, Chicago 1940
Ross I J K 1985 Trends in repetition strain injury statistics in NSW. The Journal of Occupational Health and Safety Australia and New Zealand 1(2): 96–101
RSI Committee 1985 Interim Report of the RSI Committee. National Occupational Health and Safety Commission (1985) Australian Government Publishing Service, Canberra
Sakurai T 1977 Vibration effects on hand arm system. Pt. 2 Observations of skin temperature. Industrial Health 15: 59–66
Schroter G 1967 Zur eignung des hettinger-testes fur die erkennung der disposition zu uberlast ungeschaden. Internationale Archiv fur Gewerbepathologie und Gewerbehygine 23: 99–105
Silverstein B, Fine L, Armstrong T, Joseph B, Buchholz B, Robertson M 1985 Cumulative trauma disorders of the hand and wrist in industry. Ergonomics International: 892–894
Stevenson M G 1985 Work design guidelines for the minimisation of repetitive strain injuries. The Journal of Occupational Health and Safety Australia and New Zealand 1(2): 113–121
Taylor R, Gow C, Corbet S 1982 Repetition injury in process workers. Community Health Studies 6(1)
Thompson A R, Plewes L W, Shaw E G 1951 Peritendinitis Crepitans and simple tenosynovitis: a clinical study of 544 cases in industry. British Journal of Industrial Medicine 8: 150–160
Thompson D 1975 Ergonomic data for evaluation and specification of operating devices on components for use by the elderly. Loughborough University of Technology Institute for Consumer Ergonomics
Tichauer E R 1966 Some aspects of stress on forearm and hand in industry. Journal of Occupational Medicine 8: 63–71
The Times 1985 Keyboard women suffer stress injury in silence. 5 Sept
Velpeau A 1841 Lecons orales de clinique chilcurgical faites a l'Hopital de la charite 3: 94, Paris
Walker J 1979 Tenosynovitis. A crippling new epidemic in industry. New Doctor 13
Welch R 1971 Bursitis, tenosynovitis. In: Parmegiani L (ed) Encyclopedia of occupational health and safety, 2nd edition, Ilo, Geneva p 223–224
Welch R 1973 The measurement of physiological predisposition to tenosynovitis. Ergonomics 16: 665–668

SECTION 2

Investigative methods

6. Geographical patterns of disease, with special reference to cancer, and the search for occupational risks

M. J. Gardner

APPROACHES TO THE DETECTION OF OCCUPATIONAL EFFECTS ON ILL-HEALTH

The identification of occupational risks of disease has come from a variety of different approaches. Given the large number of occupational hazards that are known, it is probable that there are others as yet undiscovered. Multiple avenues of exploring for these are likely to be more successful then any single method. This is so particularly for effects which occur many years after the relevant work experience, and less so for acute effects which are more able to be recognised clinically within the workplace and during the appropriate employment. Examples of such opposites are mesotheliomas arising some 20 or more years after exposure to asbestos, and dermatoses which present within a short period of time.

The main means in the past of detecting occupational health hazards has been through astute clinical observation, and this is likely to continue to be so in the future. In recent times the same can be said of the first suggestions that working in the furniture industry might predispose to nasal adenocarcinoma and that exposure to vinyl chloride monomer may cause angiosarcoma of the liver. This way of identifying occupational health risks is most likely to be successful for acute irritant, toxic, traumatic or allergic effects, and for causes of conditions with a high risk relative to the background rate.

Such clinically observed suggestions and discoveries are often followed by more formal studies to confirm the findings, or to measure the actual size of the risk or the exposure-response relationship. Epidemiological techniques, such as prevalance surveys, case-control and cohort studies, contribute to these approaches for the assessment and quantification of occupational health hazards. They will be used sometimes to search for the causal exposure: what, for example, in the working environment is responsible for the excess number of cases of some disease, say occupational asthma? Other studies will be directed towards establishing whether or not some known animal carcinogen has similar undesirable ill-effects on human health. A longitudinal study would be needed to investigate the relationship between dust levels in miners and measures of lung function to establish the rate of decline, if any, of for example vital capacity with increasing cumulative exposure to respirable dust.

An alternative approach to clinical observation is to use routinely collected health statistics to search for potential occupational risks. In the past in the UK this has been possible to some extent through the industrial injury benefit scheme, but now that it has been abolished the amount of data on occupational ill-health and injuries has been reduced considerably. Replacements such as registers kept in departments of dermatology or respiratory medicine, or reporting by general practitioners, will not be so wide-ranging. Mortality statistics are also used in this way through analysing causes of death by occupational groups. This is carried out periodically around censuses in a number of countries, but has the potential disadvantages that a full occupational history is not obtained and that the occupation given at death may not be the relevant one to any particular causally-related work exposure or experience.

The situation generally in relation to the detection of occupational health risks is thus fairly unsatisfactory in that the main approach still relies heavily on sensitive clinical observation. Although such perspicacity has successfully identified a large number of effects, nonetheless it will not operate with uniform high effectiveness and additional approaches are needed. The possibility discussed in this chapter is of examining the geographical distribution of disease (particularly within countries) to search for clues to investigate in further detail so as to identify occupationally-related problems. This requires the existence of some systems for collecting, fully, information on disease events. In most countries mortality data will be available, in others cancer registration information and hospital morbidity data. A geographical approach is, of course, not new and is unlikely to determine occupational hazards absolutely. However it could, used systematically, generate leads to be investigated on a local basis by studies of individuals, much as those following risks suggested clinically.

GEOGRAPHICAL DISEASE MAPPING AND OCCUPATIONAL RISKS DISCOVERED IN THE PAST

The history of geographical studies of disease in detecting occupational health hazards is slim, and only two instances are commonly recognised. This is a warning that the approach is unlikely to be over-productive, even though in the past searches have not been mounted in a systematic manner. The two known instances are both of carcinogenic effects, which are necessarily less likely to be identified by an occupational physician because of the likelihood of their presentation after the relevant employment has ended. Each of the two discoveries are of interest in themselves and are described briefly.

Lung cancer in fluorspar miners

During the early 1950s a high level of mortality from lung cancer was noted among the male population of St Lawrence, a small community on the Burin Peninsula of Newfoundland (de Villiers & Windish, 1964). Comparisons with the mortality experience of a control community in the same geographical region and with the population of the rest of Newfoundland estimated an excess of lung cancer in St Lawrence of about 29-fold among men, but not women.

St Lawrence had been principally an isolated fishing town until the 1920s when the price of salt-cured fish began to drop. In 1929 the Grand Banks earthquake disrupted the fishing grounds and all the fishing equipment was destroyed by the ensuing tidal wave. At this time development of the local fluorspar deposits began, and mining displaced fishing as the principal male occupation.

A detailed analysis of the working histories of the lung cancer cases and all other members of the workforce of the mines since they opened was undertaken. This showed that 28 of the 29 lung cancer cases were underground workers, who formed a much smaller proportion of the overall employees. Various constituents of the rock and ore were discussed as potential carcinogens, but the conclusion was that 'exposure to radon and its daughters appears to be the most important single hazard' (de Villiers & Windish, 1964). This was suggested both by known lung cancer excesses in uranium miners (Lorenz, 1944) and by measured radiation levels.

Nasal cancer in boot and shoe manufacturers

In 1965 Hadfield and MacBeth, ENT surgeons in High Wycombe and Oxford, first suggested that adenocarcinoma of the nose and nasal sinuses might be related to work in the furniture industry in England (MacBeth, 1965). The manufacture of tables, chairs, cabinets and such items is concentrated in the area of Buckinghamshire around High Wycombe where the hardwood forests of the Chiltern Hills produced a plentiful supply of suitable timber. This suggestion was confirmed by a detailed study carried out in the area (Acheson et al, 1968), and at the same time the geographical distribution of nasal cancers in the Oxford Cancer Registry Region (which includes Buckinghamshire) was examined. This revealed a focus of high incidence in Northamptonshire, an area outside the furniture-making locality but within the geographical boundaries of the Oxford registry (Acheson et al, 1970). When the occupational histories of the Northamptonshire cases were looked into, a concentration among workers exposed to dusty conditions in the manufacture and repair of boots and shoes was found. Again, the excess was mainly due to adenocarcinomas, but was also found for squamous cell tumours, and has been confirmed in other parts of England and Wales (Acheson et al, 1981a) and in Italy (Cecchi et al, 1980). So this discovery was a product of serendipity, and offers the potential for further clues to be unearthed by a wider search in the future of the sources of ill-health statistics on a geographical basis.

GEOGRAPHICAL DISEASE MAPPING AND KNOWN OCCUPATIONAL RISKS—THE PRESENT STATE

The use of maps of disease as a method of searching for causes, as well as describing the distribution of disease and pointing to areas of high risk as possible localities for concentrating prevention activities, received a great impetus from the publication in the mid 1970s of a cancer atlas for the USA (Mason et al, 1975). This was followed by an atlas showing mortality patterns from a selection of other non-malignant causes (Mason et al, 1981), and by atlases of disease mortality in a number of other countries—Germany (Becker et al, 1984), China (Chinese Academy of Medical Sciences, 1981), Japan (Shigematsu, 1982), England and Wales (Gardner et al, 1983, 1984), Italy (Capocaccia et al, 1984), among others. Earlier atlases had been produced decades ago, for example, Stocks (1936, 1937, 1939) and Howe (1970), but had not been exploited in the same detailed way as has happened recently, particularly in the USA. These studies have served to indicate that for a number of occupationally-related diseases (in particular, cancers of various sites) the geographical pattern is such that the occupational associations are apparent. A number of these will now be described for the USA and England and Wales.

Known risks in the USA

One of the earliest maps from the USA cancer atlas which was seen to mirror clearly a major occupational cause was that of bladder cancer in men (Hoover et al, 1975).

Clusters of areas with elevated mortality correlated with areas of occupational exposures already linked with this tumour, in particular, there was a concentration of areas with raised rates in New Jersey. In this state a high proportion of the employed male population worked in the chemical industry, including the manufacture of organic chemicals known to have the potential to cause bladder cancer in man. In the USA overall, local area death rates showed high correlations with the proportions of the local male population employed in the manufacture of dyes, dye intermediates and organic pigments (Hoover & Fraumeni, 1975). These branches of the industry, of course, entailed exposures to high levels of beta-naphthylamine and benzidene, both of which have been demonstrated in studies of individuals to be causes of bladder cancer (Case et al, 1954).

The geographical pattern of lung cancer mortality demonstrated a concentration of areas with high rates along the southern and eastern coast, primarily among white men (Blot & Fraumeni, 1976). Detailed case-control studies in some of these coastal areas, aimed at identifying responsible factors, showed an excess of lung cancer among workers employed in shipyards, particularly those in operation during World War II (Blot et al, 1978, 1980, 1982). It is thought that this excess is related to exposures to asbestos, which was used extensively in the insulation of ships during construction and repair. This example is covered in more detail later in the chapter.

Known risks in England and Wales

In a similar manner to the map for bladder cancer among men in the USA, that for England and Wales showed concentrations of areas with high mortality rates (Gardner et al, 1982b, 1983). The main location of these areas was in parts of northern and central England where there had been major production activities in the dyestuffs and rubber industries.

The geographical distribution of areas with raised mortality from cancer of the nose, nasal cavities, middle ear and accessory sinuses in men also indicated clearly known risks (Gardner et al, 1982b). Thus the two main areas associated with the furniture industry were shown, and one of the areas in the Northamptonshire boot and shoe industry. Five areas only with markedly high rates were identified, and three of these were explicable on the basis of known occupational risks.

Although, in contrast to the USA, there is no important geographical relationship between lung cancer mortality rates in England and Wales and the distribution of the past asbestos industry, there is a striking association in the maps for pleural and peritoneal mesothelioma in each sex (Gardner et al, 1982a, 1985). Areas with high usage of asbestos, particularly crocidolite, such as in ship-building and repairing and in naval dockyards, showed high death rates for men as was expected (Elmes & Wade, 1965; Rossiter & Coles, 1980). For women, the areas with high mesothelioma rates were mainly those where filters containing crocidolite for wartime gas masks had either been manufactured or the masks assembled, again these were known risks (Newhouse & Berry, 1979; Jones et al, 1980; Acheson et al, 1982; Morgan & Holmes, 1982; Wignall & Fox, 1982). These maps suggest strongly that asbestos, amphibole mainly, has been the only numerically important cause of mesothelioma in England and Wales, since areas where asbestos was not used do not show raised death rates (Acheson, 1984).

In addition, mortality rates from pneumoconioses in men are aligned geographically with mining areas of the country (Gardner et al, 1984). A breakdown of overall pneumoconioses into sub-divisions (such as coal miner's pneumoconiosis, silicosis and asbestosis) shows relationships with the distribution of coal mining, slate quarrying, tin mining and other occupational exposures to silica dusts, and the asbestos industry (Gardner et al, 1986).

Comments on the geography of these known occupational risks

The above examples have established that an examination of the geographical distribution of some diseases shows clearly the associated occupational risks. What is not obvious is whether the maps, in the absence of knowledge about the causes producing their patterns, would have led to establishing new occupational hazards such as in dyestuffs, asbestos, etc. This, of course, is the most important question in relation to their projected future use, producing pertinent hypotheses leading to the identification of new occupational risks. Before looking at ways of exploring the geographical patterns for clues to causes, a few comments on the described relationships can be made.

1. For more common diseases, such as lung or bladder cancer, diagnosis and death certification is unlikely to be much influenced, if at all, by knowledge of the person's occupation, as these conditions are produced also by other non-occupational causes. However, for an individual to be diagnosed and classified as having pneumoconiosis a known relevant occupational exposure is usually required. Thus, the geographical distribution of the disease is constrained to certain industries and geographical areas by nature of the causes already known. Therefore, in this situation an examination of the geographical distribution of the disease cannot really lead to new causes being identified, but can describe the locations of high risk areas and indicate the relative sizes of the risks.

2. Although an appropriate occupational history of working with an established cause is not paramount for some diseases, it may influence to a greater or lesser degree the diagnosis finally reached. Such an example could be the case for mesothelioma, where there may be a greater recognition of the condition and tendency to certify it in cases where an occupational history of working with asbestos is known and/or the local area is one where past asbestos usage was high and mesothelioma relatively common. In such instances, lack of awareness of the possibility in other patients and in other geographical areas may bias the certification rates downward (Wright et al, 1984).

3. An interesting feature of the maps which have been discussed is that, although the diseases shown are the result of long-term rather than acute effects, the geographical distributions of the associated occupations and industries are preserved. Given that in most cases first exposure would have occurred many years past, and that some persons would have changed jobs and moved house, the ability to see the occupational relationship is of relevance. If this had not been the case it would have suggested that the geographical approach was inefficient, because migration (which has increased and will probably continue to increase) would destroy any pattern. However, the

majority of population movement in the past has been only local, and long-distance change of job and place of residence is still relatively unusual.

NON-OCCUPATIONAL FACTORS IN THE GEOGRAPHICAL DISTRIBUTION OF DISEASE

There are many non-occupational determinants of disease which have the potential to modify geographical differences, and must be considered (Gardner, 1984). These can conveniently be grouped into three categories as follows.

First, there are factors in the general environment which, by and large, tend to affect all persons living within particular areas. These would include climate (such as humidity, sunshine and temperature), geochemistry, water and air quality, and other such locally determined factors. Examples of suggested disease relationships are exposure to sunshine and the development of malignant melanoma of the skin (Swerdlow, 1979; Armstrong, 1984) which is apparent in the south/north gradient in the cancer atlases for both the USA (Mason et al, 1975) and for England and Wales (Gardner et al, 1983) and for which an occupational component of sunlight exposure is suggested as important for melanomas of the head, face and neck (Beral & Robinson, 1981); environmental nitrates and gastric cancer, for which the evidence is really quite weak (Fraser, 1985); and the softness of drinking water and heart disease, which has been observed for many years within a number of countries but the causal mechanism, if any, is still unknown (Comstock, 1986; Pocock et al, 1986).

Secondly, there are personal environmental factors, which because they relate to individual habits and behaviour will differentially affect persons living in the same geographical areas. Such personal factors would include diet, cigarette smoking, and other individually determined influences on health. In addition any resultant effects of the home residential environment, such as respiratory illnesses, would be included.

Thirdly, genetics and the potential predisposition of some individuals to certain diseases have to be considered.

As well as considering non-occupational causes as influences in the geographical pattern of disease, there are additional items related to the quality of the mortality or morbidity data themselves that need to be considered. Thus, the following should be assessed in any study where possible.

1. As alluded to earlier, the quality of the medical information input needs to be evaluated to provide guidance on whether diagnostic and certification practices in different geographical areas produce valid and comparable data.

2. The influence, if any, of the use of mortality rather then morbidity data on the geographical distribution of disease. In particular, if treatment, and hence survival, is better in some areas than others, this will modify the disease patterns. This will occur in such a way that mortality rates will not be good comparative indices of disease frequencies in different areas.

3. The location of residence at the time of death may be different from that of usual home residence. This can occur when sick people move to caring institutions or to live with relatives. Although this is unlikely to generally produce misleading disease rates, an exceptional example where a small area's cancer mortality was doubled due

to the presence of a home for the terminally ill was given by Gardner & Winter (1984a).

4. Another problem which can occur is that the size and demographic structure of the population living in particular geographical areas may not be well estimated. Thus, in most countries local population figures are only available in detail by age and sex at censuses which are usually 10 years apart. Inter-censal estimates can be grossly incorrect if areas are either increasing or decreasing rapidly in population size.

GEOGRAPHICAL DISEASE MAPPING AND THE SEARCH FOR UNKNOWN OCCUPATIONAL RISKS

Looking at the geographical distribution of disease has a long history (Alderson, 1982), but its systematic exploitation for suggesting clues to environmental, including occupational, determinants has not received very much attention. This can be either on an international scale, within countries such as examining disease patterns between counties, within smaller areas such as wards or constituencies, or even at an individual home level, since the example of Snow should not be forgotten (Snow, 1849). I shall focus mainly on using any of the first three area scales, and the approach outlined here would not be appropriate for the last type of geographical investigation. The larger areas will be less homogeneous internally in terms of their occupational characteristics and hence less suitable than smaller areas; on the other hand, the smaller areas will suffer from lack of stability of estimated disease rates due to smaller numbers.

Current and future strategies

One of the first research groups to set up detailed geographical studies was the Environmental Epidemiology Branch of the National Cancer Institute in the USA. Their springboard was the cancer mortality atlas for the USA mentioned earlier, which was published in the mid 1970s (Mason et al, 1975). The need to explain and understand the patterns shown in this atlas, and later those for non-whites and for other (non-cancer) causes of death, led to a major expansion of activity in this area within that research group. The range of approaches envisaged and which have been taken are covered in one of their earlier papers on the subject (Blot et al, 1979a), which indicates lines of enquiry for turning geographical maps into factors in disease causation. A similar approach was later outlined by the Medical Research Council Environmental Epidemiology Unit (Acheson, 1981; Acheson et al, 1981b).

One of the initial investigations carried out with the US data was an examination of the geographical relationship between nasal cancer mortality and work in the furniture industry. This was undertaken to see whether the finding in the UK described earlier was relevant in another setting, where different processes and woods may be used. The answer was that there were high death rates among white males in USA counties with a heavy concentration of furniture manufacturing industries (Brinton et al, 1976), and a later examination of death certificates in such counties in North Carolina showed a four-fold excess risk among those whose usual trade, as recorded on their death certificate, was furniture manufacturing (Brinton et al, 1977). A more recent case-control study obtaining occupational histories from subjects or their next of kin and histological data on the tumours showed an increased risk of adenocarcinoma in the furniture industry in a similar area of the USA (Brinton et al, 1984). These studies therefore demonstrated that a known occupational risk could be discovered in routinely collected mortality and census data. Because of this it would be imprudent to suggest that a more in-depth study of inter-relationships between disease and occupation on a geographical scale would not generate hypotheses about aetiological factors which would eventually prove to be substantiated.

The example in the previous paragraph outlines some of the proposed steps which are now listed in more detail. This is largely based on the suggestions from the USA (Blot et al, 1979a; Blot & Fraumeni, 1982), with some supplementation.

1. The starting points are the geographical maps of disease obtained from routine data, and now available in many countries in atlas form.

2. These can be scanned, with a general background of known disease causes and of the environmental geography of a country or area (in particular, the location of occupations and industries) to see whether there are obvious suggestions from the contrasts, in particular, between areas with low and high disease rates.

3. As a supplement, the geographical distribution of occupation and industry can be mapped in the same area format as the mortality rates. This is usually easily possible now with relevant data on computer tape and appropriate software available. The costs are low at least for in-house maps. These enable visual comparisons of the maps of disease and of occupation for assessment of inter-relationships, and is a sensible addition to one's own notion of how industry is distributed (as it is for disease). In addition, with a suitable data-base, maps can be produced indicating geographical areas where the workforce has high potential exposure to specific chemical, physical and biological hazards, as has been done for formaldehyde in the USA (Frazier et al, 1983).

4. A relevant next step for the areas with high disease rates, particularly if any occupational relationship is suggested by 2 or 3 above, is to examine the occupation/industry information given on the death certificate for persons contributing to the high rates. The comparison is of usual occupations, which are commonly the ones recorded, for persons dying from the disease of interest with the usual occupations of control persons dying in the same areas from other causes.

5. Because the usual occupation is not always going to be the important or relevant one, a case-control study of new cases of the disease may be carried out to obtain life-time occupational and industrial working histories to enable all jobs to be examined, rather than the one, often chosen by a relative, reported to the registrar at the time of death. Again, these studies will be more appropriate in selected areas with high disease rates, but in this instance will require field work to collect the relevant information from suitable informants.

6. Correlation studies of a formal nature, by contrast to the visual approach of 2 and 3 above, can be carried out if relevant geographical data on the distribution of occupation and industry are available. These can usually be found in censuses, either national censuses at 10- or 5-yearly intervals, or from statistics that are collected

and produced by industrial organisations themselves. If similar data on possible confounding factors, such as climate, air pollution, housing conditions, etc. are obtainable these should be taken into account. With this approach it is possible to examine a large number of relationships and should be considered as a hypothesis-generating exercise. Any forthcoming ideas should be tested in more direct studies, such as case-control or cohort studies, focussed particularly on the suggested occupations.

7. Although implicit somewhat in 2 above, the formal analysis of whether areas with high disease rates appear in geographical clusters, as opposed to being randomly spread, may be relevant to considerations of occupational causes. This does not necessarily have to be so. An effectively random scatter of areas with raised rates could still have an occupational cause if persons in the relevant occupation are similarly haphazardly scattered around the country.

Example from the USA

The example, outlined earlier, of the finding of excessive lung cancer mortality rates among men on the eastern and southern seaboard of the USA is illuminating. The pattern was unexpected and to seek explanations the rates were correlated with different local indices, including measures of urbanisation, socio-economic status, income levels and industrial activity. Correlations with the paper and petrochemical industries were found, and because the latter were prominent in Florida/Georgia and Texas/Louisiana respectively it suggested that an industrial component may exist (Blot & Fraumeni, 1976).

A study in Georgia, looking at occupations on death certificates of lung cancer cases and controls (persons dying from other causes), indicated that the excess related to the wood-paper industry was limited to residents of the rural coastal countries and not found inland (Harrington et al, 1978). A more detailed case-control study of newly diagnosed lung cancer cases obtaining lifetime residential, occupational and smoking histories uncovered a lung cancer risk associated with employment in shipyards during World War II (Blot et al, 1978). This relationship with early employment was not apparent from the usual occupation as recorded on the death certificate, indicating an acknowledged weakness of this approximate approach.

An examination of areas throughout the USA with large shipyards during World War II subsequently showed high rates for lung and laryngeal cancers, and indications of raised levels of oropharyngeal, oesophageal and stomach cancers also (Blot et al, 1979b). This suggested, from known evidence of the risks from asbestos exposure, that exposure to asbestos in the shipyards was responsible for the excesses. It is known that asbestos, particularly amphibole, was used in large quantities for insulation purposes in naval shipyards during the last war. Thus, an important past occupational determinant of lung cancer mortality was found, not a new one, but one operating on an unsuspected scale and in an unsuspected place, despite the fact that cigarette smoking is the major cause of lung cancer in the USA.

Forthcoming atlases for the USA based on mortality data will include maps for the decades 1950–1959, 1960–1969 and 1970–1979 separately. These will enable the evaluation of any time-related changes in the geographical distribution of diseases, for example the three maps for lung cancer in men show that inland eastern areas have increasing relative mortality which will require explanation (Pickle & Mason, 1986).

Examination of the changing geographical patterns over time may well provide more useful clues to disease causation than static maps, especially when changing occupational and exposure distributions are studied alongside.

Examples from England and Wales

1. The discoveries of the association between employment in the furniture industry and adenocarcinoma of the nasal cavities by careful clinical observation and between employment in the boot and shoe industry and nasal cancer by serendipity have been mentioned earlier. It is interesting that, had these two relationships not been found 15–20 years ago, a systematic correlational analysis of geographical nasal cancer mortality rates and the distribution of industry in England and Wales would have suggested looking at these industries (Gardner & Winter, 1984b; Gardner, 1984). The furniture industry was found to have the highest relationship of 207 different industrial categories with nasal cancer, and would have pointed to an in-depth investigation of the industry, such as the cohort study that was carried out by Acheson and his colleagues (Rang & Acheson, 1981; Acheson et al, 1984). It is possible that the relationship is highest because of diagnostic bias exaggerating rates in furniture workers due to knowledge of the association, although most doubtful, and it is debatable whether the statistical correlation would have emerged and led to the discovery of the cause without prior knowledge of the situation. The finding itself is now well established having been replicated in a number of other countries, as well as the USA including Italy (Battista et al, 1983) and Scandinavia (Hernberg et al, 1983).

An alternative approach, of carrying out a local case-control study, is being adopted in East London where an area with relatively high nasal cancer rates was found unexpectedly. This is based in adjacent localities with high rates for women as well as men (Gardner et al, 1982b), and is being carried out with the co-operation of a number of local hospitals in the district. The area has some history of furniture and of boot and shoe industries, but it seems unlikely to be able to explain the raised mortality among women. Associations with employment in the tailoring industry were suggested by the correlational analysis (Gardner et al, 1982b; Gardner & Winter, 1984b), and this industry among all others will be looked at for an explanation. Occupational and industrial information is being collected from hospital records, death certificates, cancer registries, and also from the 1939 Register which forms the basis of the National Health Service Central Register at Southport. The inability to interview the known cases because most of them had died, and because newly diagnosed cases are uncommon due to the relative rarity of the tumour, made it necessary to use routinely recorded information, and the occupation/industry in 1939 may be particularly appropriate for cases diagnosed in the 1970s and 1980s. However, the study is lacking full life-time occupational and industrial histories.

An analysis from occupations recorded on death certificates alone has been published for this same part of London (Baxter & McDowall, 1986), the study arising from the same observation of high nasal cancer mortality in the area. The results did not suggest an association with any branch of industry other than furniture, but the data were limited to men. The occupational history of women is more difficult to ascertain from routine records due to their general lack of regular employment. Results of the case-control study of nasal cancer in North Carolina and Virginia which have recently been published (Brinton et al, 1985) take further the suggested relationship with the textile and clothing industries in showing an elevated risk of adenomacarcinoma in both men and women workers.

2. Higher rates of bladder cancer for men were seen in areas with higher concentrations of the chemical industry among the 1366 Local Authority areas which were used for the cancer atlas in England and Wales. A formal correlational analysis between the rates and the proportion of local men employed in the industry confirmed the visual impression from the map. A subdivision of the chemical industry overall into its components showed a high association of bladder cancer with the dyes and pigments sub-category, which is as would be expected from the known risks due to past exposures to beta-naphthylamine in particular. Other components, such as soaps and detegents and synthetic resins, showed lesser relationships and some none at all.

When correlations with other cancers were examined, it was found that there was a stronger association between bladder cancer rates and dyes and pigments than for other major cancers, and also that the relationships of the other cancers with soaps and detergents and with synthetic resins were as high or higher than for bladder cancer. The lack of specificity in the other industries, as well as the small numerical levels of the correlation coefficients, is counter-suggestive of a particular occupational risk. This is an additional method of assessing the importance of any particular disease and industry association found from geographical studies where many correlations are investigated.

ADVANTAGES AND DISADVANTAGES OF A GEOGRAPHICAL APPROACH TO IDENTIFYING OCCUPATIONAL RISKS

The growing availability of maps showing the geographical distribution of disease, even if mainly for mortality, makes an examination of this approach possible in a wider way than before. That it could be successful in identifying occupational hazards is indicated by the examples which have been given. The distribution of occupations and industries within a country is one environmental factor which could influence the appearance of disease maps. Information on the geographical spread of occupations and industries are available from routine data sources such as censuses, and may possibly be obtained for relevant periods in the past. Thus, for cancer mortality occupational information by area from 10, 20, 30, ... years ago would be more relevant. Occupational data available in this way would usually be recorded on census forms by the individuals themselves, rather than those on death certificates or cancer registrations which are commonly given by relatives or other proxy informants.

The geographical approach is of course indirect, in the sense that it looks at rates and relationships in populations rather than in individuals. Thus a correlation between a disease and an occupation on an area basis needs, as discussed earlier, studies at the individual level to advance further the association in terms of any direct causal effect. Migration may be a difficulty, although examples have been given where this has not obscured relationships. However, it could bias them depending upon any selective migration of the healthy or sick, and will probably have a larger effect in the future with increasing mobility of populations. There will also be a dilution effect on any relationships because occupational groups are usually small sub-sections of local populations. This effect will be greater for common diseases with confounding causes than for diseases without non-occupational causes.

In summary, geographical studies of occupational risks have indicated a potential value that makes their exploration in parallel with other methods useful. A consistency of findings with, for example, analyses based on occupational data from individuals, such as the Occupational Mortality Decennial Supplement in England and Wales (Office of Population Censuses and Surveys 1978, 1986), would add to the need to investigate further. This type of comparison would limit the potential number of false associations, in a causal sense, which might be followed up if the geographical studies alone were used for hypothesis-generating independently of any other source of information.

REFERENCES

Acheson E D 1981 Towards a strategy for the detection of industrial carcinogens. British Journal of Cancer 44: 321–331

Acheson E D 1984 Some geographical patterns by cancer site from the Atlas of cancer mortality in England and Wales, 1968–78. In: Gardner M J (ed) Maps and Cancer. Medical Research Council Environmental Epidemiology Unit, Southampton, p 25–27

Acheson E D, Cowdell R H, Hadfield E H, MacBeth R G 1968 Nasal cancer in woodworkers in the furniture industry. British Medical Journal ii: 587–596

Acheson E D, Cowdell R H, Jolles B 1970 Nasal cancer in the Northamptonshire boot and shoe industry. British Medical Journal ii: 385–393

Acheson E D, Cowdell R H, Rang E H 1981a Nasal cancer in England and Wales: an occupational survey. British Journal of Industrial Medicine 38: 218–224

Acheson E D, Gardner M J, Pippard E C, Grime L P 1982 Mortality of two groups of women who manufactured gas masks from chrysotile and crocidolite asbestos: a 40-year follow-up. British Journal of Industrial Medicine 39: 344–348

Acheson E D, Gardner M J, Winter P D 1981b Towards a strategy for the identification of occupational carcinogens in England and Wales — a preliminary report. In: Peto R Schneidermann M (eds) Banbury Report 9: Quantification of occupational cancer. Cold Spring Harbor Laboratory, New York, p 591–598

Acheson E D, Pippard E C, Winter P D 1984 The mortality of English furniture makers. Scandinavian Journal of Work Environment and Health 10: 211–217

Alderson M R 1982 The geographical distribution of cancer. Journal of the Royal College of Physicians of London 16: 245–251

Armstrong B K 1984 Melanoma of the skin. British Medical Bulletin 40: 346–350

Battista G, Cavallucci F, Comba P, Quercia A, Vindigni C, Sartorelli E 1983 A case-referent study on nasal cancer and exposure to wood dust in the province of Siena, Italy. Scandinavian Journal of Work Environment and Health 9: 25–29

Baxter P J, McDowall M E 1986 Occupation and cancer in London: an investigation into nasal and bladder cancer using the Cancer Atlas. British Journal of Industrial Medicine 43: 44–49

Becker N, Frentzel-Beyme R, Wagner G 1984 Atlas of cancer mortality of the Federal Republic of Germany. Springer, Berlin

Beral V, Robinson N 1981 The relationship of malignant melanoma, basal and squamous skin cancers to indoor and outdoor work. British Journal of Cancer 44: 886–891

Blot W J, Davies J E, Morris Brown L, Nordwall C W, Buiatti E, Ng A, Fraumeni J F 1982 Occupation and the high risk of lung cancer in northeast Florida. Cancer 50: 364–371

Blot W J, Fraumeni J F 1976 Geographic patterns of lung cancer: industrial correlations. American Journal of Epidemiology 103: 539–550

Blot W J, Fraumeni J F 1982 Geographical epidemiology of cancer in the United States. In: Schottenfeld D, Fraumeni J F (eds) Cancer epidemiology and prevention. Saunders, Philadelphia, p 179–193

Blot W J, Fraumeni J F, Mason T J, Hoover R N 1979a Developing clues to environmental cancer: a stepwise approach with the use of cancer mortality data. Environmental Health Perspectives 32: 53–58

Blot W J, Harrington J M, Toledo A, Hoover R, Heath C W, Fraumeni J F 1978 Lung cancer after employment in shipyards during World War II. New England Journal of Medicine 299: 620–624

Blot W J, Morris L E, Stroube R, Tagnon I, Fraumeni J F 1980 Lung and laryngeal cancers in relation to shipyard employment in coastal Virginia. Journal of the National Cancer Institute 65: 571–575

Blot W J, Stone B J, Fraumeni J F, Morris L E 1979b Cancer mortality in US counties with shipyard industries during World War II. Environmental Research 18: 281–290

Brinton L A, Blot W J, Stone B J, Fraumeni J F 1977 A death certificate analysis of nasal cancer among furniture workers in North Carolina. Cancer Research 37: 3473–3474

Brinton L A, Blot W J, Becker J A, Winn D M, Patterson Browder J, Farmer J C, Fraumeni J F 1984 A case-control study of cancers of the nasal cavity and paranasal sinuses. American Journal of Epidemiology 119: 896–906

Brinton L A, Blot W J, Fraumeni J F 1985 Nasal cancer in the textile and clothing industries. British Journal of Industrial Medicine 42: 469–474

Brinton L A, Stone B J, Blot W J, Fraumeni J F 1976 Nasal cancer in US furniture industry counties. Lancet ii: 628

Capocaccia R, Farchi G, Mariotti S, Verdecchia A, Angeli A, Morganti P, Panichelli Fucci M L 1984 La mortalita in Italia nel periodo 1970–79. Istituto Superiore di Sanita, Rome

Case R A M, Hosker M E, McDonald D B, Pearson J T 1954 Tumours of the urinary bladder in workmen engaged in the manufacture and use of certain dyestuff intermediates in the British chemical industry. British Journal of Industrial Medicine 11: 75–104

Cecchi F, Buiatti E, Kriebel D, Nastasi L, Santucci M 1980 Adenocarcinoma of the nose and paranasal sinuses in shoemakers and woodworkers in the province of Florence, Italy (1963–77). British Journal of Industrial Medicine 37: 222–225

Chinese Academy of Medical Sciences 1981 Atlas of cancer mortality in the People's Republic of China. China Map Press, Beijing

Comstock G W 1986 Studies into cardiovascular disease and water quality in the United States and Canada. In: Thornton I (ed) International Symposium on Geochemistry and Health. Science Reviews, London

de Villiers A J, Windish J P 1964 Lung cancer in a fluorspar mining community 1. Radiation, dust and mortality experience. British Journal of Industrial Medicine 21: 94–109

Elmes P, Wade O L 1965 Relationships between exposure to asbestos and pleural malignancy in Belfast. Annals of the New York Academy of Sciences 132: 549–557

Fraser P 1985 Nitrates: epidemiological evidence. In: Wald N J, Doll R (eds) Interpretation of negative epidemiological evidence for carcinogenicity. IARC Scientific Publications 65, Lyon, p 183–194

Frazier T M, Lalich N R, Pedersen D H 1983 Uses of computer-generated maps in occupational hazard and mortality surveillance. Scandinavian Journal of Work Environment and Health 9: 148–154

Gardner M J 1984 Mapping cancer mortality in England and Wales. British Medical Bulletin 40: 320–328

Gardner M J, Acheson E D, Winter P D 1982a Mortality from mesothelioma of the pleura during 1968–78 in England and Wales. British Journal of Cancer 46: 81–88

Gardner M J, Jones R D, Pippard E C, Saitoh N 1985 Mesothelioma of the peritoneum during 1967–82 in England and Wales. British Journal of Cancer 51: 121–126

Gardner M J, Winter P D 1984a Mapping small area cancer mortality: a residential coding story. Journal of Epidemiology and Community Health 38: 81–84

Gardner M J, Winter P D 1984b Extensions to a technique for relating mortality and environment—exemplified by nasal cancer and industry. Scandinavian Journal of Work Environment and Health 10: 219–223

Gardner M J, Winter P D, Acheson E D 1982b Variations in cancer mortality among local authority areas in England and Wales: relations with environmental factors and search for causes. British Medical Journal 284: 784–787

Gardner M J, Winter P D, Barker D J P 1984 Atlas of mortality from selected diseases in England and Wales 1968–78. Wiley, Chichester

Gardner M J, Winter P D, Pannett B, Powell C A 1986 The geographical distribution of mortality from pneumoconioses and related industries in England and Wales. Annals of Occupational Hygiene, in press

Gardner M J, Winter P D, Taylor C P, Acheson E D 1983 Atlas of cancer mortality in England and Wales 1968–78. Wiley, Chichester

Harrington J M, Blot W J, Hoover R N, Housworth W J, Heath C W, Fraumeni J F 1978 Lung cancer in coastal Georgia: a death certificate analysis of occupation. Journal of the National Cancer Institute 60: 295–298

Hernberg S, Collan Y, Degerth R, Englund A, Engzell U, Kuosma E, Mutanen P, Nordlinder H, Sand Hansen H, Schutlz-Larsen K, Sogaard H, Westerholm P 1983 Nasal cancer and occupational exposures: preliminary report of a joint Nordic case-referent study. Scandinavian Journal of Work Environment and Health 9: 208–213

Hoover R, Mason T J, McKay F W, Fraumeni J F 1975 Geographic patterns of cancer mortality in the United States. In: Fraumeni J F (ed) Persons at high risk of cancer: an approach to cancer etiology and control. Academic Press, New York, p 343–360

Hoover R N, Fraumeni J F 1975 Cancer mortality in US counties with chemical industries. Environmental Research 9: 196–207

Howe G M 1970 National atlas of disease mortality in the United Kingdom, 2nd edn. Nelson, London

Jones J S P, Smith P G, Pooley F D, Berry G, Sawle G W, Madeley R J, Wignall B K, Aggarwal

A 1980 The consequences of exposure to asbestos dust in a wartime gas-mask factory. In: Wagner J C (ed) Biological effects of mineral fibres. IARC Scientific Publications 30, Lyon 2: 637–653
Lorenz E 1944 Radioactivity and lung cancer: a critical review of lung cancer in the mines of Schneeberg and Joachinstal. Journal of the National Cancer Institute 5: 1–15
MacBeth R G 1965 Malignant disease of the paranasal sinuses. Journal of Laryngology and Otology 79: 592–612
Mason T J, Fraumeni J F, Hoover R, Blot W J 1981 An atlas of mortality from selected diseases. NIH Publication No. 81-2397. US Government Printing Office, Washington
Mason T J, McKay F W, Hoover R, Blot W J, Fraumeni J F 1975 Atlas of cancer mortality for US counties: 1950–1969. DHEW Publication (NIH) 75-780. US Government Printing Office, Washington
Morgan A, Holmes A 1982 Concentrations and characteristics of amphibole fibres in the lungs of workers exposed to crocidolite in the British gas-mask factories, and elsewhere, during the Second World War. British Journal of Industrial Medicine 39: 62–69
Newhouse M L, Berry G 1979 Patterns of mortality in asbestos factory workers in London. Annals of the New York Academy of Sciences 330: 53–60
Office of Population Censuses and Surveys 1978 Occupational Mortality Decennial Supplement for England and Wales 1970–72 (Series DS 1). Her Majesty's Stationery Office, London
Office of Population Censuses and Surveys 1986 Occupational Mortality Decennial Supplement for Great Britain 1979, 80, 82, 83. Her Majesty's Stationery Office, London
Pickle L W, Mason T J 1986 Mapping cancer mortality in the United States. In: Thornton I (ed) International Symposium on Geochemistry and Health. Science Reviews, London
Pocock S J, Shaper A G, Powell P B, Packham R 1986 The British regional heart study: cardiovascular diseases and water quality. In: Thornton I (ed) International Symposium on Geochemistry and Health. Science Reviews, London
Rang E H, Acheson E D 1981 Cancer in furniture workers. International Journal of Epidemiology 10: 253–261
Rossiter C E, Coles R M 1980 H M Dockyard, Devonport: 1947 mortality study. In: Wagner J C (ed) Biological effects of mineral fibres. IARC Scientific Publications 30, Lyon 2: 713–721
Shigematsu I 1982 National atlas of major disease mortalities for cities, towns and villages in Japan 1969–1978. Japan Health Promotion Foundation, Tokyo
Snow J 1849 On the mode of communication of cholera. Churchill, London
Stocks P 1936 Distribution in England and Wales of cancer of various organs. British Empire Cancer Campaign Annual Report 13: 239–280
Stocks P 1937 Distribution in England and Wales of cancer of various organs. British Empire Cancer Campaign Annual Report 14: 198–223
Stocks P 1939 Distribution in England and Wales of cancer of various organs. British Empire Cancer Campaign Annual Report 16: 308–343
Swerdlow A J 1979 Incidence of malignant melanoma of the skin in England and Wales and its relationship to sunshine. British Medical Journal ii: 1324–1327
Wignall B K, Fox A J 1982 Mortality of female gas mask assemblers. British Journal of Industrial Medicine 39: 34–38
Wright W E, Sherwin R P, Dickson E A, Bernstein L, Fromm J B, Henderson B E 1984 Malignant mesothelioma: incidence, asbestos exposure, and reclassification of histopathology. British Journal of Industrial Medicine 41: 39–45

7. Pooling strategies for data from occupational epidemiological studies

R. Frentzel-Beyme

THE NEED FOR RECOMMENDATIONS AND STANDARDS FOR POOLING DATA OF DIFFERENT ORIGINS

Epidemiologic research has to face so many dilemmas which stem from its observational character that the importance of combining the evidence derived from different studies is rising. One of the epidemiologic arguments for a causal association is the combined evidence from several studies (although results from several studies may be wrong). Therefore, it becomes apparent that some formalisation of the process of combining data from different studies carried out independently, i.e. without a common protocol, is needed.

This task has been taken up by several authors, and at present a Working Group of the ICOH Scientific Committee on Occupational Epidemiology is engaged in a search for guidelines, recommended methods and practical directions for the application of such methods or pooling strategies.

In recent years, attempts also have been made to pool individual studies, such as clinical trials, the results of which sometimes appear to be conflicting (Lewis, 1985). Informal pooling of data is already a common practice, if only by a summation or overview of positive or negative effects found in the individual studies taken together. Attempts to formalise this process have been made but yielded more difficulties than success. As stated by Lewis, three steps can be taken to proceed in a favourable direction.

Firstly, to make the objectives of combining the evidence sufficiently clear.

Secondly, to check 'poolability' in that design of the various studies must be consistent in certain respects arising from these objectives.

Thirdly, the relevant data should be collected and reported similarly. Keeping this in mind, it should be possible to combine the evidence from a series of studies of differing design to enhance sensitivity. This encouraging experience should also guide the work concerning epidemiological occupational studies.

In the course of the Symposium *Living in a Chemical World* which was held by the Collegium Ramazzini in Bologna in 1985, one discussion remark after the presentation of a risk assessment of acrylonitrile addressed the need to pool data from different epidemiological studies instead of qualifying each study in an attempt to dismiss insufficient information (Mast et al, 1986). This need to pool data from studies mostly of sufficient quality carried out all over the world arises more and more frequently and indeed in the case of acrylonitrile has already been dealt with by the Working Group (Frentzel-Beyme & Goldsmith, 1986).

Generally, the epidemiological cohort study has a non-experimental design and, therefore, depends on available cohort size, mostly too small to reach conclusion of sufficient certainty. But not only are occupationally exposed populations usually bound to be of small size (i.e. limited by work and factory requirements), in many cases the results of a series of such studies may also be inconsistent.[1]

Therefore, one is faced with the problem of combining data from studies in order to reach a plausible conclusion. Problems in making such a synthesis are seen both in general and also in terms of statistical foundations, and this applies all the more if recommendations for pooling are to be given.

It is the objective of this chapter to evaluate ways and strategies of pooling data obtained from several different occupational cohorts. Under the assumption that a pooled data set contains more information than separate reports, the desirable qualities of data sets for such a combination are examined and some examples are given. It is in the nature of such an overview that it cannot be definitive or conclusive, given the many ongoing processes of trial and error. Likewise, next to cohort studies, other instances of pooling cannot be considered at once; thus studies with other types of design, such as case-referent studies, will not be considered in this preliminary chapter.

RECENT ADVANCES

Goldsmith and Beeser suggested some approaches and guidelines for pooling data from occupational epidemiological studies in a recent publication (Goldsmith & Beeser, 1984). Epidemiologic studies taken separately can rarely provide convincing evidence for proper identification of a hazard or incontrovertible proof for a risk assessment. Yet, since epidemiology's role is to provide a basis for decisions to protect workers, for relatively grave effects such as the occurrence of excess cancer, the optimal approach should possess a high probability of detecting a true risk at the earliest time. Combining data from several similar (and comparable?) studies is becoming a preferred approach. Some strategies for this are compared by Goldsmith and Beeser. One important point made is that the loss of material (for pooling) due to the so called default strategy preventing publication impedes an earlier recognition of hazards. If, however, pooling were more prevalent, such insufficient data would be more readily published and could contribute to respective endeavours.

In a study of reproductive outcomes among anaesthetists, the authors demonstrated the potential bias of the usual summation of data on the exposed and reference populations. Large differences in response rates of the reference populations in the UK (67.8%) compared with the USA (45%) influenced the results and required adjustment.

Although this is a problem of poolability, it led to considerations of how to distinguish a fallacious result from a significant finding. In an example of pooling of data from studies on occupational groups exposed to manmade and naturally-occurring mineral fibres, it was shown that some 'general pooling strategies are needed which encourage smaller plants to obtain and transmit valid and well characterised data for pooling and collective interpretation'. This clearly addresses the differences in hazard between large plants and smaller ones due to differing exposure conditions,

[1] Rosenthal (1978) has summarised and provided guidelines for the use of a variety of methods for combining results of independent studies, particularly in psychological research.

plant layouts and record keeping systems. In addition to the methodologic issues, the need for recommendations for storage or listing of data sets for possible pooling becomes evident.

Moreover, it is obvious that bias due to a simple summation of observed events across the various age groups could possibly be avoided by first estimating the probability that each pair of age-specific (observed/expected) values could have occurred as a random process, and then pooling the resulting probabilities. For this purpose, the authors claim the need of access to the detailed data sets and call attention to the method of Fisher to obtain the joint probabilities of observing a set of such pairs of values. The approach, if based on probability estimates for each set can accommodate sets of varying sizes, without the large ones being unduly dominant.

Finally, besides the general considerations given to the poolability of results from different epidemiological studies, there are the necessary instruments for measuring an effect, such as SMR, being under scrutiny, and arising problems instigate the search for better strategies. Keeping in mind that an average conceals more than it reveals (and the reduction of data in the sense of compressing information is basically intended by pooling), the typical SMR is in fact an average. This particular problem will be addressed in the section Methodological Aspects.

In his thorough review of the available methodology, Nurminen (1985) concentrated on the aspect of pooled and stratified analyses of two rates, paying attention to the poolability of results, inconsistency of pooled estimators and the prevailing analytic practices for stratified data. Besides the very good principal evaluation of the state of the art, the limitations of the statistics have also been worked out. It was shown that some additional research seems necessary before making any recommendations.

In view of the needs described earlier not only the theory but also pragmatic approaches as to the use of available information for a demonstration of the effect of pooling strategies will be pursued in the following section.

METHODOLOGICAL ASPECTS

One definition given by Nurminen may be used here in dealing with all possible kinds of epidemiological data: when an aggregation of summary data of different origin is carried out directly (without reference to the original population segment of the study base) the strategy is referred to as pooling. As to the conditions for this according to Cochran (1954) the procedure of pooling is only legitimate if the probability of an illness occurring can be assumed equal. As formulated by Miettinen & Cook (1981) the pooled analysis is sufficiently founded if the stratification factor (in the study base) is not predictive of the illness outcome or is unrelated to the exposure under study. Clearly this addresses confounders, and the issues of confounding and of limitations of any combination of data are considered by Nurminen in some detail.

Pooling has to aim for the best estimate of an effect. The results of cohort studies are usually expressed by standardised mortality ratios, observed deaths (O) divided by expected deaths (E), where the Es are considered as being without sampling error, if derived from national, regional or even local total populations. The examples given later will all refer to this simplest set of data, although pooling strategies are needed likewise for data including measured levels of exposures or of body burdens, or even data on non-lethal conditions such as chromosome changes or sperm counts. For each pair of O and Es the probability is determined that a value as great as or greater than the observed one would be obtained from a distribution defined by the expected values (as the mean *and* the variance of a Poisson distribution). For the purpose of pooling such probabilities, the Fisher relationship[1] is: If the probabilities P_1, $P_2, P_3 \ldots P_{-n}$ are pooled, then:

$-2\,\Sigma\,\text{Lnp}\,(P_1)$ has a χ^2 distribution with 2n degrees of freedom.

The exact probability of χ^2 that large or larger can be obtained from either the tables (e.g. Goldsmith & Beeser, 1984), or by use of programmable calculators (Liddell, 1985; Rothman & Boice, 1979).

It seems to be important to pool only findings which are essentially homogeneous. For this reason the Mantel-Haenszel χ^2 pooling procedure has been given preference by many experts, including Nurminen, who reviewed the whole matter extensively. As to the pooling of probabilities, the following considerations are important.

1. Each P has to be one-sided before pooling;
2. Each test must be carried out on the same side;
3. After pooling, the pooled value of P should be turned into a two-tailed probability. It would be incorrect to use a one-tailed procedure (ignoring the side) for pooling purposes; this would cause all probabilities to be less than 0.5 and would falsely enhance the significance of the finding (Liddell, 1984).

Yet the procedure seems far from being unanimously accepted, in particular because of limitations demonstrated by an example (next section).

Some other concepts have been developed during the search for feasible and acceptable methods. These include the extension of the simple pooling of data at face value towards a Bayesian approach of judging the impact of each sequential set of data. This is in essence the judgment of the likelihood that a given data set will arrive at the risk ratio agreed upon as being fixed, whereas successively available data sets are examined for their likelihood ratio (LR).

Assuming this strategy and that the Poisson distribution assumption holds as a basis for establishing LRs for each data set, the LR is the probability that O cases of a disease are observed, given that E × R are expected, divided by the probability of observing O events, given that E are expected. The LR is then:

$$LR = R^{O}_{e}{}^{-E(R-1)}.$$

An application to data and demonstration of this approach was given by Goldsmith & Wendel (1986).

Recently, Gaffey (1985) proposed pooling by ranking results by power of the different studies, i.e. looking at the power which depends on several circumstances such as exposure, and ranking studies by increasing order of the power from the least powerful to the most powerful. The evaluation by the Fisher test involves the product of all the P values, whereas a test for the rank correlation would take the individual

[1] Fisher has earlier pointed out that 'when a number of quite independent tests of significance have been made . . . although a few or none can be claimed individually as significant, yet the aggregate gives an impression that the probabilities are on the whole lower than . . . by chance. It is sometimes desired to take account only of these probabilities, and not of the detailed composition of the data from which they are derived, which may be of very different kinds, to obtain a single test of the significance of the aggregate, based on the product of the probabilities individually observed'. He also introduced the basic mode of dealing with the problem (Fisher 1936, p 104).

probabilities with their property as a function of sample size into account. Armstrong (1986) in this context stated that, for power considerations, small-size studies may show the relative (un-)importance of a single contribution to the overall estimate, whereas testing for trends and ranking may appreciate such effects.

APPLICATION OF POOLING

Some examples are provided to demonstrate some of the issues in pooling (rather than to identify as ideal studies or procedures still under consideration). Data on formaldehyde-exposed persons were summarised by Gibbs et al (1985) and included six mortality studies among chemical and garment workers and eight studies of pathologists, anatomists and morticians. This overview of absolute numbers did not consider probability pooling, and studies of some different backgrounds revealed no deaths from nasal cancer (versus three expected), little evidence for cancer induction at contact sites, but an excess of brain cancer, leukaemia (in professional groups) and of colon cancer. In dealing with deficiencies of the published information in regard to pooling, the failure of reporting important background information (only one study linked exposures to job) and inadequacies of study design were brought forward, by which in-depth pooling seemed to have been impeded. As one result of their effort, the authors name a few tasks of an international committee which is supposed to become helpful in making suggestions providing guidelines for a better comparability of on-going studies and for a minimum reporting for future publications. This conclusion of a pooling effort with a substantial set of data at least led to a clearer understanding of inherent problems often underestimated.

Another example is the pooling of epidemiological data of persons with the same exposure to acrylonitrile from 10 different studies. The review mentioned before included an evaluation of the credibility and limitations of the majority of them on the grounds of some requirement allegedly put forward by the US Environment Protection Agency, which also evaluated studies, so far, however, without any intention of pooling (Mast et al, 1986). However, mixed exposure, lack of accurate exposure estimates, short follow-up periods, cohorts' young average age and small cohort size were criticised and accused of restricting reliability. Thus, only one study was found acceptable to the reviewers but naturally too small in size to be ultimately conclusive (O'Berg, 1980).[1]

And yet, when pooling the data of the five more sizeable studies, a remarkable consistency of findings enhances the impression of a human risk (Table 7.1). This is valid at least for lung cancer, whereas problems arose for pooling other sites without expected values being given, such as brain tumours, lymphatic neoplasms and Hodgkin's disease.

Although of different origin, all studies on acrylonitrile were likely candidates for pooling because:

a) all studies were carried out according to the same principle (follow-up studies on mortality);
b) the figures of interest in each study were small;
c) some remarkable cancer deaths had occurred in more than one study;
d) an excess of several cancers was indicated but still liable to the influence of chance.

[1] The same may apply to the recent update by the same author (O'Berg M, Chen J L, Burke C A et al 1985 Journal of Occupational Medicine 27: 835–840).

Table 7.1 Total and lung cancer deaths (observed and expected) with SMRs and probabilities in five cohort studies of acrylonitrile workers

	Cancer				Lung cancer			
Study	0	E*	SMR	p†	0	E*	SMR	p†
O'Berg	25	20.5	122	0.161	8	2.6	308	0.003
Werner	21	18.6	113	0.281	9	7.6	118	0.293
Thiess	27	20.5	132	0.081	11	5.65	195	0.022
Kiesselbach	20	20.4	98	0.521	6	6.9	87	0.611
Delzell	22	17.9	123	0.167	9	5.9	153	0.11
Fisher Pooled p	(χ^2 = 16.1, 10 df = 0.048				(χ^2 = 27.1, 10 df) < 0.003			
Total	115	97.9	117	0.045‡ 0.049§ 0.049‖	43	28.65	150	A 0.006‡ B 0.007§ C 0.007‖

* Expected values from the male population national rates (exc study I).
† Poisson cumulative probability of an observed value as great or greater; the P value exceeds 0.5, if the observed value is smaller than the expected .
‡ Miettinen exact one-sided p.
§ Fisher one-sided p.
‖ Approximate p value.

Thus, in terms of poolability, the cohorts were reasonably comparable, diagnostic procedures of all events being based on the same principles (namely on information from death certificates for all of them, even if in two studies additional information on the morbidity was available). The total numbers and expected cancer deaths per study pointed to the relatively homogeneous nature of five sizeable studies, also placing some confidence in the poolability. Table 7.1 shows the results from summary pooling, and Fisher's pooled p for comparison (Delzell & Monson, 1982; Frentzel-Beyme & Goldsmith, 1986; Kiesselbach et al, 1979; O'Berg, 1980; Thiess et al, 1980; Werner & Carter, 1981).

The unimpressive gain of information by using the pooled p does not outweigh the principal difficulty given by the basically non homogeneous results, i.e. the SMRs of one study being below 100 for total and for lung cancer. Looking at the age distribution of the entire cohort and of the total cancer deaths, serious doubts arise as to the completeness of follow-up and analysis of this study (Kiesselbach et al, 1979). Still, would it be practical to dismiss sizeable information from one study where a certain exposure during acrylonitrile production prevailed?

Looking closer at this study, the overall total (a pooled figure expressed as the SMR) hides some important information, such as the high cancer mortality before the age of 60 years. If one had used the same approach of pooling the age-specific rates as before, a pooled p given in Table 7.2 would result, indicating the highest SMRs in earlier age groups (although very small numbers) and the influence of the reverse finding in others, in this case the highest age-group. Obviously, in this cohort a premature mortality from cancer exists, and with higher figures in the cohort (e.g. twice the size) a statistically significant result could already be obtained (Table 7.2.). The uniform approach to the non homogeneous data is, however, questionable (Liddell, 1984; Mantel & Ederer, 1985). Reservations concern not only the negative direction of outcomes in pooling but also issues of colinearity, loss of power because the

Table 7.2 Age specific observed and expected deaths from cancer in a cohort exposed to acrytonitrile

Age group	Observed	Expected*	SMR	p†
<30	1	0.2	500	0.193
30–44	4	2	200	0.143
45–59	11	6.9	159	0.092
⩾60	4	11.3	35	0.996
Total	20	20.4	98	0.564

Fisher pooled p ($\chi^2 = 12.6$, 8 df) = 0.127

When duplicating all observed and expected values:

	<30	30–44	45–59	⩾60
p	0.062	0.051	0.026	0.999

Fisher pooled p ($\chi^2 = 18.9$, 8 DF) = 0.016

* Expected values from the male population of Northrhine-Westphalia (8 million male population).
† Poisson cumulative probability of an observed value as great or greater; the P value exceeds 0.5 when the observed value is smaller than the expected.

directions of the stratum-specific differences are not taken into account. For this reason (and other limitations) the ingenious method is not an omnibus test (Nurminen, 1985), and as a remedy a stratified analysis is required.

Another set of data from occupational mortality pooling based on data compiled by Dubrow & Wegman (1983) has been presented by Goldsmith in the example of larynx cancer among painters. Ten different studies showed SMRs in the range of 58–205, with p values ranging from 0.00066 to 0.99. The summation strategy (189 observed/171.6 expected, SMR 110, P = 0.10) could not demonstrate a clear excess, whereas probability-pooling (Fisher) resulted in χ^2 values of 33.3 (raw) and 30.1 (adjusted), with 20 df and p values of 0.031 and 0.068 (Table 7.3). Depending on the purity of thinking, the latter approach would provide a significant result, although based on non-homogeneous data, whereas the former (and usual) one would show a result far from significant, nevertheless consistent with an excess risk.

Table 7.3 Larynx cancer (observed and expected) with SMRs and probabilities in 10 cohort studies of painters

Study number		O	E	SMR	p
I	(MA)	10	4.88	205	0.028
II	(Wash)	5	7.04	71	0.831
III	(CA)	10	9.1	111	0.414
IV	(UK)	16	12.6	127	0.201
V	(US)	28	14.0	200	0.00066
VI	(LA)	10	9.52	105	0.481
VII	(RP)	10	10.2	98	0.567
VIII	(UK)	14	24.14	58	0.99
IX	(UK)	36	32.14	112	0.27
X	(UK)	50	48.08	104	0.41
Summation-strategy		189	171.6	110	0.10

Fisher probability pooling
Raw χ^E 33.3, 20 df $P = 0.031$
Adjusted χ^2 30.13, 20 df $P = 0.068$

From: Dubrow and Wegman (1983) modified by Goldsmith.

RECOMMENDATIONS

Common to all proposed approaches is that minimum requirements have to be met. Thus, recommendations have to concentrate on poolability issues first. These include guidelines for future data collections with a view to the application of pooling methods to be chosen according to the characteristics of the data. The guidelines recommended by an interagency Regulatory Liaison Group showed that even with existing well-established standards the need was felt to formalise the minimum requirements for the conduct, documentation and publication of such standard studies (Epidemiology Work Group, 1981). It will be the declared task of the Working Group on Pooling Strategies to pursue all directions described herein in order to find acceptable and recommendable methods of combining scattered information, individually not sufficiently strong. It is, however, necessary to develop new avenues of providing access to basic data for further use in pooling. The examples shown present only a few aspects of inherent problems in finding simple, useful and recommendable procedures. The challenge remains with the scientific community to provide the necessary contributions and improvements.

ACKNOWLEDGEMENT

The critical comments of Sheila Zahm-Hoar, Barry Miller, Douglas F. K. Liddell, J. R. Goldsmith and Markku Nurminen are gratefully acknowledged; similarly the entire project would not be possible without the selfless input by members of the Working Group on Pooling Strategies.

REFERENCES

Armstrong B 1986 Choosing boundaries for exposure groups: An arbitrary element in analyses of epidemiological data? Abstract, Fourth Internatl. Symposium Epidemiology in Occupational Health, Como. 10–13 September 1985. La Medicina del Lavoro 77: 1

Cochran W G 1954 Some methods for strengthening the common χ^2-test. Biometrics 10: 417–451

Delzell E, Monson R R 1982 Mortality among rubber workers: VI. Men with potential exposure to acrylonitrile. Journal of Occupational Medicine 24: 767–769

Dubrow R, Wegman D H 1983 Setting priorities for occupational cancer research and control: Synthesis of the results of occupational disease surveillance studies. Journal of the National Cancer Institute 71: 1123–1142

Epidemiology Work Group of the Interagency Regulatory Liaison Group 1981 Guidelines for documentation of epidemiologic studies. American Journal of Epidemiology 114: 609–618

Fisher L R 1936 Statistical methods for research workers, 6th edn.

Frentzel-Beyme R, Goldsmith J R 1986 Application of pooling strategies for data from occupational epidemiological studies of acrylonitrile. Abstract. Fourth International Symposium Epidemiology in Occupational Health, Como, 10–13 September 1985. La Medicina del Lavoro 76: 6

Gaffey W 1985 Ranking results by power of a study. Working Group on Pooling Strategies. Meeting in Como, Italy, 9 September 1985

Gibbs G W, Levine R J, Blair A E 1985 Ad hoc committee on formaldehyde. Working Group Meeting on Pooling Strategies, Como, Italy, 9 September 1985

Goldsmith J R, Beeser S 1984 Strategies for pooling data in occupational epidemiological studies. Annals of the Academy of Medicine 13 (Suppl): 297–307

Goldsmith J R, Wendel J G 1986 A Bayesian strategy for pooling data from occupational epidemiological studies. Abstract, La Medicina del Lavoro 77: 1

Kiesselbach N, Korallus U, Lange H-J, Neiss A, Zwingers T 1979 Acrylonitrile — epidemiological study — BAYER 1977. Zentralblatt für Arbeitsmedizin 10: 258–259

Lewis J A 1985 Combining the evidence from clinical trials. Abstract, 6th Annual Meeting of the International Society for Clinical Biostatistics, Düsseldorf, 15–20 September 1985

Liddell F D K 1984 Simple exact analysis of the standardised mortality ratio. Journal of Epidemiology and Community Health 38: 85–88

Liddell F D K 1985 Personal communication
Mantel N, Ederer F 1985 Exact limits for the ratio of two SMR values. Letter to Journal of Epidemiology and Community Health 39: 367–368
Mast R W, Strother D E, Kraska R C, Franko V 1986 Acrylonitrile as a carcinogen: Research needs for better risk assessment. Annals of the New York Academy of Science (in press)
Miettinen O S, Cook E F 1981 Confounding: Essence and detection. American Journal of Epidemiology 114: 593–603
Nurminen M 1985 On pooled and stratified analysis of two rates. Paper presented at a Working Group Meeting. Fourth International Symposium in Occupational Health, Como, 10–13 September 1985
O'Berg M T 1980 Epidemiologic study of workers exposed to acrylonitrile. Journal of Occupational Medicine 22: 245–252
Rosenthal R 1978 Combining results of independent studies. Psychological Bulletin 85: 183–193
Rothman K, Boice J B 1979 Epidemiologic analysis with a programmable calculator. Washington: US Department Health, Education and Welfare NIH Publication 79–1649
Thiess A M, Frentzel-Beyme R, Link R, Wild H 1980 Mortalitätsstudie bei Chemiefacharbeitern verschiedener Produktionsbetriebe mit Exposition auch gegenüber Acrylnitrile. Zentralblatt für Arbeitsmedizin 30: 259–267
Werner J B, Carter J T 1981 Mortality of United Kingdom acrylonitrile polymerisation workers. British Journal of Industrial Medicine 38: 247–253

8. Toxicity testing for industrial materials

P. A. Martin L. S. Levy

INTRODUCTION

Under UK law Notification of New Substances Regulations 1982 came into force on 26 November 1982. It embodied the notification provisions of EC Directive 67/548/EEC as amended by Directive 79/831/EEC relating to:

a) its provisions concerning the classification, packaging and labelling of dangerous substances;
b) the introduction into the EEC of a common notification scheme for new substances.

The latter objective is aimed at ensuring that the potential of any new substance to cause harm to Man or the environment is adequately assessed prior to its marketing within the EEC.

A new substance is defined as:

> a chemical element or compound, in its natural state or as produced by industry, and includes any additives such as stabilisers, inhibitors and impurities that are part of the final marketable substance.

However, compliance with the Regulations is not retrospective, and so an Inventory of existing substances marketed within the EEC and covering the period January 1971–September 1981 was carried out. A new substance would therefore be one which does not appear on this Inventory. Thus, any manufacturer or importer of a new substance must supply the Directive, via a competent authority (in the UK, the Health and Safety Executive) with such data as required by its notification provisions.

The type of information that is required depends on the tonnes produced per annum and includes basic physico-chemical properties of the material such as structural/empirical formulae, solubility (in both water and fat), vapour pressure and also details pertaining to flammability and explosive properties. In addition there is a requirement for a whole range of toxicological tests to assess the new substance's biological impact. Again, depending on the quantity of the material supplied per annum (1 tonne or above) the requirements of the Directive demand certain toxicity testing, both acute and chronic studies, unless the material is a substance expressly excluded, or covered by the limited announcement procedures (HSE Booklet, HS(R)14). The rationale behind these tests is described and discussed later in this chapter.

CHEMICAL INTERACTIONS

Most of the testing that has been done to date has been conducted in an artificial environment, that is an experimental laboratory, and the studies have been concerned with exposure to a single agent. However, in the real world it is rare for industrial exposure to be to only one chemical at a time. Humans are exposed simultaneously to a variety of chemicals including occupational and environmental chemicals as well as food additives, pesticides and drugs (Parke, 1983). Thus, the possibility of exposure to a large number of chemicals creates an increasing necessity for the consideration of their interactive effects. A number of terms have been utilised to describe the principal types of chemical interactions relevant to toxicologic studies (Shy et al, 1974):

Additive reaction

This is the effect produced when two chemicals are given in combination and the effect produced is equal to the sum of the effects of the individual components. For example, when two organo-phosphate insecticides are given together the resulting cholinesterase inhibition is usually additive.

Synergism

This is the situation in which the combined effect of two chemicals is much greater than the sum of the effect of each component alone. For example, carbon tetrachloride and ethanol are both hepatotoxins but given in combination the resulting liver damage is far greater than that of the mathematical sum of their individual effects. Fortunately synergistic toxic effects appear to be very limited in Man.

Potentiation

This is the situation where one substance, which does not have a toxic effect, in combination with another chemical increases the toxicity of the latter. For example, isopropanol which is not hepatotoxic, when given with carbon tetrachloride, substantially increases the liver damage due to the latter.

Antagonism

This is the effect produced by two chemicals which when given together is less than the effect of the single most active component. For example, zinc reduces the toxic effects of cadmium exposure in experimental studies.

Whatever the pattern of joint action, information on the sequence of this action (that is, comparison of the pattern of effect when factor A is known to precede factor B and the reverse situation) would contribute to an understanding of the mechanism(s) involved. Thus, it can be seen that chemicals may mutually facilitate, or, more probably, compete with each other for absorption, tissue distribution, metabolic detoxification and activation, reaction with specific receptors and target organs, and excretion. However, with current techniques and the present state of knowledge there is no ideal way of approaching the problem of assessing the toxicity of exposure of mixtures of chemicals to humans. With this in mind and unless evidence is available to the contrary, the toxic effects of a mixture of chemicals must be presumed to be at least equivalent to the effect which would be produced by simply adding the toxicities of each of the components of the mixture at the levels at which each is present (Bridges & Hubbard, 1982). This is a pragmatic, rather than a scientific approach and the rationalisation will have implications for industry in the area of Occupational Exposure

Limits (OELs) which are presently used as a guide for acceptable upper levels of airborne contaminants in the workplace.

TOXICITY TESTING AIMS AND PROCEDURE

The previous discussions have outlined some of the problems encountered when trying to assess the risk of harm from single or multiple exposures to one or more materials in the workplace. Obviously it is not possible to prescribe a definitive series of evaluative tests that will guarantee total information about all possible toxic effects under all conditions, nor take into account possible interactions between different substances. Instead, the Notification Scheme prescribes a compromise package of tests, the information from which will allow a reasonable judgement to be made on the likely toxicity of that material under most conditions of use as well as providing sufficient information that can be used for classification and labelling purposes.

The set of tests advocated by the HSE and contained within an Approved Code of Practice (ACOP, 1982) are not particularly novel and in general, are closely modelled on those which have been developed over recent years by the Organisation for Economic Cooperation and Development (OECD) and accepted within the EEC. In this section, we will attempt to outline the tests, show how they relate to industrial materials and where applicable, highlight the limitations of those tests in human risk evaluation.

The tests have been placed into categories, although these categories are essentially for administrative convenience rather than indicative of a precise scientific segregation. The groups are set out as follows:

1. Investigation of acute toxic effects.
2. Investigation of skin and eye irritation and sensitisation.
3. Investigation of sub-acute toxic effects.
4. Screening tests for mutagenic and potential carcinogenic activity.

Information gained from these four groups of tests comprise a basic package of toxicological information which form the minimum data source from which an informed judgement may be made on likely human risk. However, should it be decided that further information be required, the ACOP sets out a series of additional test methods which may be appropriate. These comprise tests for effects upon reproduction, investigations for sub-chronic or long term toxicity (including the use of non-rodent species), further mutagenicity tests, carcinogenicity studies, further acute and sub-acute toxicity studies and investigations into toxicokinetics. Clearly, some justification is required for embarking upon these additional tests as some are at early stages of development whilst others are extremely expensive. The justification for the use of these additional tests will be illustrated later.

ACUTE TOXIC EFFECTS

The type of tests advocated under this heading are those which are looking for the total adverse effects, within a short space of time (minutes, hours) following a single administration of the test substance. The substance can be administered by one or more of the three standard routes of administration: orally (e.g. by lavage), percutaneously by placing the substance in contact with the skin and by inhalation. Normally, the choice of route(s) will depend on the physical property of the substance and likely use of that substance. Normally, the oral route plus one other is used but for gases, vapours and particulate materials the inhalation route is preferred.

Although in theory quite a lot of information can be gained from such investigations, these tests are best known for the derivation of LD_{50} (median lethal dose) or LC_{50} value (median lethal concentration). These terms are used to describe a statistically derived single dose or concentration that can be expected to cause death in 50% of dosed animals. In the case of LD_{50} it is expressed in terms of weight of test substance per unit weight of test animal (mg/kg) and for LC_{50} value, it is expressed as weight of test substance per standard volume of air (mg/1). Such tests use rodents, usually rats, of both sexes at several dose levels in order that a dose-response relationship can be established. Over the last few years this test has attracted much attention and criticism because of its relative scientific crudeness and its unpleasant public image (Brown, 1981). Clearly, regardless of the arguments, we do need some broad estimation of the likely risk associated with a single, perhaps accidental, high exposure to a substance so that appropriate precautions can be made in its handling, transport and usage. Even if you end up with categories of very toxic, toxic and harmful based on mg/kg body weight, then at least you have some useful information. The danger can be that there is a temptation to express the *total* range of toxicity of a material, based on this single parameter.

INVESTIGATION OF SKIN AND EYE IRRITATION AND SENSITISATION

It is extremely important to assess whether or not a material is likely to harm the skin or eyes. However, in practical terms, these harmful effects will depend on many factors such as solvents, the nature and concentration of the test material, the duration of contact with the target tissue. Also, the nature and degree of the inflammatory responses can vary with these factors from very mild to severe, which can make classification difficult. For these reasons it is necessary to have fairly well defined protocols for such effects which can, where possible, act as a model for likely human exposure. As an example, the test system may use either single or repeated application of test substance. Skin irritancy tests must also be able to distinguish between those pathological processes which are directly inflammatory and those which are mediated by an immunological process. These latter allergic skin sensitisation reactions are complex and difficult to mimic in animal models, but such tests are available. The kind of test that may elicit these responses are described in the ACOP as below.

Skin irritation

Briefly, the test most commonly used for skin irritation involves the single application of the test substance to the shaven skin of several test animals, each animal acting as its own control. After a specified interval, the site of application is examined and graded for specified effects. As a minimum, usually three rabbits are used and 0.5 ml or 0.5 g of the test substance are applied to a small area of skin, which is then covered with a gauze patch for a 4-hour exposure period. At the end of this time, the skin is cleaned and observations usually made from 30 minutes to 72 hours thereafter. Observations, with a scale of 0–4 are made of erythema and oedema and also other features that may be present. In this way it is possible to grade a wide range of industrial substances into those which are likely to be human skin irritants and grade them from mild to severe. Considering the wide range of handling processes that takes place in industry it is clearly important to have such information so that appropriate labelling and warning be given and suitable protection afforded.

It should be noted that the document clearly states that any substance that is strongly acidic or alkaline does not require testing in this way as skin irritancy can be assumed to be a property of that material.

Eye irritation

This, in the public and scientific view, is probably the most distasteful and controversial of the toxicological tests that is used and recommended in this basic package of tests. Briefly, the test substance is applied in a single dose to one of the eyes in each of three rabbits. Usually for liquids 0.1 ml is used, or for solids a volume of 0.1 ml or weight approximately 0.1 g. The other eye of the animal acts as a control. Thereafter, the degree of irritation is evaluated and graded at specified intervals. In addition, the duration of observation allows for evaluation of reversibility or irreversibility of the observed effects, usually 3 weeks is considered sufficient to allow for these latter observations. The effects of applying the substance to the eye are observed at intervals of 1, 24, 48 and 72 hours and if no effects are seen in the cornea, conjunctivae or iris, then the material is assumed non-irritant to the eye. The grading of the ocular lesions is fairly detailed and complex and best facilitated by ophthalmic equipment such as binocular loupe, biomicroscope and fluorescein. The observations of ocular lesions for the corneal changes are essentially degrees of opacity and range from 0–4. Zero indicates no ulceration or opacity and 4 indicates an opaque cornea with the iris not discernible through the opacity. Observation of the iris can range from 0–2 where 0 indicates normal and 2 indicates no reaction to light, haemorrhage or gross destruction. Conjunctival changes relate to redness and swelling of the lids and/or nictitating membrane. In the case of the former, the scale of 0–3 ranges from normal blood vessels to diffuse redness and for the latter, the scale 0–4 ranges from no swelling to full swelling with eyelids more than half closed.

Despite the obvious unpleasantness associated with this test, it is important to have as much knowledge as possible on this aspect of toxicology because of the known existing high risk of contact of substances with workers' eyes. The rabbit's eye is very sensitive and probably a good model for the human eye. What must be emphasised is that this test is not required if the substance has already been shown to be a skin irritant. In this case it may be assumed to be an eye irritant. The argument for the continued use of this test is that the human eye is an extremely sensitive organ and it cannot be assumed that a non-skin irritant will not harm the eye.

Skin sensitisation

Skin sensitisation reactions are becoming more and more recognised in relation to occupational skin disorders. From a toxicological point of view however, any response which involves a complex and species-specific immune mechanism makes the development of a reliable animal model very difficult. There are about six methods that have been developed, all using the guinea-pig. The reason for this is that this animal's immune system more closely resembles that of humans than any other rodent. However, it must be emphasised that no single test can adequately identify all substances that have the potential for sensitising human skin and thus no single test will be relevant for all industrial materials. Factors which will dictate the choice of method will include the ability of the test substance to penetrate the skin and likely human skin contact. The method that is described in the ACOP is the one which is currently accepted as being most reliable. It is known as the guinea-pig maximisation test (GPMT). In this system, following initial exposure to the test material (the induction period), the guinea-pigs are subjected after the last induction exposure to the test substance to determine whether or not a state of hypersensitivity has been induced. Sensitisation is determined by examining the skin reaction, as previously described, to the challenge exposure.

A lot of skill is required in performing and evaluating the results of this test. The GPMT system, as with many of the tests used to attempt to induce sensitisation in guinea-pigs, relies on the use of an adjuvant, a material which is injected alongside the test material and is able to potentiate antigenic activity. It is essential to use adequate controls because the adjuvant, usually Freund's Complete Adjuvant (FCA), is able to induce generalised changes in the guinea-pig including increases in immunoglobulins, erythrocyte sedimentation rates, macrophage number and other immunologically related effects all of which predispose the animals to generate inflammatory responses to amounts of the test material innocuous to normal, unactivated animals. Naturally, these irritant responses could be misinterpreted as evidence of allergy.

In the ACOP, it is suggested that 20 treated and 10 control animals are used. The test consists of two phases, the induction phase and the challenge phase. In the induction phase, intradermal injections of the test material are given to the shaven skin in the shoulder region with and without the FCA in a standardised way. The controls are similarly treated but of course, without the test material. On day seven, topical application of test material with a suitable dressing to permit a 48-hour exposure is made. On day 21, the challenge is made topically and held in place with a suitable dressing for 24 hours and the observations of skin reaction made over the next 2 days. A second challenge can be made. Naturally, the challenge dose should not evoke a direct irritant effect upon the skin.

Although the model is quite complex, it is considered to be quite a good predictor of human response but, because of immunological drift, has to be checked regularly against known strong and moderate sensitisers to reduce the risk of the model giving false negative results.

INVESTIGATION OF SUB-ACUTE TOXIC EFFECTS

The previous tests have been mainly concerned with single, perhaps high dose, exposures. By contrast, the purpose of investigating sub-acute toxicity is to try to establish harmful effects of repeated exposure to a substance, often at lower doses than those which cause acute effects. The test advocated in the ACOP suggests administering the test material to the animals regularly, usually daily for at least 28 days. In theory, such tests can identify the organs and tissues injured by exposure and enable the detection of possible early cumulative toxicity and even the estimation of the 'no observed effect level'.

The route(s) of administration should be the most appropriate having regard to the nature of the material, its known acute toxicity and the way people are likely to be exposed to it. Usually, oral or inhalation routes are preferred. A range of doses are used, the highest of which elicits a clear toxic, but not lethal response.

The animals used in this kind of study are usually rats of both sexes, 10 to a group. There are at least three dose levels plus controls used and detailed observations are made during the 28-day study period. At the termination of the study, a very detailed examination takes place including haematology, clinical biochemistry, careful gross necropsy and histopathological examination of most organs and tissues. Most procedures apply whether the route of administration is via the oral route, the dermal route or the inhalation route.

The term 'limit test' is used to describe a 28-day study performed in accordance with the above outline, at one dose level of 1000 mg/kg body weight/day, or a higher dose level related to possible human exposure where this is known. If no evidence of a toxic effect is seen then further testing may not be considered necessary.

Although experience over many years has shown that this kind of limited duration repeated dosage test gives good information so that a reasonable prediction of the possible effects repeated exposure in humans may be made, the limitations of such a test must be appreciated. Not only may the harmful effects in man differ from those seen in animals, but the dose at which they occur as well as the dose-rate dependency may vary greatly. For this reason, in trying to design such a test and in the interpretation and extrapolation of the results to humans, one must take into account all available knowledge of the compound, its analogous structures and all other known toxicology.

TESTS FOR MUTAGENICITY AND POTENTIAL CARCINOGENICITY

Over the last 15 or 20 years a variety of tests have been developed which attempt to assess the ability of a test material to cause transmittible genetic changes. These latter mutational changes are thought by most scientific authorities to be the fundamental first step in the process of carcinogenesis. Thus, any chemical or substance which is able to evoke such changes must be suspect as a possible cancer-inducing agent. It must, however, always be borne in mind that cancer is a multistage disease and that mutation of the genetic material is probably only one, albeit crucial, stage in a series of events needed for cancer to be caused. For this reason such tests for mutagenic activity are useful as pre-screens for potential carcinogenicity. However, as there are over 100 such tests, it is inappropriate for the ACOP to describe or even advocate exactly which of these tests should be used. What is suggested is that for an initial screen of mutagenicity (hence potential carcinogenic activity) information should be obtained from two categories of biological end-points, namely gene mutation and chromosomal aberration. For the former of these two end-points, the ACOP suggests that a test on the production of gene (point) mutations in prokaryotic cells such as *Salmonella typhimurium* or *Escherichia coli* is acceptable depending on the nature of the chemical, and for the latter, a test on the production of chromosomal aberrations in mammalian cells grown in vitro, or in vivo systems such as the micronucleus test or metaphase analysis of bone marrow cells may be suitable. The reason for virtually insisting on at least two tests with different biological end-points relates to the fact that not all carcinogens, let alone mutagens, operate by the same mechanisms and thus two systems will decrease the chance of ending up with a false negative.

Point mutation in bacteria

The protocols for using bacterial point mutation systems in *Salmonella typhimurium* or *Escherichia coli* have and are being continually developed. Essentially, both systems employ the use of a reverse mutation system in which there is an amino acid dependence in the tester strains of bacteria used. The biological end-point measured is the ability of the test chemical to cause base substitutions or frameshift mutation in the genome of the organisms which is phenotypically expressed by the increase in the growth of colonies which have mutated backwards to the specific amino acid interdependence.

In practice, a range of tester strains of each of these bacteria have been developed which are sensitive to different kinds of chemicals. Technically, the bacteria are exposed to the test chemicals with and without the presence of a metabolising system (to activate any organic procarcinogens to their carcinogenic state). After a suitable period of incubation the revertant colonies are counted and compared with the number of spontaneous revertants in an untreated and/or solvent control. Different dose levels and positive controls are included to both test the sensitivity of the system and to make sure any positive result can be substantiated by a dose effect curve.

Chromosomal aberrations in mammalian cells

Although there are a wide range of tests available for estimating chromosomal aberrations, the ACOP outline three only, which are generally accepted to be useful indications of likely human harm. These are briefly outlined below:

In vitro mammalian cytogenetic test

The principle of this test system is the detection, as a biological end-point of structural chromosomal changes in either established or primary cell cultures following exposure to the test chemical. The protocol involves the use of positive and negative control cell cultures with and without a metabolising system (a liver enzyme activation mixture obtained from a post-mitochondrial fraction) at least at three different dose levels. Following the test treatment, the cell cultures are treated with a spindle inhibitor such as colchicine to accumulate cells in a metaphase-like stage of mitosis. Cells are harvested at appropriate times, preparations are stained and cells in metaphase are analysed for chromosomal abnormalities.

In vivo mammalian bone marrow cytogenetic test

Again, this system tests for the presence of structural chromosomal aberrations but, here, usually evaluated in the first post treatment mitoses. With chemical mutagens, most induced aberrations are of the chromatid type. The method employs bone marrow cells of mammals which are exposed to the test chemical by an appropriate route and are then killed at sequential intervals. Prior to being killed, animals are treated with a spindle inhibitor such as colchicine to accumulate cells in a metaphase-like stage of mitosis. Chromosomal aberrations are analysed microscopically from stained cells.

Micronucleus test

This assay detects chromosomal damage or damage of the mitotic apparatus by an increase in micronuclei in the polychromatic erythrocytes of treated animals versus controls. These micronuclei are in fact chromosomal fragments or even whole chromosomes lagging in mitosis. When erythroblasts mature into erythrocytes, the main nucleus is expelled while the micronucleus may be retained in the cytoplasm. The test employs young polychromatic erythrocytes in the bone marrow of mice, usually, which have been exposed to the test material by the appropriate exposure route. Bone marrow is extracted, smear preparations made and stained, and polychromatic erythrocytes scored for the presence of micronuclei under the microscope. The ratio of polychromatic to normochromatic erythrocytes can thus be established.

The above four groups of tests comprise a basic package of toxicological systems which are intended to produce a compromise of toxicological information; which in most circumstances will allow a reasonably well balanced judgement to be made of the likely harm that may be encountered with human exposure. However, when it is considered that further toxicological information is reasonably required for an adequate evaluation of the potential risks created by the material and its use, the Health and Safety Executive may require additional studies to be conducted. The following extra studies are outlined below.

INVESTIGATIONS OF EFFECTS ON REPRODUCTION

There are now a series of known examples of the effect of exposure of industrial chemicals on the reproductive capacity of both men and women as well as reproductive outcome. There might be other examples, as yet uncovered, but evidence is hard to come by of harmful reproductive outcome from human populations whether it be loss of libido through to fetal malformations which may not present until adulthood. Thus, the difficulties of investigating human reproductive effects are compelling reasons why it is important to consider the use of toxicological tests for such hazards if there is any reason to suspect a material of creating such a risk. It must be stressed that it is rarely possible to predict the potential of a substance to damage reproductive processes from its chemical structure or the results of other toxicity studies (acute or repeated dose toxicity studies); nor is there a single toxicological system for reproductive effects. Reproduction is a complex, multistage process and this is reflected in the number of tests which could be used to assess the possible effects at each of these stages. In general, however, it is possible to initially limit toxicity testing to two key areas, namely fertility and teratogenicity.

Fertility

Usually rats are used in numbers sufficient to produce 20 litters at dose level. Both sexes are dosed with the test chemical, by the appropriate route, quite some time prior to mating. During the post-natal development each litter is carefully observed until weaning or even later, and gross autopsy takes place thereafter. In this way, information can be gathered at each dose level on a number of matings, litter size and number of stillborn animals in each litter, presence and type of any abnormality in pups, viability of pups at various times and weight gain of pups. This, together with the autopsy results, can give a good idea of any likely effects on fertility.

Teratogenicity

These tests usually employ rabbits or rats and involve the administration of the test substance in graduated doses for at least that part of the pregnancy covering the period of organogenesis to several groups of pregnant experimental animals, one dose being used per group. Shortly before delivery the animal is killed and the uterine contents examined. In this way embryo and fetotoxicity are examined. Other investigations are possible, including multigeneration studies, but these would normally have to be justified in the light of the above tests and other information or likely risks to humans.

INVESTIGATIONS OF SUB-CHRONIC TOXICITY

Studies for the repeated exposure to test materials which lasted around 28 days were described earlier and it was suggested that the early changes associated with chronic toxicity could often be detected at this stage (hepatotoxicity, nephrotoxicity, etc). However, experience has shown that certain effects such as ocular toxicity and peripheral neuropathy may take many months to make themselves obvious. This also applies to materials that have a long biological half-life. Thus, in some circumstances it may be desirable to carry out studies, again usually on rats, along the lines of the 28-day repeated dose study but for a period between 3 months and 1 year. These can be by inhalational, dermal or oral routes. It is important to realise that this will not elicit a carcinogenic response and thus cannot be used to prove or disprove carcinogenicity. Examples of why such a subchronic study might be appropriate are indicated in the APOC, these include repeated exposure in humans via a route that is likely to lead to significant exposure, experimental evidence to suggest slow excretion with consequent tissue accumulation, irreversible or unusual toxic effects, or an unidentified no-effect level, seen in the 28-day study.

FURTHER TESTS FOR MUTAGENICITY

It was indicated earlier that there is a large number of screening tests for mutagenicity and potential carcinogenicity available, and that the APOC only suggested a limited number in the basic package of toxicity tests. Clearly, it may be appropriate for others for these systems to be applied.

CARCINOGENICITY STUDIES

The reasons for undertaking a full carcinogenicity study requires justification. They are extremely costly, involve a large number of rodents and tie up skilled personnel and valuable resources for a long time. Nevertheless, because of the severity of the harmful outcome, there are circumstances in which the use of such studies is appropriate. The reasons may include the following.

- (a) Large population exposed.
- (b) The material is similar in structure or metabolic activity to a known carcinogen.
- (c) There are positive results from one or more of the screening tests for mutagenic or potential carcinogenic activity.
- (d) The material is subject to tissue accumulation.
- (e) There are human case studies, epidemiology or animal data, which suggest a cancer risk.

A number of bodies have suggested guidelines for the conduct of carcinogenicity studies and notably those used by the National Cancer Institute (NCI) of the USA and the National Toxicology Programme (NTP) are those which contain most of the key elements that make them acceptable for both regulatory and scientific purposes. In essence, they suggest that animals receive repeated daily doses of the test material (by oral or inhalational routes) for 5–7 days per week for the greater part of their life span. For mice this is 18–24 months and for rats 24–30 months.

At least three dose levels should be used with the highest dose eliciting some signs of minimal toxicity (slight depression in body weight gain). At least 100 animals (50 females and 50 males) are used at each dose level and it is sometimes considered appropriate to use two species, rats and mice.

It is not appropriate here to discuss the details of the protocols for such studies, but clearly, the control and analysis of such investigations must be extremely thorough and exhaustive so that the possibility of a false negative is eliminated. Interpretation of the results of such studies is fraught with difficulties, even with well conducted studies, especially when the results of such an animal investigation is at variance with well conducted epidemiological investigations. For this reason it is often useful to go to authoritative bodies such as the International Agency for Research on Cancer (IARC) and their Monograph Series for evaluations on specific substances, where the quality of both animal and human investigations has to be assessed by independent experts.

TOXICOKINETICS

In the pharmaceutical and drugs industry the term pharmacokinetics is used to cover the study of the rate of absorption, metabolism, distribution and excretion of compounds. Toxicokinetics is the term applied for equivalent studies but into non-pharmaceutical products. Such invesigations can be invaluable in helping to understand mechanisms, dose-rate effects and formation of active metabolites of industrial materials. These are especially important where differences in metabolic handling is indicated from other toxicological investigations or the material in question has functional groups that may give cause for concern. Ideally, such studies should be carried out on all materials to help our general understanding of toxic mechanisms and bioavailability of industrial materials, but in many cases they are hard to justify in the simple determination of the presence or absence of toxic effects.

ALTERNATIVES TO ANIMAL TESTING

The recent increasing public demonstrations and activities of certain pressure groups regarding the use of living animals in the testing of materials has led both scientists and regulatory bodies to investigate the potential of numerous alternative, non-animal testing systems for their ability to predict likely toxicological effects on Man and the environment.

The major argument proposed by the anti-vivisectionist lobby is that 'alternatives are available which are faster, cheaper, more accurate, more reliable, and, above all, safer' (BUAV Leaflet, 1978). This view is simplistic and erroneous and in many instances misleading. Whilst many toxicologists would agree that certain non-animal testing systems are available, which give information on toxicological events within the system being used, in no way is this information the same as that obtained from, for instance, the LD_{50} test. What the non-animal test does provide is information of a corroborative nature pertaining to the potential of a tested material to cause deleterious changes which, when viewed with other information gained from other toxicity tests, may indicate that the tested material represents a threat to Man and the environment.

Hence, it would be more correct to refer to these non-animal toxicity tests as complementary or ancillary methods (Balls & Clothier, 1983).

Below are listed some of the currently available alternative methods of toxicity testing:

(a) computer-aided predictive models;
(b) use of lower organisms: algae, fungi, bacteria and invertebrates;
(c) use of cell/tissue fractions;
(d) cells in culture, including human cells and organ cultures;
(e) human volunteers.

Most of the above methods already have applications in biochemical research including, experimental pharmacology and toxicology. However, they have not been evaluated as to their potential as replacements for established animal toxicity testing procedures.

There can be little doubt that these alternative methods would contribute a significant amount of data concerning specific mechanisms of toxicity at the cellular level, and initially they may be used as a pre-screen for new materials before animal testing. This would instantly result in a reduction in animal utilisation since many new chemicals, with high toxicity would be removed at this pre-screen stage.

Computer modelling

Computers can be used in the theoretical modelling for prediction of toxicity of new materials. This is based on chemical structure and biological activity of related classes of chemicals (Eschenroeder et al, 1980). The usefulness of the method relies upon co-operation between producers, testing agencies and universities regarding the issue of confidentiality of toxicological data of substances manufactured, tested and utilised in research projects. Provided the data are available, the selection of an appropriate chemical class control compound and the most appropriate test will enable a judgement to be made as to the likelihood of a toxic effect. The computer can be programmed to select the most appropriate test or tests since, not all assays have the same response to a particular chemical class. It is obvious that the more data that is available concerning structure-activity relationships in a particular class of materials the greater will be the potential of this method. The potential gain therefore of a valid, cost-effective predictive model for toxicity evaluation with the reduction of loss of some putative products (which at present arises from toxicological data of some dubious relevance with respect to hazard) may provide the incentive required by the agencies concerned to achieve this collaboration and reduce the indiscriminate and wasteful screening of today.

Lower organisms

The reduction of mortality due to major infectious diseases in developed nations, with the concomitant increase in life expectancy, coincided with a measurable increase in the incidence of cancer. Conventional toxicity studies for carcinogenicity involved lifetime studies with laboratory animals and it was apparent that with the increasing production of new chemicals, and the paucity of toxicity data pertaining specifically to carcinogenic risk for the thousands of chemicals in every day use, the development of a rapid screening method for the detection of carcinogenic potential was crucial (Hollstein et al, 1979). These tests, mentioned in the ACOP, commonly referred to as short-term tests, rely on the presumption that there is a relationship between DNA damage and chemical carcinogenesis. Hollstein et al (1979) produced a list of over 100 different short-term tests the majority of which utilised bacteria. Bacterial mutation tests are based on the somatic mutation theory of carcinogenesis. These tests identify those chemicals that:

(i) react with;
(ii) modify;
(iii) induce the repair of;
(iv) interrupt the replication of, or otherwise alter the structure, function or subsequent fate of DNA.

These chemicals are often referred to as genotoxic agents.

The Ames test (Ames et al, 1975a,b) is generally recognised as being the primary short-term screening method for the detection of potentially carcinogenic chemicals, with an accuracy of 80–90%. However, there is a problem of extrapolation of the results to Man. A positive result indicates that the chemical has the potential to react with DNA and cause a mutation; that is, the chemical is a mutagen. Not all mutagens are carcinogens, indeed neoplasia is probably only seen in vertebrates. Carcinogenesis is a multi-stage process and these short-term tests are indicators of the ability of a chemical to initiate the process. Overt neoplasia also requires that the initiated cell(s) (clonal, in origin) undergo some secondary progression or promotion in order to develop into a frank tumour (Cairns, 1981).

Sub-cellular fractionation

It has been known for a number of years that exposure to certain chemicals results in some degree of reaction and ultimately binding with cellular macromolecules. However, it was not until the studies of Brodie et al (1971), using radiolabels and cellular fractionation techniques, that it was fully appreciated that the type of binding involved was covalent bonding, and that this had implications concerning the persistence of a chemical within the cell. These and other studies led to the hypothesis that a chemical, or its metabolite(s) might undergo two types of reaction with cellular macromolecules; one resulting in possible beneficial effects and the other in a toxic effect (Gorrod, 1981). Fig. 8.1 illustrates this point.

This hypothesis allows for the development of a pre-screen for toxicity testing. Further work needs to be carried out to evaluate for instance, whether a divided dose would provide a better correlation of toxicity with covalent binding as opposed to a single dose, the time lag for optimum binding and indeed the type of macromolecular binding; to proteins, lipids or nucleic acids. It should also be noted that the magnitude of the observed toxic effect will depend upon not only the initial interaction but also on the persistence of the damage. This in turn will be related to the tenacity of the binding and the efficiency of the body's repair mechanisms. At very high exposure levels the repair mechanisms, and hence the return of the tissue's

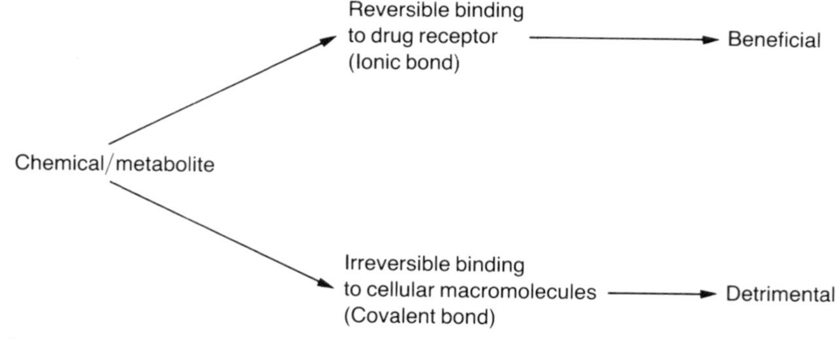

Fig. 8.1

is probably the rate-limiting factor. This latter point is interesting to bear in mind when one considers the relevance of extrapolation of animal studies (where unusually high doses are given) to Man (where exposures are proportionally lower).

The major advantage of these methods is that information on target organ toxicity and possible mechanisms of action can be obtained in relative isolation from the complexities of other influences: immunological, humoral and neural.

Cells in culture

The possibility of utilising cell or tissue cultures as a method of testing agents for toxicity has been postulated for some years, although it was not until the 1970s (Iype, 1971; Williams & Gunn, 1974) that certain technical problems were overcome.

Cell cultures are usually established by the enzymatic dissociation of the cells of a tissue, followed by their distribution in a suitable growth medium. Initial or primary cultures can be subcultured, or passaged, giving rise to a secondary culture or an established cell line, and certain ones may give rise to permanent, potentially immortal, cell lines. The primary cell line retains many of the characteristics of the parent cells, chromosomal numbers and biochemical properties, whereas an established cell line invariably has different chromosome numbers and a loss of specialised biochemical characteristics.

Progress in the development of cell culture techniques for toxicity testing has lagged behind that of carcinogenicity screening (Rees, 1980). The major reason for this is the diversity of primary insults that the numerous chemicals evoke on the cells in culture, usually resulting in the death of the cell and also because of the specificity of certain chemicals, or groups of related chemicals, for a particular cell type. Established cell lines, which are readily available, were tested in earlier studies without much success since many of the specialised properties of the parent cells had become modified and hence the response to the toxic action of the chemical did not always correlate with that of the known toxic effect in vivo.

Development of the method with primary cultures is potentially more rewarding, although interpretation of the in vitro effects must be done with some care. The concentration of potentially toxic chemicals must be controlled since there must be a level at which most, if not all chemicals, will become cytotoxic and this will then be a non-specific effect which may bear no relationship to the in vivo situation. Nevertheless, perhaps with a small range of primary cultures it may be possible to offer a reasonable prediction of degree of toxicity of a substance. A chemical with low toxicity would be more difficult to evaluate with this technique due to the high concentrations that might be required in order to elicit an effect.

One example of a cell in culture which may find increasing use in assessing the mechanisms of action of new chemicals is the hepatocyte. In man the hepatocyte accounts for some 85% of the liver cell population. The liver is frequently the site of chemically mediated toxicity, and this may be because:

(i) it contains the highest concentration of the mixed function oxidases, the enzymes primarily responsible for detoxification of xenobiotic materials;
(ii) being the largest organ in the body, the liver has the greatest capacity to bind chemicals (non specifically) and hence to remove them, albeit transiently, from the circulating blood. This in effect renders them less harmful.

A recent review by Bridges (1981) indicates that if certain technical problems, relating to improved viability of the isolated cultures, can be overcome, then the hepatocyte may be a very useful tool in both detection of carcinogens and also in ascertaining mechanisms of toxicity in order to be able to predict toxicity of new, but physico-chemically related compounds.

Human guinea-pigs

There are obvious ethical problems involved in obtaining most forms of toxicological data from man in vivo and also from human tissues, especially those which might be provided from spare embryos. At the present time certain new drugs are tested, under rigid conditions, on volunteers for financial reward and also on terminally ill patients where there is a theoretical possibility of some beneficial effect. If extrapolation from animal/non-animal systems continues to be questioned then these problems of ethics regarding human studies must be approached and discussed anew.

CONCLUSIONS

The foregoing discussions have outlined the current approach adopted in the UK for gathering toxicological data, using animal models, for helping to assess likely human risk from industrial materials and also a number of alternative methods which may find increasing use in toxicity testing by a contribution to the understanding of specific mechanisms of action of chemicals. What these later alternatives will not provide at present is an answer to the question, 'How much of this substance is toxic to humans?'

These alternatives would serve as a pre-screen for new materials thereby resulting in a decrease in some of the acute animal toxicity trials. Toxicity testing using these new methods will thus be seen to be more humane but the battery of tests required may not provide the same sort of essential, relevant data, and most certainly will not be more rapid, cheap or more readily obtainable than the acute toxicity dat using animals. Many of these new methods require very sophisticated equipmen and a higher degree of technical expertise than those required for example in th observation of animals in an LD_{50} test. Nevertheless it is certain that the numbe of animals currently used in routine testing programmes has been, and will continu to be, reduced if regulators and producers critically assessed the suitability of th sort of data provided by some of these outmoded procedures and were given rigorousl validated alternate, non-animal bioassays in which there was a good correlation betwee the in vitro results and the in vivo experience. This correlation between in vitr and in vivo results has been shown by a number of workers (Smith et al, 1963 Ekwall, 1983), and so it would appear that simple in vitro tests may be used fo predicting acute toxicity of certain types of chemicals.

REFERENCES

A guide to the notification of new substances regulations 1982 Health & Safety Series, Booklet HS(R)14. Health & Safety Executive, HMSO, London

Ames B N, Kammen Ho, Yamasaki E 1975b Hair dyes are mutagenic: identification of a variety of mutagenic ingredients. Proceedings of the National Academy of Science USA 72: 2423–2427

Ames B N, McKann J, Yamasaki E 1975a Methods for detecting carcinogens and mutagens with the Salmonella mammalian microsome mutagenicity test. Mutation Research 31: 347–364

Approved Code of Practice 1982 Methods for the determination of toxicity. Notification of new substances regulation. Health & Safety Commission, HMSO, London

Balls M, Clothier R 1983 Differential cell and organ culture in toxicity testing. Acta Pharmacologica et Toxicologica 52: 115–137

Bridges J W 1981 The use of hepatocytes in toxicological investigations. In: Gorrod J W (ed) Testing for toxicity. Taylor & Francis, London

Bridges J W, Hubbard S A 1982 Principles, practice, problems and priorities in toxicology. In: Ward Gardner A (ed) Current approaches to occupational health Vol. 2. Wright PSG, Bristol

Brodie B B, Reid W D, Cho A K, Sipes G, Krishna G, Gillette J R 1971 Possible mechanism of liver necrosis caused by aromatic organic compounds. Proceedings of the National Academy of Science USA 68: 160–164

Brown V H K 1981 Acute toxicity testing — a critique. In: Gorrod J W (ed) Testing for toxicity. Taylor & Francis, London

BUAV Leaflet 1978 These drugs can kill you. British Union for the Abolition of Vivisection, London

Cairns J 1981 The origin of human cancers. Nature 289: 353–357

Ekwall B 1983 Correlation between cytotoxicity in vitro and LD_{50} values. Acta Pharmacologica et Toxicologica 52: 90–99

Eschenroeder A, Irvine E, Lloyd A, Tashima C, Tran K 1980 Computer simulation models for assessment of toxic substances. In: Haque R (ed) Dynamics, exposure and hazard assessment of toxic chemicals. Ann Arbor Science Publishers Inc, Ann Arbor, Michigan

Gorrod J W 1981 Covalent binding as an indicator of drug toxicity. In: Gorrod J W (ed) Testing for Toxicity. Taylor & Francis, London

Hollstein M, McKann J, Angelosanto F, Nichols W W 1979 Short-term tests for carcinogens and mutagens. Mutational Research 65: 133–226

Iype P T 1971 Cultures from adult rat liver cells I. Establishment of monolayer cell-cultures from normal liver. Journal of Cell Physiology 78: 281–288

Parke D V 1983 Biochemical mechanisms of chemical interactions. In: Health effects of combined exposures to chemicals in work and community environments. Interim Document 11. WHO, Geneva

Rees K R 1980 Cells in culture in toxicity testing: a review. Journal of the Royal Society of Medicine 73: 261–264

Shy C M, Alvarie Y, Bates D V, Frank R, Hackney J S, Horrath S M, Nadel J A 1974 Synergism or antagonism of pollutants in producing health effects. Report to US Senate Committee of Public Works, National Research Council, National Academy of Science and National Academy of Engineering, p 483–499

Smith C G, Grady J E, Northam J I 1963 Relationship between cytotoxicity in vitro and whole animal toxicity. Cancer Chemotherapy Reports 30: 9–12
Williams G M, Gunn J M 1974 Long-term cell culture of adult rat liver epithelial cells. Experimental Cell Research 89: 139–142

9. Genetic screening and monitoring for workers

P. A. Schulte W. E. Halperin

INTRODUCTION

A continuum of techniques (Halperin & Frazier, 1985) exists for preventing occupational disease. Major examples include: eliminating hazards from the workplace; containing hazards with engineering controls; protecting workers with personal protective devices such as gloves and respirators; measuring intoxicants in the environment or once absorbed in biological samples; and detecting and treating through screening, occupational disease at early stages when it is reversible or more easily treatable.

One technique, the medical screening of workers, may involve evaluating employees before they begin work. Such screening should look for physical or psychological attributes that may place an employee at high risk even in normal circumstances, such as visual impairment in a vehicle operator. Screening also involves the medical monitoring of employees for early signs or symptoms of disease. Biological monitoring involves the periodical examination of workers for either the absorption of an intoxicant or its metabolite.

The topic of this, and other reviews (Office of Technology Assessment, 1983; Omenn & Gelboin, 1984; Calabrese, 1984) is whether to add other techniques to the list above. The other preventive techniques that might be added are: evaluating workers for genetic traits that may place them at unusual risk in the work environment (genetic screening) or evaluating genetic material from workers for evidence of the adverse effects of occupational exposures (genetic monitoring).

The discussion presented here does not represent the policy of the National Institute for Occupational Safety and Health. Rather, it represents an effort to initiate discussion on this topic.

GENETIC SCREENING

For some time (Chetwynd, 1917; Haldane, 1938), researchers have speculated that genetically determined hypersusceptibility may predispose workers to occupational disease; however only recently has research demonstrated that genetically determined variability in humans may affect health (Kalow, 1984). During the Korean conflict, some soldiers, usually blacks, ingesting primaquin to prevent malaria, developed episodes of acute haemolytic anemia; it is now recognised that their anaemia resulted from a genetically controlled deficiency in glucose-6-phosphate-dehydrogenase (G6PD) (Beutler, 1978). Also in the 1950s, researchers identified a genetic variant of plasma cholinesterase in individuals receiving pharmacologic doses of succinylcholine (Kalow, 1984). In the same decade, researchers also recognised slow acetylators of pharmacologic doses of isoniazid (Hughes et al, 1954).

Early examples like those above established that genetically controlled characteristics made people unusually susceptible to the adverse consequences of therapeutic agents. These discoveries led to the field of pharmacogenetics. The related term ecogenetics encompassed a broader view:

> 'a branch of science concerned with hereditary factors that determine the pharmacological or toxicological responses to chemicals, be these drugs, food constituents, or other ingredients around us' (Kalow, 1984).

Occupational ecogenetics confines itself to the genetically conditioned effect in workers from exposure to chemicals found in the workplace. Genetic screening is the testing of individuals to determine whether they have heritable traits that would predispose them to occupational disease. We present for discussion some traits pertinent to occupational environments. Other sources (Office of Technology Assessment, 1983; Omen & Gelboin, 1984) contain more comprehensive reviews.

A REVIEW OF SELECTED GENETIC VARIANTS AND ASSOCIATED DISEASES

Red blood cell traits

Several genetic deficiencies involve identifiable red blood cell traits. Below, we discuss the glucose-6-dehydrogenase (G6PD) deficiency, sickle cell anaemia, and thalassaemia.

Gluocse-6-dehydrogenase (G6PD) deficiency

When patients have genetically-determined deficiency in the enzyme G6PD of their red blood cells, they are predisposed to acute haemolytic anaemia after they receive certain pharmaceutical drugs (Beutler, 1978). In 1963, Stockinger and Mountain speculated that workers with G6PD deficiency would be susceptible to haemolytic anaemia after they were exposed to industrial chemicals that are strong oxidising agents. A study by Szeinberg et al (1959) is the only attempt to evaluate this occurrence in workers; Szeinberg et al studied 25 workers with G6PD deficiency in a workforce of 241 workers and compared their incidence of haemolytic anaemia with rates from a general population survey. They found no difference in the rates of haemolytic anaemia. After subsequent in vitro studies, they found that G6PD deficient erythrocytes were more sensitive to 13 chemical agents that were common in certain industries such as textile-dyeing, pharmaceutical processing, munitions preparation, rubber tyre manufacturing. In another report (Linch 1974) which was based on the records of biological monitoring and medical examinations at a dye production plant, Linch maintained that G6PD deficient workers were more susceptible to chemical (i.e. aromtic nitro and amino compounds) induced cyanosis than were normal individuals. In two other studies (Larizza et al, 1960; Djerassi & Vitany, 1975) six G6PD deficient workers developed haemolytic anaemia after exposure to trinitrotoluene (TNT). A relatively large number of industrial chemicals could pose a risk of haemolytic anaemia to G6PD deficient workers, but researchers have not demonstrated this occurrence in anyone other than TNT workers.

Sickle cell trait and disease

Sickle cell anaemia results from a change in the amino acid structure of the peptide chains of haemoglobin. The resultant physical changes in haemoglobin and the red cell are most important when oxygen tension of the blood is reduced. Those individuals who are not disabled by the homozygous condition by the time they are adults will probably risk untoward reactions if they take jobs with the potential for hypoxia. Researchers have not looked into this risk, probably because the implications of hypoxia for those with sickle cell disease are well recognised, and also because the morbidity associated with sickle disease usually precludes relevant employment.

Screening for sickle cell trait, the heterozygous condition, is substantially more controversial than screening for the disease. The primary reason for detecting the asymptomatic carrier of the sickle cell trait is to counsel prospective parents (Moutulsky, 1974). Carriers of the sickle cell trait are generally asymptomatic; they have few established complications attributable to the trait (Wintrobe et al, 1974). Whether or not individuals with the trait might sickle during episodes of hypoxia, such as during surgery on in unpressurised aeroplanes is controversial.

The fear of sickling in unpressurised aeroplanes led the US Air Force to screen pilot candidates (Omenn, 1982). Recently, the policy of exclusion of such candidates was abandoned for lack of evidence that applicants with sickle cell trait demonstrated evidence of intravascular sickling when exposed to controlled hypoxic conditions in air chambers during training (Personal communication, Dr Hugh Smith, US Air Force).

Thalasaemia

The rate of haemoglobin synthesis is the primary genetic defect in thalassaemia. Haemoglobin, the oxygen-carrying pigment in red blood cells, consists of two sets of peptide chains. The inherited deficit in production of these chains can be homozygous, producing thalassaemia major, a chronic progressive anaemia that starts in childhood, or heterozygous producing thalassaemia minor, a variable, but usually mild anaemia. However, either kind of anaemia can be exacerbated by stresses such as infections, which have led researchers to question whether people with this anaemia have increased susceptibility to haematologic toxins such as lead and benzene (Calabrese, 1984).

Research on workers exposed to benzene and lead shows very limited evidence that workers with thalassaemia minor might be at an increased risk of anaemia if they are exposed to a variety of chemicals (Office of Technology Assessment, 1983; Girard et al, 1967). Hence, while thalassaemia may explain mild anaemia detected in workers, there is insufficient documentation to warrant screening of asymptomatic workers for thalassaemia minor.

Traits associated with lung diseases

Increased aryl hydrocarbon hydroxylase (AHH) activity and alpha-1-antitrypsin deficiency are traits that appear to have strong associations with lung diseases.

AHH activity

AHH is an inducible enzyme involved in the metabolism of some chemical carcinogens and associated (correlated) with the risk of lung cancer (Kellerman et al, 1973; Kouri et al, 1982). Researchers using animal studies have demonstrated increased activity in AHH after they administered a number of agents, such as polycyclic aromatic hydrocarbons PAH, insecticides, and steroids (Conney, 1976).

In a study of 50 cases of bronchogenic carcinoma, Kellerman et al (1973) found three distinct groups having low, intermediate, and high inducible AHH activities. Paigen et al (1977) found striking seasonal variability in AHH activity in 53 cancer patients, but could not differentiate the patients into three distinct groups, as Kellerman (1973) reported. When they controlled for age, cigarette smoking, hospital diet, and medication, Kouri et al (1982) found a statistically significant positive correlation between AHH activity (0.89 units versus 0.47 units) in lung cancer patients when those patients were compared with non-lung cancer patients with respiratory disease. However, whether the high AHH levels are the result of the primary lung cancer or predispose a person to lung cancer remains to be determined. Results of studies of AHH vary considerably. Some studies show a correlation between high AHH levels and risk of lung cancer and some do not. Kouri et al (1982) have identified eight sources of variation; variations in outcome result from technical difficulties in the test system used to measure AHH induction in peripheral blood lymphocytes, and limitations in the experimental design. Kellerman et al (1973), Kouri (1982) and others (Emery et al, 1978, Gahmberg et al, 1979) have conducted studies suggesting a relationship between AHH activity in certain tissues and the occurrence of lung cancer, but many features of this relationship are unknown.

Alpha-1-antitrypsin deficiency

Researchers have demonstrated that homozygous serum alpha-1-antitrypsin deficiency influences a person's susceptibility to chronic obstructuve pulmonary disease, particularly susceptibility to emphysema (Tobin et al, 1983). Homozygotes for the deficiency allele have approximately 10–15% of the normal concentration of alpha-1-antitrypsin, while heterozygotes have 55–60% of the normal concentration. The trait is codominant; both alleles are expressed in the heterozygous individual. Nearly 80% of the individuals who are homozygotes develop emphysema (Keuppers & Black, 1974). The prevalence of the homozygous trait in human populations is approximately 1/4000 to 1/8000.

The risk for individuals with heterozygous serum containing an alpha-1-antitrypsin deficiency is less clear. Researchers do not agree about whether or not heterozygous deficiency is associated with increased risk of emphysema. Earlier research (Kueppers et al, 1969; Lieberman et al, 1969) showed an increased risk of emphysema in the heterozygotes; however larger, more recent, controlled studies show no risk of emphysema for one particular allelic state, MZ heterozygotes (McDonough et al, 1979; Cole et al, 1976).

These latter studies are limited in size and hence in their abilities to detect an association of heterozygotes with emphysema, which may occur in only 10% of heterozygotes. Furthermore, available studies do not adequately address the presence or absence of coexisting factors, such as environmental exposures that may be necessary to express emphysema. In fact, because emphysema is a multicausal disease, the heterozygous state (while not predisposing in itself) may combine with environmental factors (for example, exposure to cadmium, ozone or cigarette smoke) to present an increased risk. In the USA, heterozygotes for alpha-1-antitrypsin deficiency comprise about 3% of the population; a sizeable group could be at risk if the number of people who are both heterozygotes and exposed to hazardous environmental factors are considered (Office of Technology Assessment, 1983).

Other selected metabolic polymorphisms

A variety of other genetic polymorphisms potentially can affect workers. Three of these are discussed below.

Slow acetylator phenotype

Clinical studies have shown that patients can be divided into two distinct populations according to their ability to acetylate drugs: those with the slow acetylator phenotype (recessive) and those with the fast acetylator phenotype (dominant) (Bonicke & Reif, 1953; Evans et al, 1960; Weber & Hein, 1979). Acetylation in the liver is a common pathway for the metabolism of various substances (especially carcinogenic aromatic amines).

Results from animal and human epidemiologic studies suggest that the rate at which aromatic amines are acetylated may enhance bladder cancer development (Lower et al, 1979; Glowinski et al, 1978). Cartwright et al (1982) found the slow acetylator phenotype to be present in 22 of 23 (96%) people with bladder cancer who had been exposed to aromatic amines, compared with 52 of 88 (59%) people with bladder cancer who had not been exposed. Similar results have been found in animals. Cartwright et al (1982) also found a strong tendency for slow acetylators to harbour a more invasive form of bladder cancer than fast acetylators. Thus far, the number and size of studies is too small to confirm a susceptibility to aromatic amine-induced bladder cancer in slow acetylators.

Microsomal oxidising systems

The cytochrome P-450 systems are essential for the metabolic activation and de-activation of exogenous chemicals. Within the P-450 system substrates are metabolised by oxidation; the resultant metabolites may be either more or less toxic than the parent compound.

In 1977, a landmark report identified a monogenic polymorphism, the slow metabolism of the antihypertensive drug debrisoquine, which affects many types of oxidations. This polymorphism offers what Omenn (1984) has referred to as

> 'a much needed mutant analysis of the highly complex cytochrome P-450 system so critical to the metabolic activation and inactivation of exogenous chemicals.'

Maghoub et al (1977) reported that approximately 7–9% of the population of the UK were slow metabolisers of debrisoquine. The debrisoquine polymorphisms may affect the metabolism of potent carcinogens, such as aromatic amines and aflatoxin.

Ayesh et al (1984) concluded that metabolic oxidation phenotypes may serve as genetic markers for determining individual susceptibility to lung cancer. Researchers estimate that heterozygous individuals who display an intermediate oxidation capability represent 50% of the population; homozygous-recessive individuals comprise 6% of the population (Idle & Smith, 1979). Hence, quite a large number of workers are potentially affected; the number of occupational chemicals that might be influenced by a defect in carbon oxidation is also quite large.

Paraoxonase deficiency

Paraoxinase is a hydrolytic enzyme found in normal plasma that inactivates the pesticide, parathion, as well as its normal metabolite, paraoxon. A single gene codes for low activity for the enzyme (Geldmacher von Mallinckrodt et al, 1973; Playfer et al, 1976). Paraoxonase activity is constant within a given individual but the activity varies considerably among individuals. Individuals with the low activity allele may be more likely to suffer the acute effects of parathion exposure, such as the inhibition of acetyltransferase at neuromuscular junctions.

An estimated 50% of the population is homozygous for the low activity allele; those with the low-activity allele exhibit from one-third to one-sixth the activity of those homozygous for the high activity allele (Geldmacher von Mallinckrodt, 1978). If this hypothesis of risk is confirmed, people with low activity alleles would be at increased risk of parathion toxicity.

GENETIC MONITORING

Cytogenetic indicators

Ultimately the object of monitoring genetic material is to predict the risk of disease caused by genetic damage. Results from animal studies show an empirical association between chromosomal damage and mutagenic/carcinogenic agents; these results show that researchers may use chemically-induced chromosome aberrations and sister chromatid exchanges as indicators of exposure to carcinogenic agents (Preston et al, 1981). It has not been unequivocally determined whether these indicators of exposure are predictors of disease risk. However, there is strong evidence that chromosomal aberrations and sister chromatid exchanges induced by chemical agents may be used as reasonable indicators of carcinogenic risk.

Studies of occupational populations have revealed strong associations between exposures and chromosomal effects (Office of Technology Assessment, 1983). These exposures include: ionising radiation (Evans et al, 1979), arsenic (Nordenson et al, 1978), benzene (Forni et al, 1971), epichlorohydrin (Sram et al, 1980), ethylene oxide (Garry et al, 1979), styrene (Andersson et al, 1980), and vinyl chloride (Purchase et al, 1978). The studies on ionising radiation are the most extensive; they have shown increased incidence of chromosomal aberrations even at occupationally acceptable levels. These aberrations persist months after acute exposure (Evans et al, 1979).

While there is substantial evidence that populations at increased risk for occupational cancer have increased prevalence of cytogenetic abnormalities, the clinical significance for individual workers of positive occupational cytogenic studies is unclear. No occupational studies have followed individuals with positive findings for any chromosomal abnormality through time to observe unusual rates of disease. Furthermore only gross chromosomal damage can be detected by analysing chromosomal aberrations and sister chromatid exchanges. The absence of aberrations does not exclude the existence of other types of DNA damage caused by occupational carcinogens (Mitelman & Pero, 1979). Therefore, it may prove more useful in evaluating individual workers to address the impact of occupational carcinogens by studying chromosomal aberrations in relation to other parameters for evaluating DNA damage.

New genetic monitoring techniques are likely to enhance the capabilities to detect precisely genetic damage. One such technique allows for the measurement of specific locus mutation rates in cultured human lymphocytes using the 6-thioguanine resistant lymphocytes. These techniques have been applied to cancer chemotherapy patients and radiation exposed persons with positive results (Albertini & deMars, 1973; deMars, 1974).

Effects on sperm

Studies are performed on germ cells in order to estimate the genetic consequences of occupational exposures to workers (Preston, 1982). Traditionally, most studies of chromosomal adnormalities have used cultured peripheral lymphocytes. This approach has limitations with regard to sensitivity and the ability to extrapolate to germ cells and subsequently to the genetic consequences of such germ cell exposures (Preston, 1982). A manifestation of these genetic consequences may be an adverse reproductive effect. There are substantial impediments to detecting adverse reproductive health effects by assessing macroscopic events such as birth rates, congenital abnormalities and the like among worker populations (Ratcliffe, 1985). Often only large frequencies of these types of effects, occurring in close temporal proximity to exposures, will be determinable by these assessments. More subtle effects, with longer latent periods, require more sensitive techniques such as the analysis of sperm, for their determination.

The potential for occupational exposure to have an adverse effect on sperm was proven convincingly when a 1977 study demonstrated that workers producing Dibromochloropropane (DBCP) had markedly reduced sperm counts and a decrease in offspring (Whorton et al, 1979). Although DBCP is unusual in the severity of its reproductive effects, studies of other chemical agents or physical factors have demonstrated or suggested that these can affect sperm production or quality in humans (Steeno, 1983). Also, toxicologic studies on laboratory animals have shown that other agents now in use affect animal testes; their findings are without confirmation in man.

It may be useful to test semen quality in order to determine the relationship between adverse reproductive effect and recent (months rather than years) exposures. Unfortunately, sperm count varies greatly from man to man, and from time to time, depending upon such factors as abstinence, fever, alcohol consumption, and so forth (Ratcliffe, 1985). Studies that depend on sperm count in relationship to occupational exposure would have to be sufficiently large, and sufficiently detailed in order to detect any but the most extreme effects.

Some studies that have shown other physical characteristics of sperm result from occupational exposure (Ratcliffe, 1985); the characteristics include: atypical shape, teratospermia (Wyrobek et al, 1981), abnormal fluorescence of Y bodies signifying chromosomal nondisjunction (Kapp & Jacobson, 1980), and abnormal sperm motility. In general, these techniques await further evaluation in research studies to determine their value. It remains for further research studies to evaluate whether measuring sperm production, quality, and physical characteristics represents a valuable way to assess risk.

Oncogenes

According to Omenn (1984b):

> 'Perhaps the most exciting prospect in genetic toxicology will be the development of assays to measure the activation and inhibition of activation of human cancer cell genes.'

These genes are known as oncogenes and they are contained in every human cell. Cellular oncogenes are genes that are probably involved in regulating cell growth. These oncogenes originate from normal sequences of DNA and are conserved with great fidelity during evolution. Carcinogens may convert normal genes (proto-oncogenes) into pathologic genes by inducing higher levels of normal gene products or inducing structurally aberrant gene products (Slamon et al, 1984; Hunter, 1984). Future studies will have to elucidate the actual role of cellular oncogenes in the development of cancer, but recent experiments have demonstrated that the transcriptions of c-myc, c-fos, c-Ras^{Ha}, and c-ras^{Ki} were higher in malignant tissue than they were in corresponding normal tissue (Slamon et al, 1984; Hunter, 1984). Research has not shown whether this finding is merely associated with malignancy or whether it indicates some causal factors. If the latter is true, the ability to monitor for products of oncogenes may provide scientists with the opportunity to evaluate the potential for certain gene-environment interactions to develop carcinogenesis in worker populations.

Noncytogenetic indicators

A variety of noncytogenetic monitoring techniques can be used to assess exposure to mutagenic agents or damage to genetic material. Noncytogenetic monitoring includes tests for the presence of mutagens in body fluids or tests for DNA damage resulting from the presence of mutagens. These tests are not yet established as reliable techniques for human populations. However, human studies (Yamasaki & Ames, 1977; Eisenstadt et al, 1982) have demonstrated that different mutation assays, such as the Ames assay, are simple and inexpensive for detecting human exposures; in some cases, such assays can lead to semi-quantitative estimates of individual exposure. Noncytogenetic monitoring techniques are used because mutagenicity tests may predict a chemical's carcinogenic potential (generally the main concern in occupational settings), and a strong correlation exists between carcinogenicity and mutagenicity (Bartsch et al, 1982). Huessner et al (1985) compared aluminium reduction plant workers exposed to coal tar pitch volatiles to those not exposed; they reported statistically significant higher rate of positive urines in the exposed workers. Results were controlled for cigarette smoking.

Determining mutagenesis in humans is difficult. Consequently, most determinations rely on extrapolations from test systems in nonhuman and often nonmammalian organisms. Mutation assays have four other limitations:

1. studies emphasising occupational exposures contain the potential for confounding by other sources of mutagenic exposures (e.g. drugs, alcohol, smoking, cosmetics, ambient air and water;
2. urine assays detect short-term rather than cumulative exposure;
3. the Ames assay is insensitive to certain classes of chemicals;
4. measuring the mutagenicity of urine is not necessarily a good substitute for concentration of active mutagen at the target tissue (Perera, 1984).

Researchers have found that assays for carcinogen-DNA binding are promising as biological dosimeters because of their biological relevance, their ability to provide chemical-specific data, their sensitivity, and their applicability to various worker populations (Perera, 1984; Eherenberg, 1979). However, such assays have two limitations:

1. all electrophilic carcinogens or their metabolites form complex spectrums of DNA adducts; identifying critical adducts is a difficult analytical challenge;
2. the rate at which adducts are removed can vary substantially from one cell type to another. A chemical that interacts with DNA in one species is likely to interct in others, but a variety of mitigating factors prohibit generalisations.

Although specific techniques are not yet established, researchers increasingly use these techniques to detect exposure to mutagens. Eherenberg et al (1979) have demonstrated that workers exposed to ethylene oxide had alkylated haemoglobin molecules and that there was a dose-response relationship. Research can only use adduction determination for those substances, such as ethylene oxide, that react with blood macromolecules either directly or after metabolism, such as does 4-aminobiophenyl. Adduction with blood macromolecules may occur (Garner, 1985) for many carcinogens that have extremely short lived reactive metabolites.

An excellent indicator of genotoxic and carcinogenic effects of occupational chemicals (Pero et al, 1982) is measuring chemically-induced DNA repair as unscheduled DNA synthesis. Measuring DNA repair is useful because the alteration of DNA appears to be a key step in carcinogenesis inititation and in germ cell mutations that are responsible for heritable genetic changes. It is possible to assess induced repair in many tissues, thus it provides information on target-organ specificity. However, this technique is unlikely to indicate previous chronic exposure; the technique would not work well with non-genetically controlled (epigenetic) mechanisms that do not induce excision repair (Butterworth et al, 1982).

The greatest deficiency in various noncytogenetic tests is the lack of baseline data in normal populations. Without good estimates of the range of responses in unexposed populations, the data from test populations will be difficult to interpret (Office of Technology Assessment, 1983).

REQUIREMENTS TO DEMONSTRATE EFFECTIVENESS OF TECHNIQUES

One should evaluate a screening test in terms of its sensitivity, specificity, and predictive value. We can use these terms to discuss the ability of one test as it compares with some other test, or of one test in relationship to disease. For instance, using a simple physical measure of red cells (Pearson et al, 1973) to determine the heterozygote thalassaemia may be sensitive, specific, and predictive as compared with electrophoretic means. However, thalassaemia minor and occupational clinical disease are hardly related, making use in the occupational setting of the physical measures of red cells appear ineffective at present. Sensitivity is the probability of the test being positive in those with the condition sought. Specificity is the probability of the test being negative in those free of the condition. False positives in people without disease lead to the test being nonspecific. Predictive value is the probability that those who tested positive have the condition, rather than being falsely positive because the test is not perfectly specific. The predictive value of a test is substantially related to the specificity of the test and the prevalence of the condition sought (Vecchio, 1966). The interrelationship of these operating characteristics and the limitations in achieving a large predictive value can be seen in the following example: in a test for a genotype frequency of 1%, even if the sensitivity and specificity are each 99%, the predictive value is only 50% (Office of Technology Assessment, 1983). This is not generally an adequate predictive value upon which to base a screening programme.

Khoury et al (1985) have developed general formulas to predict the usefulness of any genetic screening programme in terms of sensitivity, specificity, and predictive value. The use of these formulas depends on having reliable estimates of the frequency of the marker in the population of interest; the probability of developing the disease subsequent to the time of the screening programme; and the strength of the association between the marker and the disease. Based on the arithmetic relationships derived by Khoury et al (1985) it can be shown that with very large relative risks, sensitivity and positive predictive value are affected by the relative magnitude (frequency) of disease and genetic marker frequencies. However when the genetic marker is less frequent than the disease, the positive predictive value increases with increasing relative risk but the sensitivity remains low.

> 'When the genetic marker is more frequent than the disease, sensitivity increases with increasing relative risk but the positive predictive value remains low. When marker and disease frequencies are equal, both positive predictive value and sensitivity increase with increasing relative risks . . .' (Khoury et al, 1985).

Demonstration that a test is effective requires that beyond an evaluation of the operating characteristics of the test, it must be shown that the test leads to the prevention of occupational disease in practice. For a screening test to be effective in practice (Sackett & Holland, 1975; Halperin & Frazier, 1985) there are other considerations:

1. the disease to be prevented should be important;
2. the disease should be preventable;
3. follow-up care should be planned for those who participate in screening programmes;
4. professionals who use the test should be skilled in interpreting the test;
5. a screening test should be targeted to specific job risks;
6. action levels should be decided upon before the screening programme starts;
7. the screening programme should be acceptable to the workforce;
8. the goals of the screening programme should be specified.

RESEARCH NEEDS

With almost no exception, researchers have not adequately shown the practical relevance of genetic polymorphism to the health of workers. While it might be expected that industrial chemicals with similar properties to pharmacological agents may induce similar diseases in workers, because workers are usually exposed to lower than pharmacological doses, and perhaps because of a dearth of research interest in this area, we have little evidence pro or con that genetic polymorphism places workers at risk.

Two types of research are indicated. First, workers with clinical illnesses (such as haemolytic anaemia, bladder cancer, and so forth) should be carefully compared with coworkers without such disease to establish the presence or absence of predisposing genetic abnormalities. An example of this approach is the comparison of dye workers with bladder cancer with those without for slow or fast acetylation (Cartwright et al, 1982). Alterntively, a more demanding but informative approach would be to screen large populations for genetic variants and follow these populations into the future with careful documentation of chemical exposure and illness.

Research needs in the area of genetic monitoring are somewhat different. First, researchers should undertake careful cross-sectional studies that attempt to establish whether exposure to certain chemicals is associated with the genetic outcome of interest. These studies must be done at various levels and durations of exposure to establish dose-response relationships and control for individual variation (Legator, 1982). If these relationships are established, then genetic monitoring may serve as a kind of biological dosimeter for populations, if not for individuals. We may find so much individual variation that it clouds any association between effect observed in an individual and effect in a population. Second, research should move towards an understanding of the prognostic implications for individuals or populations with altered genetic material. For this purpose, researchers should document and follow populations whose genetic abnormalities can be documented now.

METHODOLOGIC DIRECTIONS IN GENETIC TESTING RESEARCH

Recombinant-DNA-based tests

Over the last 10 years, researchers have developed recombinant-DNA techniques that, in theory, will detect all genetic diseases and many genetic predispositions to diseases. Four advances in this technology make it possible to identify previously inaccessible genes:

1. the development of restriction fragment length polymorphisms (RFLP);
2. the development of monoclonal antibodies;
3. the development of direct gene probes;
4. the ability to sequence libraries of recombinant DNA-derived genes (Lappe, 1984). Examples of the use of some of these advances follow.

Researchers identified the molecular basis for alpha-1-antitrypsin deficiency and the responsible gene through using specially-synthesised DNA sequences that mimic the region responsible for the defect (Kidd et al, 1983; Lappe, 1984).

Monoclonal antibodies have been used to identify specific human leukocyte antigens (HLAs) with greater precision than the existing methodologies. For example, one monoclonal antibody has been used to detect an antigen (HLA-B27) associated with ankylosing spondylitis, an arthritis-like condition that may be linked to an individual's capacity for certain types of work (Lappe, 1984).

Restriction enzymes have allowed researchers to identify almost 30 different oncogenes associated with a variety of cancers (See discussion of oncogenes). The specificity of this new technology is apparent in the oncogene for bladder cancer, a malignancy with well-documented occupational risk factors. Using recombinant-DNA techniques, researchers have identified a single base substitution that might be responsible for the altered nature of the bladder cancer oncogene (Taparowsky et al, 1982).

The use of DNA probes as genetic markers has great potential to aid in identifying the extent to which persons with particular genetic conditions may be at increased risk to environmental mutagens/carcinogens. This technique offers a clear advantage over current approaches which are constrained to use genetic probabilities to estimate the likelihood of genetic relationships possessing a trait (Hill, 1985).

The specificity allowed by these newer technologies is still some years away. At present, genetic disorders are only able to be detected if a gene is discovered near the actual risk-generating gene. For example, genes have been detected for conditions such as atherosclerosis, angiopathy and Huntington's disease. In these conditions, the markers identified were too imprecise for researchers to use in identifying who is really at risk. However, researchers can predict, to a greater or (lesser) likelihood the occurrence of potentially disabling conditions, Life insurance and workers' compensation underwriters, and membership directors of health plans will be likely to seek such data (Lappe, 1984).

Gene-environment interaction

The primary goal of genetic testing research is to determine the strength and degree of association between a genetic marker (or predisposition) and a disease. Of the approximately 2000 genetic diseases identified to humans, only a few diseases have been studied to assess whether (and to what extent) genetic conditions may predispose persons to environmentally-induced disease. The fundamental concepts that allow us to determine relationships between genetic markers and disease have already been established. In fact, the field of pharmacogenetics has extended those concepts to cover the link between genetic predispositions and disease given exposure to drugs. There is a close parallel between that work and work to be done with industrial chemicals.

Too often genetic variables have been excluded from multivariate models for evaluating disease risk factors because surveys have gathered nothing more than a broad question on family history (Skolnick, 1984). Before genetic testing can be implemented researchers must determine the association between genetic markers and disease, with control or adjustment for physiological, demographic, and cultural, as well as environmental, variables. Traditional epidemiological methods, such as the case-control and cohort studies with analyses by multiple logistic regression, have been useful in evaluating these associations (Skolnick, 1984).

In the past, most genetic research has focused on rare and apparently inherited disorders. But the emphasis has begun to shift to include, among other interests, common diseases, related to life-style, and environmental and occupational conditions (Schull & Weiss, 1980). Such diseases have multifactorial aetiologies.

Research on gene-environment interactions is now beginning to be a more frequent topic of study. A recent conference, sponsored by the United States Council on Environmental Quality, on long term environmental research concluded that of the 2000 gentically identifiable human diseases 50 have been suggested as enhancing the risk of affected individuals to environmental agents but that for the most part whether and to what extent such genetic conditions may predispose persons to environmentally-induced disease is not known. The conference recommended that for research purposes registries of high risk genetic groups be developed (Hill, 1985).

King et al (1984) reviewed five approaches for disentangling genetic and environmental influences on disease distribution:

1. twin studies;
2. adoption studies;
3. path analysis;
4. analysis of the cultural transmission of disease risk factors;

5. studies of association between specific genotypes and diseases.

Path analysis is a statistical method derived from multiple regression and it is used to evaluate associations with health conditions in relatives. Such analysis is particularly valuable in assessing genetic and environmental influences.

One problem in assessing gene-environment interactions is that different types of data are appropriate for evaluating the specific components of the interaction. The genetic component is best determined with data from pedigree studies; the environmental component is best determined with data from case-control and cohort studies (Skolnick, 1984).

Skolnick (1984) believes a solution to this dilemma might be to map a gene, and then study a cohort of individuals with the susceptible genotype and a set of controls so that gene-environment interactions can be clearly defined. Using genetic markers from DNA cloning appears to be a useful way to map genes. In the past, the gene products or proteins were the only tools available for investigation of genes. Now with DNA sequences themselves available for analysis we can ascertain the complexity of the underlying nature of genetic defects. The development of technologies to probe processes on a molecular level has helped increase our understanding of the biochemical components of disease and of genetic defects. Epidemiological methods must follow suit in this regard.

The role of various newly emerging laboratory procedures in epidemiology has been described by the term molecular epidemiology (Perera, 1984). The term molecular epidemiology was first used in research on viral diseases. It has more recently been applied in the context of chronic diseases (Higginson, 1967; Lower et al, 1979; Perera & Weinstein, 1982) as have the terms biochemical and metabolic epidemiology (Wynder & Reddy, 1974). Lower et al (1979) interpreted the term to mean

> 'molecular level observations demonstrating differentials in the specificity and functional capacity of host-mediated activation and detoxification processes. . . .'

These processes are analogically and spatio-temporally consistent with both cellular- and organism-level observations of risk. This term or mode of study includes evaluating genotypic expressions, such as the slow acetylator phenotype, and ABO blood groups, as well as oncogenes, mutagens in urine, DNA adducts and restriction fragment length polymorphisms. Molecular level markers and endpoints should be useful to support more productive research and to provide better candidates for genetic testing.

ETHICAL AND LEGAL ISSUES

Ethical and legal issues represent the greatest controversies over genetic testing. Any consideration of applying genetic screening tests raises troubling ethical issues because employability has traditionally been based on ability to perform essential job duties, not on projected wellness over a working lifetime (Lappe, 1983). Genetic screening tests that identify varied susceptibility in worker populations are based on individual differences and, hence, are pointedly anti-egalitarian (Murray, 1983). Such tests directly conflict with the meritocratic ideal that is a cornerstone of democratic societies. The controversy is over how to maintain this meritocratic ideal, in light of genetic diversity and of the acknowledged role of genetic factors in disease.

Purposes of genetic screening and monitoring

Genetic screening and monitoring is not in itself seen as inherently unethical, but ethical and legal problems arise from the purposes, uses, implications, and effects of genetic testing. There are at least four general purposes for genetic tests: diagnosis, research, information and job placement (Murray, 1983). The suitability of genetic screening for each of these purposes depends on the degree of certainty research has shown on the links among a genetic anomaly, the potential exposure, and the possible resultant disease. Currently, if the link among these three elements is strong and the risk is high, many concerns about genetic testing would be more easily resolved. In reality, the links are tenuous and involve only a few genetic defects; the risks, both relative and absolute, are often unremarkable.

However, the role of genetic predispositions in occupational disease is not always trivial. In some cases, genetic predispositions may turn out to be more significant as we learn more. At this time, researchers are still uncertain about the linkage of genetic traits to disease, after occupational exposure (Office of Technology Assessment, 1983). Without demonstrating effectiveness and careful weighing of ethical impacts, genetic screening for the diagnosis of disease or for excluding people from certain work is unwarranted, nor would this data be a useful basis for workers to consider employment situations. At this time the only viable purpose for genetic screening is research (Dabney, 1981). Research in this field is the greatest need; the general ethical opinion is that, given proper controls, research is warranted (Office of Technology Assessment, 1983).

Guidelines for developing genetic tests

Lappe (1983) has examined the ethical aspects of genetic screening and proposed a set of criteria that would permit researchers to develop orderly tests. Lappe originally proposed criteria for genetic screening in a medical context (Lappe et al, 1972), but they provide guidance in the occupational context as well. The criteria include the following:

1. an attainable purpose;
2. workforce participation;
3. equal access and/or random participation;
4. adequate testing procedures;
5. absence of compulsion;
6. informed consent;
7. protection of subjects;
8. access to information;
9. provision of counselling/follow-up;
10. understandable relationship between tests and therapy, if any therapy results;
11. protection of the right to privacy.

Genetic monitoring creates fewer ethical problems than genetic screening. In principle, genetic monitoring is based upon the effects of an exposure and, in that sense, genetic monitoring is similar to most other biological monitoring. Unfortunately, in practice, the link between cytogenetic or non-cytogenetic endpoints and the risk of disease is not well established (Dabney, 1981). Therefore, personnel actions that are based on the results of genetic monitoring are unclear. However, a presumed risk could warrant environmental precautions.

Genetic testing and the law

The legal issues in genetic testing have not yet been adequately addressed in the law. There is a series of unanswered questions about these issues. The most important questions in this area may be summarized as:

1. how are genetically susceptible individuals to be identified;
2. once such individuals are identified, what may or must an employer do (Rothstein, 1983).

Without legal safeguards genetic screening tests could undermine occupational safety and health legislation by shifting emphasis from controlling hazardous environments to excluding hypersusceptible workers (Diamond, 1983).

One approach to these questions is exemplified by the Council of European Communities which requires that a history of G6PD deficiency and thalassaemia be considered by the examining physician in the placement and provision of medical surveillance for workers exposed to lead (Commission of the European Communities, 1982).

Regulatory activity in the USA has not directly approached the issue of how genetically susceptible individuals are to be identified and what should happen after sensitive individuals have been identified. However, this latter issue has a parallel in attempts to regulate any employment effects that follow the medical examinations of workers. The Occupational Safety and Health Administration (OSHA) efforts involve used 'medical removal protection' and 'rate retention' in the lead standard (29 CFR 1910.1025K) to safeguard pregnant women and fetuses (Rothstein, 1983). 'Medical removal protection' is the reassignment to a new job of an employee who shows symptoms of the adverse effects of exposure until it is medically advisable for the employee to return. If the new safe job is at a lower rate of pay, 'rate retention' would require employers to maintain wage and benefit levels during the period of medical removal. Medical removal protection and rate retention have been upheld in the courts for lead (United Steelworkers of America versus Marshall 1980), but the Supreme Court in the cotton dust case (American Textile Manufacturers versus Donovan 1981) cautioned that

> 'the Act in no way authorises OSHA to repair general unfairness to employees that is related to the achievement of health and safety goals. . . .'

It is still an unanswered question whether medical removal protection and rate retention will be applied to genetically-tested workers.

Because genetic testing, especially genetic screening, deals with individual differences, the way workers who exhibit certain characteristics are treated may be unfair. Can protection for these workers be found in anti-discrimination laws, such as: the Rehabilitation Act of 1973, which prohibits discrimination on the basis of handicap; the Civil Rights Act of 1964, which prohibits discrimination based on race, colour, religion, sex, or national origin; and the Federal Fair Employment Practices Act? There is still a question of whether individuals who have a genetic susceptibility or altered genetic material come within the definition of 'handicapped'. Even if the definition is inclusive, the law lacks an adequate remedy. There is no private right to legal action, and individuals who claim discrimination must file a complaint with the Department of Labor. More effective remedies may be found in the handicap discrimination laws 41 states have, but the issue is still controversial.

Some genetic markers, such as G6PD deficiency, thalassaemia trait, and sickle cell trait, have a disparate impact along racial and ethnic lines; therefore, do employment practices that preclude affected individuals violate Title VII of the Civil Rights Act of 1964 (Rothstein 1983)? So far, not very many cases have involved this interpretation. To comply with the Federal Fair Employment Practices Act an employer must demonstrate:

1. that the genetic factor being sought in screening is a bona fide qualification for adequate job performance;
2. that the genetic factor sought be related to the ability to conduct safely the duties of the job;
3. that the test used be valid and reliable (Lappe, 1983).

Will other areas of law, especially workers compensation law, also have a bearing on the employment rights of workers who receive genetic tests? An entire body of workers compensation law has developed around whether susceptible workers are entitled to compensation (Rothstein, 1983). The predictive value of the genetic screening test would be key in any deliberations on whether susceptible workers should be compensated.

The additional issues of an employee's right-to-know and the confidentiality of medical or privileged information must, also, be addressed within the context of genetic screening. At present, though no federal legislation deals specifically with the right of hypersusceptible workers to be informed of test results or told of potential work place hazards which may pose them increased risk, both the Occupational Safety and Health Act and the Toxic Substance Control Act may be relevant (Diamond, 1983). Also, many states have promulgated right-to-known legislation, which while not specific in addressing genetic screening, may be applicable in many instances. The protection of medical information from unauthorised disclosure is generally excepted. However, due to possible familial links, medical ethics may dictate the release of relevant genetic screening information to family members (Diamond, 1983).

The ethical and legal controversies surrounding genetic testing are based on the conflict between maintaining equal employment opportunities without sacrificing worker health or economic efficiency. Two possible dangers are that companies will base their policies on scant or inconclusive data and that genetic screening procedures will be used indiscriminately and result in harsh consequences for workers at risk (Rothstein 1983). Also, the application of genetic screening in employment practice poses the risk that focus of responsibility for occupational disease will shift from the employer to employee (Diamond, 1983).

CONCLUSION

The use of genetic traits to identify workers with increased susceptibility to disease, and the use of genetic material as a sensitive early warning system for exposure or disease are highly attractive techniques which have the potential to be added to the continuum of prevention for occupational disease. The data, however, do not support the routine application of genetic screening tests at this time. Until genetic screening tests are more accurate, reliable, and valid in predicting risk of disease they should be considered as experimental and used only in research.

Monitoring changes in genetic material has a firmer foundation for implementation. There have been well-documented studies that support the use of various genetic changes (chromosomal aberations; sister chromatid exchanges) and related changes (DNA adducts; unscheduled DNA synthesis; and mutagens in biological fluids) as indicators of exposure to carcinogenic and mutagenic agents. Although these changes are reasonable indications of carcinogenic risk, this has not been unequivocally determined. Changes in genetic material are not always adverse health effects in themselves and this limitation must be considered when evaluating data from genetic monitoring studies.

There is great potential for the techniques of genetic screening and monitoring to be abused, to be discriminatory or to shift the blame for disease risk from hazardous environmental conditions to workers themselves. The ethical and legal problems must be guarded against during the inevitable attempts to release the great promise that the evaluations of genetic traits and material holds for prevention of occupational disease.

REFERENCES

Albertini R J, deMars R 1973 Somatic cell mutation: detection and quantification of X-ray-induced mutation in cultured diploid human fibroblasts. Mutation Research 18: 199–224

American Textile Manufacturers Institute, Inc et al v Donovan 1981 US 490 USLW 4720, 16 June

Andersson H C, Tranberg E A, Uggla A H, Zeherberg G 1980 Chromosomal aberrations and sister-chromatid exchanges in lymphocytes of men occupationally exposed to styrene in a plastic-boat factory. Mutation Research 73: 387–401

Ayesh, Idle J R, Ritchie J C, Crothers M J, Aetzel M R 1984 Metabolic oxidation phenotype, as markers for susceptibility to lung cancer. Nature 312: 169–170

Bartsch H, Tomatis L, Mullavillec 1982 Qualitative and quantitative comparisons between mutagenic and carcinogenic activities in chemicals. In: Heddle J A (ed) Mutagenicity: new horizons in genetic toxicology, Academic Press, London, p 35–67

Beutler E 1978 Hemolytic anemias in disorders of red cell metabolism. Plenum, New York

Bonicke R, Reif W 1953 Enzymatische inakfivierung von isonicotin square hydrasid in messchlichen and tierischen organisms. Arch Exp Pathol Pharmacol 220: 321–333

Butterworth B E, Doolittle D J, Working P K, Strom S C, Jirtle R L, Michealopoulos G 1982 Chemically-induced DNA repair in rodent and human cells. In: Bridges B A, Butterworth B E, Weinstein I B (eds) Indicators of genotoxic exposure, Banbury Report 13, p 101–112

Calabrese E J 1984 Ecogenetics: genetic variation in susceptibility to environmental agents. John Wiley, New York

Cartwright R A, Glashan R W, Rogers H J, Ahmad L A, Baraham-Hall D, Higgins E, Kahn M A 1982 Role of N-acetyltransferase phenotypes in bladder carcinogenesis: pharmacoepidemiological approach to bladder cancer. Lancet ii: 842–846

Chetwynd V 1917 In discussion on the origin, symptoms, pathology, and treatment and prophylaxis of toxic jaundice observed in munitions workers. Proceedings of the Royal Society of Medicine 10:6

Cole R B, Nevin N C, Blundell G, Merret J D, McDonald J R, Johnston W P 1976 Relation of alpha-1-antitrypsin phenotype to the performance of pulmonary function tests and to the prevalence of respiratory illness in a working population. Thorax 31: 149–157

Commission of the European Communities 1982 Directive 82/605/EEC. Official Journal of European Communities L 247/12

Conney A H 1976 Pharmacological implications of microsomal enzyme induction. Pharmacological Review 19: 3317–366

Dabney B J 1981 The role of human genetic monitoring in the workplace. Journal of Occupational Medicine 23: 626–631

deMars R 1974 Resistance of cultured human fibroblasts and other cells to purine and pyrimidine analogues in relation to mutagenesis detection. Mutation Research 24: 335–364

Diamond A L 1983 Genetic testing in employment situations. Journal of Legal Medicine 4: 231–256

Djerassi L S S, Vitanyl 1975 Haemolytic episode in G6PD deficient workers exposed to TNT. British Journal of Industrial Medicine 32: 54–58
Eherenberg L 1979 Risk assessment of ethylene oxide and other components. In: Assessing chemical mutagens: the risk to humans. Banbury Report 1, Cold Spring Harbor Laboratory, p 157–190
Eisenstadt E, Kado N V, Putzrath R M 1982 Detection of mutagens in body fluids. In: Bridges P A, Butterworth B E, Weinstein I B, (eds) Banbury Report 13, Cold Spring Harbor Laboratory, p 338
Emery A E M, Danford N, Anand R, Duncan W, Paton L 1978 Aryl hydrocarbon hydroxylase inducibility in patients with lung cancer. Lancet iii: 470–471
Evans D A P, Manley K A, McKusick V A 1960 Genetic control of isoniazid metabolism in man. British Medical Journal 2: 485–491
Evans H J, Buckton K E, Hamilton G E, Carothers A 1979 Radiation-induced chromosome aberrations in nuclear-dockyard workers. Nature 277: 531–534
Forni A, Pacifico E, Limonta A 1971 Chromosome studies in workers exposed to benzene or toluene or both. Archives of Environmental Health 22: 373–378
Gahmberg C G, Sekki A, Kosunen T U, Holsti L R, Mekele O 1979 Induction of aryl hydrocarbon hydroxylase activity and pulmonary carcinoma. International Journal of Cancer 23: 302–305
Garner R C 1985 Assessment of carcinogen exposure in man. Carcinogenesis 6: 1071–78
Garry V F, Hozier J, Jacobs D, Wade R L, Gray D G 1979 Ethylene oxide; evidence of human chromosomal effects. Environmental Mutagenesis 1: 375–382
Geldmacher von Mallinckrodt M 1978 Polymorphism of human serum paraoxonase. Human Genetics Suppl 1: 65–68
Geldmacher von Mallinckrodt M, Lindorf H H, Petenyi M, Flugel M, Fisher Th, Hillet Th 1973 Genetish determinierter polymorphismus der menschlichen serum paraoxonase. Humangenetik 17: 331
Girard K P, Mallein M L, Jouvenceau A, Tolot F, Revol L, Bourett J 1967 Etude de la sensibilite aux toxiques industriels des porteurs du trait thalassemique. Le Journal de Medicine de Lyon 48: 1113–1126
Glowinski I B, Radtke H E, Weber W W 1978 Genetic variation in N-acetylation form of carcinogenic arylamines by human and rabbit liver. Molecular Pharmacology 14: 940–949
Haldane J B S 1938 Heredity and politics. London, George Allen and Unwin, p 179–180
Halperin W E, Frazier T M 1985 Surveillance for the effects of workplace exposure. Annual Review of Public Health 6: 419–432
Heussner J C, Ward Jr J M, Legator M S 1985 Genetic monitoring of aluminum workers exposed to coal tar pitch volatiles. Mutation Research 155: 143–155
Higginson J 1967 The role of the pathologist in environmental medicine and public health. American Journal of Pathology 86: 460
Hill A 1985 Report on long-term environmental research and development. Council on Environmental Quality, Washington DC
Hughes H B, Biehl J P, Jones A P, Schmidt L H 1954 Metabolism of isoniazid in man as related to the occurrence of peripheral neuritis. American Review of Tuberculosis 70: 266–273
Hunter T 1984 Oncogenes and proto-oncogenes: how do they differ? Journal of the National Cancer Institute 73: 773–785
Idle J R, Smith R L 1979 Polymorphisms of oxidation at carbon centers of drugs and their clinical significance. Drug Metabolism Review 9: 301–317
Kalow W A 1984 Pharmacologist looks at ecogenetics. In: Omenn G S, Gelboin H V (eds) Genetic variability in responses to chemical exposure. Banbury Report 16. Cold Spring Harbor Laboratory, p 15–35
Kapp R W, Jacobson C B 1980 Analysis of human spermatozoa for Y chromosomal nondisjunction. Teratogenesis, Carcinogenesis, and Mutagenesis 1: 193–211
Kellerman G, Shaw C R, Luyten-Kellerman M 1973 Aryl hydrocarbons hydroxylase inducibility and bronchogenic carcinoma. New England Journal of Medicine 289: 934–937
Khoury M J, Newill C A, Chase G A 1985 Epidemiologic evaluation of screening for risk factors; application to genetic screening. American Journal of Public Health 75: 1204–1208
Kidd V J, Wallace R B, Hakura K, Woo S L C 1983 Alpha-l-antitrypsin deficiency detection by direct analysis of the mutation in the gene. Nature 304: 230–234
King M C, Lee G M, Spinner N B, Thompson G, Wrensch M R 1984 Genetic epidemiology. Annual Review of Public Health 5: 1–52
Kouri R E, McKinney C E, Siomiany D J, Snodgrass D R, Wray N P, McLemore T L 1982 Positive correlation between high aryl hydrocarbon hydroxylase activity and primary lung cancer as analyzed in cryropreserved lymphocytes 42: 5030–5037
Kueppers F, Fallat R, Larson R K 1969 Obstructive lung disease and alpha-l-antitrypsin in deficiency gene heterozygosity. Science 165: 899
Kueppers F, Black L F 1974 Antitrypsin and its deficiency. American Review of Respiratory Disease 110: 176–194

Lappe M 1983 Ethical issues in testing for differential sensitivity to occupational hazards. Journal of Occupational Medicine 25: 797–808
Lappe M 1984 Broken code. Sierra Club Books, San Francisco
Lappe M, Gustafson J Roblin R et al 1972 Ethical and social aspects of screening for genetic disease. New England Journal of Medicine 206: 1129–1132
Larizza P, Brunetti P, Grignani F 1960 Anemie emolitche enzimopeniche. Haematologica 45: 1–90, 129–212
Legator M S 1982 Approaches to worker safety: recognizing the sensitive individual and genetic monitoring. Annals of the American Conference of Governmental Industrial Hygienists 3: 29–34
Lieberman J, Mittman C, Schneider A S 1969 Screening for homozygous and heterozygous alpha-l-antitrypsin deficiency. Journal of the Americal Medical Association 2010: 2055
Linch A L 1974 Biological monitoring for industrial exposure to cyanogenic aromatic nitro and amino compounds. American Industrial Hygiene Association Journal 35: 426–432
Lower G M Jr, Nilsson T, Nelson C E, Wolf H, Gamsky T E, Bryan G T 1979 N-acetytransferase phenotype and risk in urinary bladder cancer: approaches in molecular epidemiology. (Preliminary results in Sweden and Denmark.) Environmental Health Perspectives 29: 71–79
Maghoub A, Idle J R, Dring L D, Lancaster R, Smith R L 1977 The polymorphic hydroxylation of debrisoquin in man. Lancet ii: 584–586
McDonough D J, Nathan S P, Knudson R J, Lebowitz M D 1979 Assessment of alpha-l-antitrypsin deficiency heterozygous as a risk factor in the etiology of emphysema. Journal of Clinical Investigations 63: 299–309
McKusick V 1983 Mendelian inheritance in man. Johns Hopkins Press, Baltimore
Mitelman F, Pero R W 1979 Indicators of DNA damage determined by a combined analysis of chromosome breakage, sister chromatid exchange, aryl hydrocarbon hydroxylase, DNA repair synthesis, and carcinogen binding. In: Berg K (ed) Genetic damage in man caused by environmental agents. New York, Academic Press, p 364
Motulsky A G 1974 Screening for sickle cell hemoglobinopathy and Thalassemia. In: Ramat B, Adam A (eds) General Polymorphisms and Diseases in Man. New York, Academic Press
Murray T H 1983 Genetic screening in the workplace: ethical issues. Journal of Occupational Medicine 25: 451–454
Nordenson I, Beckman G, Beckman L, Nordstrom S 1978 Occupational and environmental risks in and round a smelter in northern Sweden II. Chromosomal aberrations in workers exposed to arsenic. Hereditas 88: 47–50
Office of Technology Assessment 1983 The role of genetic testing in the prevention of occupational disease. Office of Technology Assessment, Congress of the United States, US Government Printing Office
Omenn G S 1982 Predictive identification of hypersusceptible individuals. Journal of Occupational Medicine 24: 369–74
Omenn G S 1984a Advances in genetics and immunology: the importance of basic research to prevention of occupational diseases. Archives of Environmental Health 39: 173–182
Omenn G S 1984b Risk assessment, pharmacogenetics and ecogenetics. In: Omenn G S, Gelboin H V (eds) Genetic variability in responses to chemical exposure. Banbury Report 16, p 11 Cold Spring Harbor Laboratory
Omenn G S, Gelboin H V (eds) 1984 Genetic variability in responses to chemical exposure. Banbury Report 16, Cold Spring Harbor Laboratory
Paigen B, Minuwade J, Ourto H L, Oargen K, Parker N B, Ward E, Hayner N T, Bross I D J, Bock F, Vincent R 1977 Distribution of aryl hydrocarbon hydroxylase inducibility in cultured human lymphocytes. Cancer Research 37: 1829–1837
Pearson H A, O'Brien R T, McIntosh S 1973 Screening for Thalassaemia trait by electronic measurement of mean corpuscular volume. New England Journal of Medicine 288: 351–353
Perera F, Weinstein I B 1982 Molecular epidemiology and carcinogen-DNA adduct detection: new approaches to studies of human cancer causation. Journal of Chronic Diseases 35: 581–600
Perera F 1984 Molecular epidemiology: a novel approach to the investigation of pollutant related chronic disease. Paper presented to the Conference on Long-term Research, Washington
Pero R W, Bryngelsson T, Hogstedt B, Akesson B 1982 Occupation and in-vitro exposure to styrene assessed by unscheduled DNA synthesis in resting human lymphocytes. Carcinogenesis 3: 681–685
Playfer J R, Eze L C, Bullen M F, Evans D A P 1976 Genetic polymorphism and inter-ethnic variability of plasma paraoxonase activity. Journal of Medical Genetics 13: 337
Preston R J et al 1981 Mammalian in vivo and in vitro cytogenetic assays: a report of the US EPA's Gene-Tox Program. Mutation Research 87: 143–188
Preston R J 1982 Chromosome aberrations in decondensed sperm DNA. In: Bridges P A, Butterworth B E, Weinsten I B (eds) Banbury Report 13, Cold Spring Harbor Laboratory, p 515–526

Purchase I F H, Richardson C K, Anderson D, Paddle G M, Adams E G F 1978 Chromosomal analysis in vinyl chloride exposed workers. Mutation Research 21: 335–340
Ratcliffe J 1985 Fertility in World Health Organization monograph on occupational hazards to reproduction. WHO Geneva (in preparation)
Rothstein M H 1983 Employee selection based on susceptibility to occupational illness, Michigan Law Review 81: 1379–1496
Sackett D, Holland W W 1975 Controversy in the detection of disease. Lancet ii: 357–359
Schull W J, Weiss K M 1980 Genetic epidemiology: four strategies. Epidemiologic Reviews 2: 1–18
Skolnick M 1984 Priority needs in the development of epidemiologic techniques that account for human physiologic and genetic variability. Paper presented at the Conference on Long-term Environmental Research and Development, Council on Environmental Quality, 11 September, Washington DC
Slamon D J, Lekernion J B, Verma I M, Cline R J 1984 Expression of cellular oncogenes in human malignancies. Science 244: 256–262
Sram R J, Sudova Z, Kuleshov N P 1980 Cytogenetic analysis of peripheral lymphocytes in workers occupationally exposed to epichlorohydrin. Mutation Research 70: 115–120
Steeno O P, Pangkahila A 1984 Occupational influences on male fertility and sexuality. Andrologia (Part 1 and 2) 16: 5–22 and 16: 93–101
Stockinger H E, Mountain J T 1963 Test for hypersusceptibility to hemolytic chemicals. Archives of Environmental Health 6: 495–502
Szeinberg A, Avindam A, Myers F, Sheba C, Ramot B 1959 A hematological survey of industrial workers with enzyme-deficient erythrocytes. Archives of Industrial Health 20: 510–516
Taparowsky E, Suard Y, Fasano O, Shimizu K. Goldfarb M, Wigler M 1982 Activation of the T24 bladder carcinoma transforming gene is linked to a single amino acid change. Nature 300: 762
Tobin M J, Cook P J L, Hutchison D C S 1983 Alpha-l-antitrypsin deficiency: the clinical and physiological features of pulmonary emphysema in subjects homozygous for P. Type Z. British Journal of Diseases of the Chest 77: 14–27
United Steelworkers of America versus Marshall, 647 F.2d 1189: 1252–1259 (DC Cir 1980)
Vecchio Thomas J 1966 Predictive value of a single diagnostic test in unselected populations. New England Journal of Medicine 271: 1171–1173
Weber W W, Hein D W 1979 Clinical pharmacokinetics of isoniazid. Clinical Pharmacokinetic 4: 401–422
Whorton D, Milby T H, Krauss, R M, Stubbs H A 1979 Testicular function in DBCP exposed pesticide workers. Journal of Occupational Medicine 21: 161–166
Wintrobe M M et al 1974 Sickle cell trait and sickle cell anemia. In: Lee G R, Boggs D R, Bithell T C, Athens J W, Foester J (eds) Clinical Hematology, Lea and Febiger, Philadelphia
Wunder E L, Reddy B S 1974 Metabolic epidemiology of colorectal cancer. Cancer 34: 801–806
Wyrobek A J, Watchmaker G, Gordon L, Wong K, Moore D, Whorton D 1981 Sperm shaper abnormalities in carbaryl-exposed employees. Environmental Health Perspectives 40: 255–265
Yamasaki E, Ames B 1977 Concentration of mutagens from urine by absorption with the nonpolar resin XAD: cigarette smokers have mutagenic urine. Proceedings of the National Academy of Science 74: 3555–3559

10. Industrial disasters: classification, investigation, and prevention

R. G. Parrish H. Falk J. M. Melius

This is the excellent foppery of the world, that, when we are sick in fortune,— often the surfeits of our own behaviour,—we make guilty of our disasters the sun, the moon, and stars, as if we were villains on necessity, fools by heavenly compulsion, knaves, thieves, and treachers by spherical predominance.... An admirable evasion of whoremaster man, to lay his goatish disposition on the charge of a star! (William Shakespeare, King Lear, I, ii).

INTRODUCTION

Disasters have occurred since the beginning of time, as evidenced in both the earth's geological record and humankind's oral and written history. Vivid accounts of the eruptions of Vesuvius and Krakatoa, the destructive floods of the Huanghe and Ganges, and the San Francisco earthquake remind us of the dynamic, often violent nature of the earth and its atmosphere.

The word disaster is derived from the Latin: *dis*-dis + *astrum*-star, literally meaning unfavourable aspect of a star (Oxford, 1966). The Oxford English Dictionary defines a disaster as 'anything that befalls of ruinous or distressing nature; a sudden or great misfortune, mishap, or misadventure; a calamity' (Compact Edition of the Oxford English Dictionary, 1971). A contemporary dictionary echoes the principal components of this definition: A disaster is 'an occurrence inflicting widespread destruction and distress' (American Heritage, 1978).

Three aspects of these definitions deserve attention. First, a disaster is usually widespread and great; it affects large numbers of people and produces a significant amount of destruction. The term should not be applied to small, limited, or isolated events (Reich, 1983). Second, most disasters are sudden events that occur over a relatively short time; most authors exclude events occurring over the course of months or years. Finally, the distressing or psychological aspect of disasters is important and should not be neglected. In fact, much of the disaster research of the 1970s concerned the sociological and psychological aspects of disasters and their aftermath and demonstrated that the stress caused by a disaster can produce health effects that persist long after the event, even in the absence of physical injury or disability (Logue et al, 1981; Lystad & Runck, 1983; McFarlane et al, 1985).

Many papers in the disaster literature include a definition of disaster, and several authors have redefined the term to suit their particular needs (Lechat, 1976; Rutherford & de Boer, 1983). These definitions are often not useful to the epidemiologist because most of them ignore the public health impact of disasters and focus instead on criteria for determining the need for postdisaster relief. For example, Rutherford & de Boer (1983) are concerned with the destruction caused by a disaster 'relative to the resources available'. Although this and other definitions may be useful for the medical assistance community, they are not appropriate for those studying the public health impact of disasters.

The definition for disaster that we will use suits the needs of the public health community and is in harmony with the historical definition of the word: a disaster 'an event that causes widespread destruction or distress and that usually occurs suddenly or over a short period of time'. Since we will discuss the epidemiological investigation and public health impact of disasters, we will limit our discussion to those disasters that cause injury or death and will exclude disasters that are principally destructive to the environment.

Another term frequently used in the disaster literature is casualty. Historically, it has been used to refer to 'a chance occurrence, an accident, esp., an unfortunate occurrence, a mishap; now, generally, a fatal or serious accident or event, a disaster (Compact Edition of the Oxford English Dictionary, 1971). A contemporary dictionary defines casualty as '1. an unfortunate accident, especially one involving loss of life. 2. one who is injured or killed in an accident' (American Heritage, 1978). Thus, contrary to current usage, casualty has referred to an unfortunate accident rather than to those affected by the accident. It has gained its current association with those affected principally in its application to those injured or killed in war. Casualty is also tied to the notion of chance. Yet, many disasters, including war and virtually all industrial disasters, are by no means the result of chance (Velimirovic, 1980). As a result of the potential for confusion in usage and the causal implications of the word casualty, we shall avoid its use in our paper, although it enjoys widespread use elsewhere.

Several different classification schemes have been developed to aid in the study of disasters. Most authors divide disasters into two major categories: natural (act of God) versus man-made (technological). Table 10.1 shows such a classification. Velimirovic (1980) adds a third category, quasi-natural, which includes events to which a combination of natural and man-made factors contribute, for example, an earthquake causing the collapse of a poorly sited man-made dam. As he points out, the distinction between natural and man-made disasters is sometimes blurred, and it may be helpful to view them as a continuum—an attitude Berren et al (1980) share.

Disasters may also be classified according to their effect on a community, their duration, their geographical extent, and their public health impact (Rutherford & de Boer, 1983). Rutherford and de Boer distinguish between a simple disaster, in which the structure of the community remains intact, and a compound disaster, in which the structure of the community is destroyed. This distinction has significant implications for the epidemiologist investigating a disaster, and we shall return to this issue in our discussion of industrial disasters. For geographical extent, they subdivide disasters into those with a radius of less than 1 kilometre (3 square kilometres), of between 1 and 10 kilometres, and of more than 10 kilometres (300 square kilometres). The public health impact is measured by the number of injured or killed. In keeping with their definition of disaster as a situation in which destruction exceeds the available resources, Rutherford and de Boer give more importance to those victims requiring hospital admission than to those killed or injured, but not requiring hospitalisation. Thus, they classify a moderate disaster as one in which there are 100 to 1000 injured or killed, or 50 to 250 requiring hospitalisation.

Table 10.1 Classification of disasters by cause

Natural	Man-made
(Act of God)	(Nonnatural)
(Act of nature)	(Technological)
	(Purposeful event perpetuated by man)
Weather	Mechanical/physical forces
Hurricane	Transportation
Flood	Aircraft crash
Forest fire (lightning)	Railroad crash
Tornado	Highway crash
	Ship collision
Geological	Collapse of structure
Avalanche/landslide	Dambreak/flood
Earthquake	Collapse of buildings, stadiums
Volcanic eruption	Mine cave-in
	War/civil disturbance
	Fire/explosion
	Industrial
	Office/hotel/apartment
	Forest fire (cigarette, arson)
	Chemical/radioactive release
	Occupational exposure
	Air pollution
	Water pollution
	Food contamination
	Improper chemical waste management

In assessing the psychological impact of disasters on individuals and communities, Berren et al (1980) suggest that three categories be added to those of cause/type and duration of disaster. They are the degree of personal impact (low versus high); potential for occurrence or recurrence (low versus high); and control over future impact, that is, the ability to prevent or control the effects of a recurrence (low versus high). They suggest that the degree of psychological impact that a disaster has can be predicted on the basis of such a scheme. Logue et al (1981) agree that these factors are important in assessing disaster effects, but they use somewhat different terminology.

INDUSTRIAL DISASTERS: INTRODUCTION

Historically, the causes of disasters have been natural, with the exception of wars and civil disturbances. Within the last 100 years, however, our increased ability to store very large amounts of chemical or potential energy has ushered in a new kind of disaster, the technological or man-made disaster. Massive storage tanks for natural gas and other highly flammable or explosive materials remind us of this new threat, and huge dams and skyscrapers have increased the potential destructiveness of earthquakes and landslides. The earthquake in Mexico City in September 1985, accompanied by the collapse of many poorly designed and poorly built skyscrapers, tragically illustrates this potential. Our ability to extract or synthesise highly flammable, explosive, or radioactive substances poses a constant threat to the people who work with them in production plants and to people living or working near the plants. This threat already has been realised in incidents in which chemical substances have injured or killed large numbers of people and have caused extensive damage to the man-made and natural environments.

Although these technological disasters have been covered extensively in the press, very little research has been done on the epidemiology of their effects. Instead, investigations have focused on the technical causes of the disasters. Little work has been done on the relationship of the affected community to industrial sites and the factors that increase the risk of injury and death. Well-designed and carefully conducted epidemiological investigations of technological and industrial disasters are essential if we are to prepare for and prevent these disasters.

INDUSTRIAL DISASTERS: DEFINITION

Industrial disasters form a small subset of the man-made disasters. For this paper, we will restrict our definition of industrial disasters to events involving fires or explosions in industrial settings *or* chemical/radioactive releases from industrial point sources. These events must occur over a short period and must result in widespread destruction or distress. Since this chapter focuses on the epidemiological investigation of disasters, we will further restrict our discussion to disasters that have an immediate public health impact, although many industrial disasters have their primary impact on the environment. Finally, we will exclude transportation-related events and the contamination of food by chemical or radioactive substances.

INDUSTRIAL DISASTERS: CAUSES

Several factors are necessary in the causal chain that leads to an industrial disaster with an immediate public health impact. (The clinical expression of the public health impact may not be immediate, although biological damage has occurred; this is particularly true when there is an exposure to radiation or a carcinogen.) The most basic elements of this chain are the presence of reactive chemicals and the exposure of a population to those chemicals or their reaction products. Table 10.2 expands this causal chain and shows how varied the specific components of an industrial disaster

Table 10.2 Factors contributing to industrial disasters

1. Storage of flammable, explosive, or toxic chemicals, including radioactive materials
2. Uncontrolled release of unreacted chemicals, chemical reaction products, or the energy from a chemical reaction
   a. Due to natural causes: earthquake, flood, hurricane, lightning
   b. Due to man-made causes: operator error, equipment failure, arson, sabotage
3. The presence of people (potential victims) near enough to the chemical or energy release to result in exposure
   a. Direct exposure
      i. Workers at industrial facility
      ii. Workers at sites adjacent to industrial facility
      iii. Community adjacent to industrial facility
   b. Indirect exposure
      i. Carriage of chemical to distant site by wind, water, etc.
      ii. Community distant to industrial facility
4. Exposure sufficient to cause serious injury or death
   a. Toxic: acute, subacute, chronic effects
      i. Radiation
      ii. Chemical
   b. Fire: burns
   c. Explosion: blast injury, burns, injury secondary to debris or falling structures

can be. Without the presence of all necessary factors, a disaster will not occur. Intervention directed at any one of the factors is sufficient to prevent the disaster. Intervention strategies may vary in different settings and can be tailored to the economic, social, and geographical factors associated with specific sites.

Table 10.3 lists some of the major industrial disasters of the past 75 years. We have divided these disasters by cause: fire/explosion versus chemical/radiation release. Most of these disasters have involved fires or explosions. The methylisocyanate (MIC) release in Bhopal, India, however, stands as an example of the public health impact that a chemical release can have, given the right setting. Table 10.3 excludes small

Table 10.3 Major industrial disasters (HMIR, 1985; Metropolitan Life, 1982)

Fire/explosion

21 September 1921. Oppau, Germany. Explosion at a nitrate manufacturing plant destroyed plant and surrounding village of Oppau. 561 dead, >1500 injured

16 April 1947. Texas City, Texas, United States. Freighter being loaded with ammonium nitrate exploded and destroyed much of Texas City. 561 dead

28 July 1948. Ludwigshafen, Federal Democratic Republic of Germany. Vapour explosion from dimethyl ether. 209 dead

5 June 1980. Port Kelang, Malaysia. Fire resulted in explosion of cylinders. 3 dead, 200 injured, 3000 evacuated

25 February 1984. Cubatao, Sao Paolo, Brazil. Gasoline leak from a pipeline exploded and burned a shantytown built over and adjacent to it. >500 dead

19 November 1984. San Juan Ixtahuapec, Mexico City, Mexico. 5 000 000 litres of liquified butane exploded at a Petroleos Mexicanos storage facility. >400 dead, 7231 injured, 700 000 evacuated

Chemical/radioactive release

10 July 1976. Seveso, Italy. Chemical reactor exploded releasing 2, 3, 7, 8-TCDD. 100 000 animals dead, 760 people evacuated, 4450 acres contaminated

19 May 1981. San Juan, Puerto Rico. Valve failure at chemical plant resulted in release of 2 tons of chlorine. 200 injured, 2000 evacuated

10 April 1982. Belle, West Virginia. Chlorine pipeline burst, releasing 29 tons. 15 injured, 1700 evacuated

3 December 1984. Bhopal, India. Release of methyl isocyanate from Union Carbide plant. >2500 dead, 100 000 injured

explosions, fires, and chemical releases without a significant public health impact. These small, nondisaster events are far more common than the large-scale industrial disasters. For example, in 1985, prompted by the Bhopal disaster, the US Environmental Protection Agency (EPA) compiled a partial list of accidents in the United States involving toxic chemicals for 1980–1985 (EPA, 1985). Although the draft report was incomplete, it contained 6928 separate incidents. Even with this remarkable number of incidents, the injuries and deaths resulting from them do not match the impact of a single major disaster such as Bhopal. However, as the developed countries continue to improve their ability to prevent major disasters and as the technological infrastructure of a country broadens and expands, the small releases assume a more important position in planning and prevention efforts.

INDUSTRIAL DISASTERS: EPIDEMIOLOGICAL INVESTIGATION

What is the purpose of conducting an epidemiological investigation after an industrial disaster? In most industrial disasters, the source, time of release, location, and nature of the hazard are known or easily determined. Consequently, the purpose of the investigation should not be to elucidate the cause of the disaster but rather to describe the extent of the damage to the human population and determine the risk factors that were important predictors of injury or death. This includes analysing the event in terms of death and injury and their causal relationship with the disaster. Investigators may need to study the potential for chronic effects and psychological stress and disability related to the disaster. This analysis should include a description of pattern, time trends, geography, and probability of exposure (Velimirovic, 1980). Once the risk factors have been defined, steps can be taken to reduce the impact of the disaster and to prevent a recurrence at the same or other facilities. The epidemiologist can also assist in evaluating the adequacy of relief efforts and the potential hazards to relief workers and in determining any continuing impact on the public health. Having conducted investigations of industrial disasters and gained insights into the important risk factors, epidemiologists can review their methods and findings and improve their approach to the study of future disasters.

The epidemiological approach taken in an industrial disaster depends on a number of factors. Most important among these are the availability of information on the population in the disaster area and the nature and extent of the disaster. Developed countries are more likely to have good record-keeping systems that allow epidemiologists to accurately determine the population at risk and calculate injury and death rates. These systems also help them assess the extent of the displacement and the health status of the population following the disaster. Three Mile Island and Bhopal, described below, contrast in this respect and show the impact that the availability of information can have on an epidemiologist's ability to conduct studies.

The nature and extent of the disaster may determine the course of (or even the epidemiologist's ability to conduct) meaningful studies. In contrast to most epidemiological investigations, disaster epidemiology occurs in the context of partial or complete disruption of a community. Therefore, the epidemiologist encounters unique problems that are a consequence of both physical destruction and social disruption. Communication systems are often disabled and may be totally destroyed. Telephone, mail and house-to-house surveys to assess health effects may not be possible. Vital records may have been damaged or destroyed or, if intact, may be difficult to locate and use; in some instances, as in Bhopal, the number of deaths overwhelms the vital records system, and many are unrecorded in official files. The population is usually displaced by the disaster, and ascertaining the public health impact on a particular neighbourhood or district may be impossible. The sheer number of victims may overwhelm hospital or health department record keeping systems. Records may be sketchy and important details omitted. For public health reasons, the dead may be buried or cremated before bodies can be positively identified or their total number accurately determined. Thus, even the most basic description of the number of injured and killed may be impossible to obtain.

These difficulties are usually more prominent in natural disasters than in industrial disasters. Natural disasters tend to be compound and affect an entire community, destroying its basic structure. In contrast, industrial disasters tend to be simple and limited in extent. The basic fabric of the surrounding community remains intact. An explosion might severely damage a plant and the neighbourhoods immediately adjacent to it, but it is less likely to destroy hospitals and government buildings at some distance than a major earthquake or extensive flooding. One exception to this is the situation in which an explosion or fire in a plant sets off fires or explosions that sweep through and completely destroy the adjacent community. This happened in the 1921 explosion in Oppau, Germany, and in the 1947 explosions that destroyed much of Texas City, Texas, in the United States.

In addition to the direct physical effects of the disaster, the postdisaster environment and relief efforts pose their own problems. These include the lack of uniform reporting systems for deaths and injuries; the unavailability of information, for legal or political reasons (the issue of liability may be raised early and may preclude access to needed information); the lack of a central coordinating council; and infighting between different relief organisations. Not suprisingly, the psychological state of the population may affect participation in surveys and the objectivity of answers to questions about relatives or friends recently killed or injured.

Given the difficulties encountered in the immediate postdisaster period, when should the epidemiologist conduct the industrial disaster investigation? Certain descriptive work is best done in the immediate postdisaster period (for example, determining the geographical extent of the destruction, observing the location of debris hurled by the explosion, and correlating these locations with injury locations). In situations involving a chemical release, environmental sampling for specific chemicals must be done without delay so that the concentrations of the toxic agents present at the time of the disaster can be assessed.

More thorough investigations may require waiting until the structure of the community is reestablished and both data and communications are again intact. In some settings this may be absolutely necessary in determining the population at risk and conducting interviews with people who had been living in the affected area but who were displaced immediately afterward. Attention must also be given to the need for long-term follow-up studies of exposed or affected persons. Protocols for long-term studies may take some time to develop, but the information collected in the postdisaster period may be essential to assembling and studying a cohort in the future.

To summarise, the basic methods of the epidemiologist do not differ markedly in disaster versus nondisaster situations. The types of studies usually undertaken are similar (that is, descriptive, case-control, cohort follow-up), and surveillance is an essential function. The identification of risk factors for exposure and disease has important scientific and preventive value. The most striking factor affecting epidemiologists is the difficulty of applying well-known or standardised techniques in the context of great destruction, public fear, communal disruption, and the unavailability of the usual infrastructure for collecting and assembling data. Immediate coordination through a central focus is essential if epidemiologists are to function closely with other workers in the postdisaster period; this is especially true when rapidly evolving technological information must be collected, assessed, digested, and interpreted by persons from diverse specialties. Also critically obvious, but nevertheless sometimes neglected by epidemiologists, is the need to coordinate closely with public health and other governmental authorities involved with medical treatment, relief, relocation, and other societal functions; it is virtually impossible for epidemiologists to be successful in a post-disaster setting if they operate on their own. Planning for how to function in a postdisaster setting therefore takes on special importance. Well-thought-out procedures and checklists for factors to be considered facilitate smooth operations; at the same time, the epidemiologist must be capable of responding to rapidly evolving and often unanticipated events. Finally, societal interest and media scrutiny are often intense: updates on activities and reporting of (interim) results must often be made frequently [for example, daily, biweekly (every other week)], again in a coordinated fashion and in a manner appropriate to the postdisaster setting.

Next, we describe several industrial (and other) disasters that illustrate these issues, the special problems they present, and how epidemiologists have dealt with them.

EXAMPLES AND LESSONS FOR THE EPIDEMIOLOGIC INVESTIGATION OF INDUSTRIAL DISASTERS

Complexities of conducting extensive epidemiologic studies in a disaster setting

Bhopal

On 3 December 1984 a Union Carbide of India, Limited, plant in Bhopal, India, released into the air over 30 tons of methyl isocyanate (MIC). The release extended, as a low-lying plume, over nearby residential areas and then dissipated. The plant produced pesticides for neighbouring agricultural areas, and MIC was a chemical intermediate in the production of carbaryl, a commonly used pesticide. The MIC was produced at the plant and stored in large quantities (15 000-gallon storage tanks) until needed for the production of carbaryl. The plant was across the street from a neighbourhood of slum dwellings and within several kilometres of a large part of old Bhopal. Starting in the middle of the night and lasting for up to 1 or 2 hours, MIC slowly spread over an area populated by several hundred thousand people, aided by a light wind blowing in the direction of the city and by an atmospheric inversion that kept the cloud close to the ground.

The circumstances of the accident made it impossible to evacuate the affected areas rapidly. These circumstances were that the release occurred in the middle of the night; the affected areas were poorly lighted; the streets were narrow and cluttered; the MIC affected the residents' vision; and communication and transportation were limited. Furthermore, the plant was near residential areas, the residents had no warning, and the local population had no knowledge of effective evacuation plans or procedures.

Estimates of the deaths ranged from less than 2000 to more than 7000, with 2000 to 2500 being the most frequent estimates. Inadequate census data, large dislocations of residents, and unrecorded deaths, multiple cremations, and unmarked graves in the rush and confusion of the immediate postdisaster period contributed to the difficulty of an accurate enumeration of deaths.

Similarly, with an estimated 15 000 people seen at the Hamidia Hospital alone on the night of 3 December, and with many thousands seen and treated elsewhere, the collection and maintenance of basic hospital charts was almost impossible for the first 24–48 hours. An estimated 50 000 to 100 000 people were affected enough to seek medical care of some type; total numbers are unavailable, and it would be extremely complex, costly, and time-consuming to assemble them after the fact. This complicates all efforts to establish representative samples of patients with particular diagnoses for the study of risk factors, the natural course of illness, or long-term effects. This is in spite of the fact that the medical response appeared to be heroic, with triage centres and treatment protocols established early and with relief supplies and medical care staff brought in as needed. In such dire, life-threatening circumstances, data of use to epidemiologists could not be the main concern.

Exposure data for this short-lived, acute toxin were also limited, and the best representations of the plume were based on estimates of the numbers of dead and the numbers and severity of ill patients in different parts of old Bhopal. Environmental data (for example, chemical measurements and meteorological information) were sparse, and anecdotal information suggested that many factors (for example, housing structure, height from ground, precautions taken or not taken by people) significantly affected individual exposure. The most precise estimates of exposure, retrospectively, would probably come from a painstaking and difficult-to-achieve (if not impossible because of the large-scale relocation of residents) recreation of attack rates by blocks and neighbourhoods. Similarly, the creation of registries of exposed or ill patients for a variety of purposes (for example, long-term health studies, notification of important health information, follow-up medical testing, or allocation of benefits) will be complicated by the factors outlined above. Limited data on the distribution of confounding factors for respiratory disease in the population (for example, smoking, air pollutants, occupational factors, tuberculosis) present another set of problems for the epidemiologist. From a purely practical standpoint, the short supply of epidemiologists, biostatisticians, trained support staff, and data-handling and communications equipment in Bhopal is a potentially remediable problem in the long-term, but it is daunting nevertheless. At the same time, adequate personnel, training, equipment, and supplies for sophisticated pulmonary follow-up of affected individuals and any appropriate control groups for clinical-epidemiological studies are equally difficult to assemble.

An unexpected, complicating circumstance was a second, temporary reevacuation of old Bhopal about 2 weeks after the catastrophic MIC release, when the start-up of the carbaryl production process to rid the plant of remaining stocks of MIC led to widespread fear, on the part of the population, of another MIC release. Finally, this disaster occurred in a setting in which it is inherently difficult to follow people long-term [no universal health or social security numbers, few systems for tracking people who leave Bhopal, no forwarding addresses from post offices, many people with the same last name (for example Singh) and so on].

The Indian Council for Medical Research (ICMR) has coordinated development of a wide-ranging and detailed series of studies to evaluate the medical effects of the disaster. Epidemiologists will be greatly interested in how the follow-up studies are conducted.

Mount St Helens

Although the eruption(s) of Mount St Helens is classified as a natural disaster in our scheme (Table 10.1), it has many similarities to an extensive environmental disaster with varied chemical exposures; it occurred in a developed country and was extensively studied, and so it may be instructive to epidemiologists (Merchant, 1986). Features of interest include the following.

Coordinated federal and state response headquartered in Vancouver, Washington, with a central Technical Information Network (TIN) releasing data and the results of the study.

Health reports issued biweekly by the Centers for Disease Control to amplify and supplement the TIN.

Extensive environmental sampling and industrial hygiene support to characterise constituents of the ash and to analyse samples for toxic elements, toxic gases, asbestos fibres, radon, crystalline silica, and other potential hazards (Olsen & Fruchter, 1986; Dollberg et al, 1986).

Epidemiologic study of deaths and survivors of the blast to identify risk factors.

Rapid initiation of an extensive hospital-based surveillance system to identify acute volcanic ash-related effects, primarily respiratory irritation and ash-related airways disease, such as bronchitis and asthma (Baxter et al, 1981).

Case-control study in heavily impacted area (Yakima, Washington) to identify risk factors for developing ash-related respiratory illness.

Identification of highly exposed groups (primarily occupationally exposed loggers) for a long-term cohort follow-up study to evaluate chronic respiratory problems (Buist et al, 1986).

Studies of the psychosocial impact of the disaster (Shore, 1986).

Extensive in vitro and animal experimentation to identify the range of hazards related to constituents of volcanic ash.

Comprehensive review of plans for dealing with future volcanic activity, with special emphasis on man-made technological factors that might exacerbate potential hazards (Baxter et al, 1982).

Spectrum of long-term epidemiological study, postdisaster: from Three Mile Island to Hiroshima/Nagasaki

Three Mile Island

Although this accident posed the threat of an extensive release of radiation and extensive destruction (Falk et al, 1982), the actual estimated radiation doses to individuals living nearby turned out to be relatively low (Fabrikant, 1981), certainly not high enough to justify long-term follow-up studies. Nevertheless, because it was believed that continuing concerns about chronic effects on the part of the public would inevitably force public health authorities to conduct studies of cancer and other effects in nearby residents, a complete census, with ample identifying and locating information, was conducted of all residents living within a 5-mile radius of the Three Mile Island nuclear reactor. Thus, detailed and scientifically sound epidemiological evaluations can be conducted at any time, if indicated.

Atom bomb survivors at Hiroshima and Nagasaki

The most extensive epidemiological follow-up studies of a man-made environmental disaster have been those of the survivors of the atom bomb blasts in Hiroshima and Nagasaki; these studies can serve as a prototype for detailed and sophisticated follow-up of large numbers of people for many decades. Originally set up by the Atom Bomb Casualty Commission (ABCC) and later transferred to the Radiation Effects Research Foundation (RERF), these studies have assembled the basic knowledge of radiation doses from the bomb blasts and the detailed data on carcinogenic and other effects. Studies have been based on mortality data from a cohort of over 100 000 persons, repeated medical examination data from a subset of about 23 000 persons, and extensive autopsy data (Schull, 1984).

Risk factor analysis, prevention efforts, and the role of the epidemiologist

Bhopal

Many factors contributed to this industrial disaster, and the major ones are evident even without detailed epidemiologic study. As described in many reports (a good review is by Bowonder et al, 1985), seven major contributing risk factors are:

1. siting of a very hazardous plant near Bhopal;
2. long-term storage of large quantities of MIC on-site;
3. failure of multiple safety systems;
   refrigeration system
   use of a spare tank
   flare tower
   vent gas scrubber
   water curtain
4. poor emergency preparedness;
5. limited or poorly enforced local health standards for hazardous chemicals;
6. difficulty of ensuring safe use of dangerous technology transferred to a new setting or developing country;
7. social, cultural, and institutional impediments to proper emergency preparedness.

Gore, Oklahoma

The recent episode in which the heating of an apparently overfilled cylinder of uranium hexafluoride led to a release that caused one death and many injuries reinforces widespread concern about safe work practices and proper training in safe procedures.

Pemex petrochemical explosion, San Juan Ixtahuapec, Mexico, 1984

Beginning at 5.35 am on 19 November 1984, a group of butane holding tanks exploded at a Pemex (the Mexican national oil company) facility in San Juan Ixtahuapec, Mexico, a poor neighbourhood about 2 kilometres north of Mexico City (Pan American Health Organisation, 1985). About 5 million litres of gas burned during and following 11 consecutive explosions that occurred over the next 75 minutes. Flames rose 1000 metres high. A 20-block area, with a radius of about 4 kilometres, was impacted. Data from the Mexico Ministry of Health, released in April 1985, placed the number injured at 7231 and the number killed at 452. An estimated 700 000 people had to be evacuated. Unofficial estimates of the number killed ranged as high as 1000–2000.

This disaster illustrates what can happen when uncontrolled urban growth occurs in an underdeveloped county, without the benefit of zoning and adequate laws to protect the community and environment (Navarro, 1984). The plant, when originally built in the 1960s, was in fields away from population centres. Over the years, however, a low income neighbourhood grew right up to the storage facility. Thus, the stage was set for the disaster. Had there been appropriate restrictions on building adjacent to the plant and had these restrictions been enforced, the public health impact of this disaster would have been substantially reduced.

Role of the epidemiologist

Well-designed and carefully conducted studies, carried out by epidemiologists, should more comprehensively define and detail many of the risk factors related to industrial disasters. Surveillance of disaster-related mortality and morbidity and review of data on the distribution of hazardous chemicals, dangerous facilities, population centres, and risk factors for serious injury should pinpoint areas for preventive action; case-control studies can explore additional, less-obvious risk factors; and long-term follow-up studies, in addition to focusing on the natural history of exposure and health effects, can track efforts to reduce risk factors and prevent adverse health effects. Several recent studies of natural disasters (Glass et al, 1980; Kilbourne et al, 1982) have highlighted how much can be learned from prompt and thorough epidemiological study of disasters whose antecedents were presumed to be well known.

PREVENTION

> And I another,
> So weary with disasters, tugged with fortune,
> That I would set my life on any chance
> To mend it or be rid on't. (William Shakespeare,
> Macbeth, III, i)

The key to reducing the public health impact of industrial disasters is to prevent these disasters. In contrast to natural disasters, industrial disasters are preventable. The location of industrial plants, their proximity to populated areas, and their methods of storing chemicals are all completely under our control. Safe plant design is one of the most effective methods of interrupting the causal chain leading to industrial disasters. This includes the use of less hazardous materials, the storage of explosive/hazardous chemicals in small amounts, and the installation and proper maintenance of adequate backup and safety systems.

The interface between the industrial plant and people should be another focus for intervention efforts. In most industrial disasters, most victims are workers at the plant. Workers should be protected from toxic hazards through the substitution of less hazardous materials; isolation from hazards, when substitution is not feasible; and engineering controls, such as ventilation, when neither substitution nor isolation is possible. Since an appreciation of potential hazards and their control is a necessity in dealing with them appropriately, adequate initial and continuing education of the plant work force is essential. This education should cover the nature and location of chemical hazards in the plant, proper plant maintenance, and safe work practices. Lapses in these areas have been important factors in the cause of many major industrial disasters.

Planners should consider the relationship of the plant to the surrounding community. Plants should be sited so as to minimise the potential for damage to the surrounding community should an explosion, fire, or chemical release occur. Ideally, plants storing or using large quantities of hazardous or flammable chemicals should be sited away from populated areas. If this is not possible, safe distances should be maintained between storage tanks, pipelines, or chemical reactors and the surrounding community. One method of achieving this is through zoning to establish safety belts around plants that prohibit housing and businesses from being built too near. In developed countries, this is done to some extent; however, zoning for safety belts continues to be a major problem in underdeveloped countries. Communities should have access to information about the nature, quantity, and location of chemicals in plants so that they can participate in decision-making related to zoning regulations.

For plants that already exist, the epidemiologist can assist in establishing surveillance systems for locations with a high potential for disaster. On the basis of studies of past industrial disasters, he or she can recommend preventive action in these settings and supply the information needed for developing remedial and regulatory measures to prevent industrial disasters.

Use of trade names is for identification only and does not constitute endorsement by the Public Health Service of the US Department of Health and Human Services.

REFERENCES

American Heritage Dictionary of the English Language 1978 Houghton Mifflin Company, Boston
Baxter P J, Ing R, Falk H et al 1980 Mount St Helens Eruptions, May 18 to June 12, 1980. Journal of the American Medical Association 246: 2585–2589
Baxter P J, Bernstein R S, Falk H, French J, Ing R 1982 Medical aspects of volcanic disasters: An outline of the hazards and emergency response measures. Disasters 6: 268–276
Berren M R, Beigel A, Ghertner S 1980 Typology for the classification of disasters. Community Health Journal 16: 103–111
Bowonder B, Kasperson J X, Kasperson R E 1985 Avoiding future Bhopals. Environment 27: 6–37
Buist A S, Martin T R, Shore J H, Butler J, Lybarger J 1986 The development of a multidisciplinary plan for evaluation of the long-term health effects of the Mount St Helens' eruptions. American Journal of Public Health 76 (suppl): 39–44
Compact Edition of the Oxford English Dictionary 1971 (sup. 1) Oxford University Press, Oxford
Dollberg D D, Bolyard M L, Smith D L 1986 Evaluation of physical health effects due to volcanic hazards: Crystalline silica in Mount St Helens volcanic ash. American Journal of Public Health 76 (suppl): 53–58
Environmental Protection Agency (EPA) 1985 Acute hazardous events database. EPA, Washington, DC
Fabrikant J I 1981 Department of Radiology, School of Medicine, Univ of CA, San Francisco. Health Physics 40: 151–161
Falk H, Caldwell G G, Stein G F 1982 Presentation of incident — Three Mile Island. Chemical and Radiation Hazards to Children. Finberg L (ed). Report of the Eighty-Fourth Ross Conference on Pediatric Research. Ross Laboratories, Columbus, Ohio p 74–78
Glass R I, Craven R B, Bregman D J, Stoll B J, Horowitz N, Kerndt P, Winkle J 1980 Injuries from the Wichita Falls Tornado: Implications for prevention. Science 207: 734–738
Hazardous Materials Intelligence Report 1985 Special Report: International safety assessed after Bhopal. Hazardous Materials Intelligence Report 11: 3–6
Kilbourne E M, Choi K, Jones S, Thacker S B 1982 The Field Investigation Team. Risk factors for heatstroke. Journal of the American Medical Association 247: 3332–3336
Lechat M F 1976 The epidemiology of disasters. Proceedings of the Royal Society of Medicine 69: 421–426
Logue J N, Melick M E, Hansen H 1981 Research issues and directions in the epidemiology of health effects of disasters. Epidemiologic Reviews 3: 140–161
Lystad M, Runck B 1983 New NIMH Center consolidates disaster research and services. Hospital and Community Psychiatry 34(9): 789–790
McFarlane A C, Wallace M, Cook P 1985 Australian research into the psychological aspects of disasters. Disasters 9: 32–34
Merchant J A 1986 Preparing for disaster. American Journal of Public Health 76 (suppl): 233–235
Metropolitan Life Foundation 1982 Catastrophic accidents — A 40-year review. Statistical Bulletin 63: 1–5
Navarro V 1984 Exportation of hazardous substances. New England Journal of Medicine 311: 546–548
Olsen K B, Fruchter J S 1986 Identification of the physical and chemical characteristics of volcanic hazards. American Journal of Public Health 76 (suppl): 45–52
Oxford Dictionary of English Etymology 1966 Oxford University Press, Oxford

Pan American Health Organization 1985 Recent events point to need for health preparedness for technological disasters. Disaster preparedness in the Americas. 1 January: 22–25
Reich M R, Spong J K 1983 Kepone. A chemical disaster in Hopewell, Virginia. International Journal of Health Services 13: 227–246
Rutherford W H, de Boer J 1983 The definition and classification of disasters. Injury 15: 10–12
Schull W J 1984 Atomic bomb survivors: Patterns of cancer risk. In: Boice J D, Fraumeni J F (eds) Radiation carcinogensis: epidemiology and biological significance. Raven Press, New York 21–36
Shore J H, Tatum E, Vollmer W M 1986 Evaluation of mental health effects of disaster. American Journal of Public Health 76 (suppl): 76–83
Velimirovic B 1980 Non-natural disasters — an epidemiological review. Disasters 4: 237–246

SECTION 3

Environmental control

11. Working in hot climates

M. Khogali M. A. Awad El Karim

HEAT PHYSIOLOGY AND THERMAL BALANCE

In many occupational situations, whether indoor or outdoor, heat exposure in addition to the added burden of metabolic heat production may well exceed the worker's physiological capacity to regulate his body temperature. This leads to adverse physiological effects and eventually heat illnesses.

Man and other homeothermic mammals maintain internal body temperature within a narrow range (36–38°C) by regulating the flow of heat produced as a result of metabolic activity in muscles and deep tissues to the body surface where it is dissipated through heat transfer mechanisms (Machle & Hatch, 1947; Hardy, 1961).

Man's core temperature exhibits a diurnal rhythm where the rectal temperature usually reaches a peak in the late afternoon or early evening and falls to its lowest level during the early morning hours. The human body is considered thermally at equilibrium and body temperature is maintained at regulated level when heat loss is equal to heat production.

Heat exchange

Heat tends to pass from places of high temperature, to places of low temperature through three channels, conduction, convection and radiation. Heat transferred by these three channels is called sensible heat. Heat transferred by evaporation is called insensible heat (Fig. 11.1).

The thermal environment which affects the human body can be exogenous and endogenous. Factors which contribute to the exogenous thermal load are:

(i) air temperature and wind speed;
(ii) relative humidity;
(iii) mean radiant temperature;
(iv) duration of exposure;
(v) clothing.

The endogenous thermal load comprises basal metabolism plus physical activity, work, of a subject in a given environment. It is important to assess metabolic heat production, metabolic rate, because it may vary from 75 kcal h^{-1} (65 W) at rest to over 600 kcal h^{-1} (517 W) during hard physical work (Passmore & Durnin, 1955). The heat generated if not dissipated will lead to a rise in body temperature. At rest when there is no heat exchange the basal metabolic rate can raise the body temperature by 1°C h^{-1}.

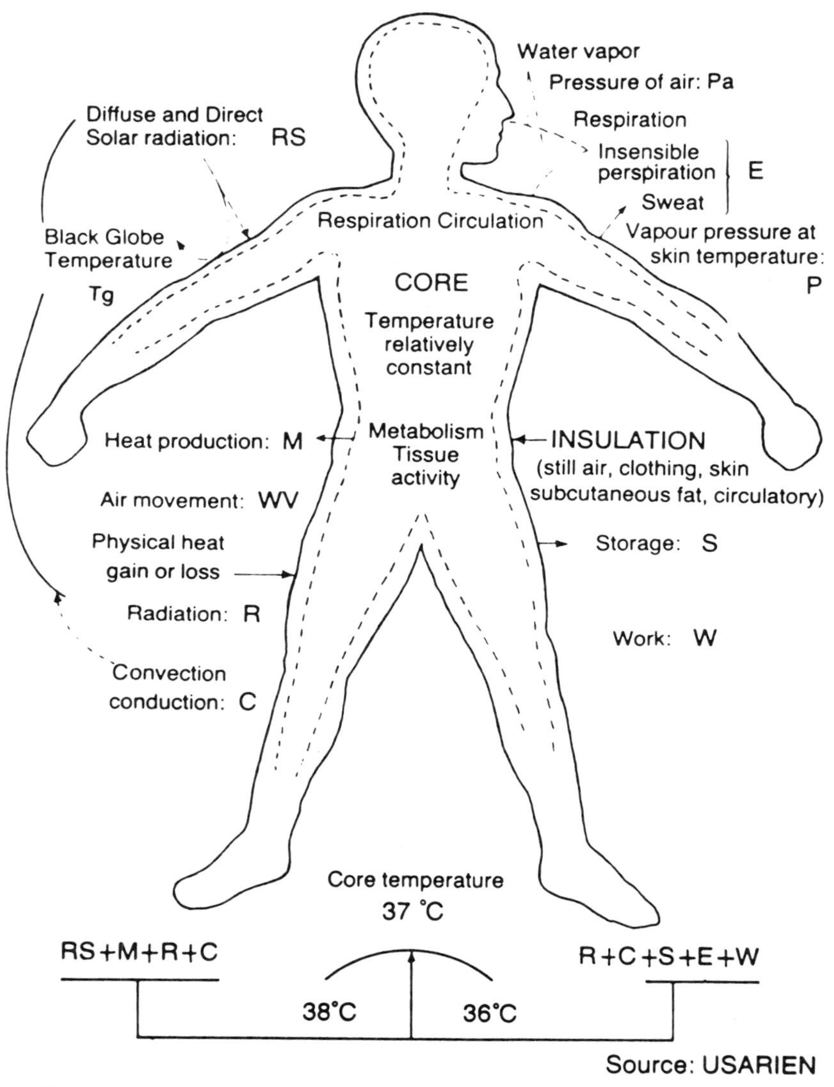

Fig. 11.1 Factors in heat exchange and balance.

Exogenous factors

AIR TEMPERATURE

If the skin temperature is lower than the surrounding air temperature the body gains heat by convection, and if it is higher the body loses heat. Natural convection is established around the body when the air movement is induced by the body heat. Normally due to different levels of air speed forced convection exceeds the influence of natural convection and it is directly proportional to the temperature difference between the skin and air and to the square root of air speed:

$$C = 8.3\,V^{0.5}\,(ts\text{-}ta)\,Wm^{-2}$$ Equation (1) (Kerslake, 1972)

where: V = air speed ms^{-1}; ts and ta = skin and air temp (°C)

Wm^{-2} = heat exchange in Watts/square metre. The convective coefficient 8.3 is for a nude man standing facing moving air.

MEAN RADIANT TEMPERATURE

Ratiation can represent heat loss or gain depending on temperature of solid surroundings. Since heat exchange by radiation is influenced by the temperature of the emitting body it is more appropriate to refer to radiant temperature as the mean radiant temperature (MRT). It is independent of air movements and, for practical purposes, is not influenced by the moisture content of the air.

$$R = EKAr\,(T^4w - T^4s)$$ Equation (2)

where: E = the emmisivity of the man; K = Stefan-Boltzman constant; A = the surface area of the man ($1.8\,m^2$ for standard man); r = the ratio of effective radiation surface to the Du Bois area (0.78 for static standing man, 0.85 for a moving man) and Tw and Ts = wall and skin temperatures in degrees absolute.

EVAPORATION

The rate of evaporation from the surface of the skin depends on the water vapour pressure gradient between sweat and surrounding air. In many industrial situations workers produce about 1 litre of sweat/h over an 8-hour work day. Evaporative cooling is impaired in environments where water vapour pressure is excessively high compared with that of the skin. Evaporative cooling is influenced by the type of clothing worn and it must be wetted to enable evaporative cooling to occur. Evaporation of one gram of sweat removes about 580 calories.

$$E = 13.7 - V^{0.5}\,(Ps - Pa)\,Wm^{-2}$$ Equation (3) (Kerslake, 1972)

where: V = air velocity in ms^{-1}; Ps and Pa = water vapour pressures in millibars of skin and air; Wm^{-2} = heat flow in watts/m^2.

CLOTHING

Clothing influences heat exchange and in hot climates distinction should be made between exposure to hot dry heat, e.g. desert environments, and hot humid environment. In humid conditions clothes impair heat loss by evaporation. Convective heat loss and radiation will also be affected, depending upon the particular characteristics of the material used. In such conditions it is desirable to have a minimum of loose clothes made of very light materials. In a hot dry environment wearing voluminous clothing allows a considerable layer of air between the clothing and the body. Heat will be lost with air stream forced from the clothing to the surrounding ambient air. For those who work under direct solar sun head gear will provide protection against solar radiation.

In industry there are a number of working situations where clothing is required to render other functions such as protecting the worker from injury and toxic hazards. While radiation and convection heat transfer is minimal across clothing fabrics, sensible heat escapes across the fabrics by conduction through the air spaces and along the fibres. Insensible heat transfer takes place by the evaporation of water vapour through the fabric. In hot environments evaporation of sweat from the surface of the fabrics aided by air movement will significantly enhance heat loss from the body. Therefore the permeability of worn clothing is of paramount importance to facilitate heat loss. The bellows effect produced by the body movement and wind pressure is also of significant importance, because direct sweat evaporation takes place through the stream of air expelled from the body.

The major problems arise in hot climates when individuals should be protected from toxic gases and vapours which can damage the skin or the eye. In such circumstances carefully designed protective garments should be considered. In industry protective clothing against heat is of two types: conditioned, the wearer is independent of the environment, and unconditioned, garments intended to extend the tolerance time. Conditioned garments are designed in such a way as to make the wearer completely independent of the surrounding environment (Crockford & Awad El Karim, 1974). They are also used for protection against toxic dusts and gases. The unconditioned garments are designed to suit various industrial situations with an objective of extending the tolerance time to excess heat exposure, e.g. radiation can be reflected back by shiny aluminium surfaces or high resistance to heat flow and can be opposed using thick layers of insulating materials.

TEMPERATURE REGULATION

Thermal balance

The total thermoregulatory response in humans is the result of both behavioural and autonomic responses. Thus, patterns of thermoregulatory responses are as follows (Cabanac, 1975).

(a) heat production through metabolic processes including shivering;
(b) vasomotor responses resulting in a modification of heat transfer between the core and skin and, in turn, of heat exchange with the environment, e.g. vasodilation in heat;
(c) evaporative heat loss;
(d) behaviour.

Thermal balance is achieved by the total thermoregulatory response in such a way as to minimise the autonomic response through behavioural activity. The heat balance equation is based on the simple thermodynamic principle of conservation of energy, at any point in time.

Rate of input − Rate of output = Rate of accumulation.

Considering the human body the equation can be written as follows:

$$\frac{dM}{dt} \pm \frac{dK}{dt} \pm \frac{dC}{dt} \pm \frac{dR}{dt} - \frac{dE}{dt} = \frac{dS}{dt} \quad \text{Equation (4)}$$

Where

$$\frac{dM}{dt} = \text{Time rate of heat production in the human body in } Wm^{-2}$$

$$\frac{dK}{dt} = \text{Time rate of heat gain or loss by conduction in } Wm^{-2}$$

$$\frac{dC}{dt} = \text{Time rate of heat gain or loss by convection in } Wm^{-2}$$

$$\frac{dR}{dt} = \text{Time rate of heat gain or loss by radiation in } Wm^{-2}$$

$$\frac{dE}{dt} = \text{Time rate of heat loss by evaporative cooling in } Wm^{-2}$$

$$\frac{dS}{dt} = \text{Time rate of heat accumulation or storage in } Wm^{-2}$$

Since conduction is a special case and under normal conditions it is negligible, dK/dt is assumed to be zero. The rate equation can simply be written as:

$$M \pm C \pm R - E = S \ldots\ldots \text{(Belding \& Hatch 1955)} \quad \text{Equation (5)}$$

At steady state, when thermal balance is achieved at different strain and discomfort levels the rate of heat storage S is equal to zero:

$$M \pm C \pm R - E = O \quad \text{Equation (6)}$$

Thermal neutrality may be defined as the total condition of man at rest when his body is in thermal equilibrium without active regulation, such as sweating or shivering. For man such a condition is achieved when the environmental temperature of a resting subject is 28–30°C. Under these conditions the heat produced by the body and removed by the environment equals 50 Wm^{-2}, the core temperature passively stabilises at near 37°C and skin temperature at about 35.5°C. In case thermal balance is disturbed, the temperature of the body tends to increase, hyperthermia, or decrease, hypothermia. A necessary condition for temperature regulation is the presence of a fixed temperature level maintained by the body's control mechanisms. Deviation from this fixed temperature is called 'load error' and as far as the body temperature is always likely to differ from the set temperature, the regulating system will operate to minimise the deviation. The heat balance equation given above describes the state of the human body in relation to the internal metabolic heat and external environment, radiation, convection and evaporation.

Mechanisms of temperature regulation

It is helpful to consider the human body in two compartments. A core or deep body tissue whose temperature is kept within a relatively narrow range controlled by thermoregulatory mechanism and a shell or superficial tissue whose temperature can be subject to greater variation. The core-shell concept is a useful mean for assisting in the evaluation of the vasomotor changes associated with temperature regulation (Burton, 1934; Hardy & Soderstrom, 1938).

The centre controlling the physiological process in temperature regulation is situated in the hypothalamus (Fig. 11.2). This thermoregulatory centre comprises two anatomically distinct subcentres: one for heat conservation by cutaneous vasoconstriction and shivering and the other for heat dissipation by cutaneous vasodilation and sweating.

Evaporation of sweat is the main line of defence against heat stress, and the second line is to increase the flow of blood to the skin. An efficient circulation is very important for efficient heat dissipation. If heat stress prevails it results in adverse changes within man leading to the different types of heat illnesses.

ACCLIMATISATION

Improved tolerance to heat develops as a result of repeated exposure to hot environments with accompanying changes in a number of physiological systems. Individuals working for the first time under heat stress, irrespective of their state of physical fitness, will develop signs of heat strain with abnormally high body temperature, increased heart rate and other signs of heat intolerance. The improved tolerance to heat gained by working in a hot environment is called heat acclimatisation. Bligh & Johnson (1978) defined acclimatisation as 'a physiological change occuring within the life time of an organism which reduces the strain caused by stressful changes in the natural climate'. They used the term adaptation to refer to genetic changes, the term acclimatisation for natural acclimatisation and acclimation for artificial acclimatisation. The major physiological responses of acclimatisation are characterised by an increased sweating rate, and lowering of heart rate and deep body temperature (Goldsmith, 1977). Acclimatisation is a gradual process, beginning with the first day and continuing at a decreasing rate for from 8 to 10 days. There is progressive improvement in the appearance and behaviour of men, changes in heart rate are dramatic, the principal reduction occuring in the first 4 or 5 days, reduction in rectal temperature 'more gradual than that of changes in circulation' reached from 7 to 10 days, and uniformal increase of sweating. Increase in rates of sweating usually begin later, from 3 to 5 days after other changes and may continue longer.

To maintain effective work performance in a hot environment replenishing of body water and salt lost is important. For a fully acclimatised worker, weighing 70 kg, sweat loss of up to 1 kg of body water can be tolerated without serious effect, however, losses equal to 1.5 kg or more during work in the heat leads to a reduction in the volume of circulating blood, increased heart rate, elevated core temperature and severe heat discomfort. Body weight loss of such magnitude should not be allowed and workers should be encouraged to replenish water every 20–30 minutes. Salt losses in sweat during work can be replaced with food intake but preferably salted water (0.1%) should be made available near the workers in hot working conditions.

Mechanisms involved in establishing acclimatisation can be very easily disturbed. Men may show some loss of acclimatisation after being away from work for a week or even after 2–3 days. Substantial loss of acclimatisation may occur because of the discontinuity of heat exposure for 2 weeks, or longer. Not only that, a fully acclimatised man may exhibit any or all of the disturbances characteristic of the unacclimatised

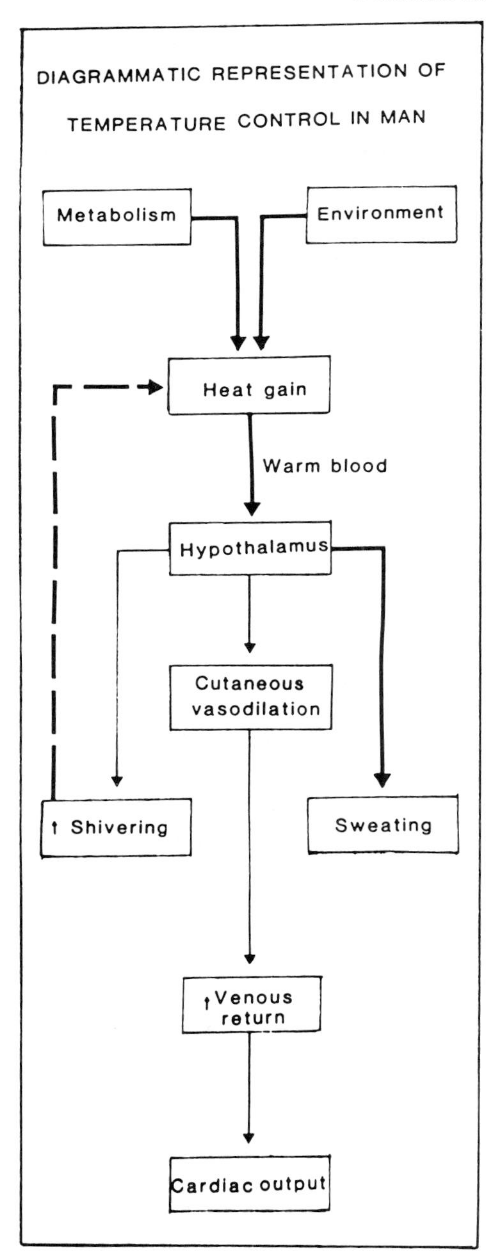

Fig. 11.2 Diagrammatic representation of temperature control in man.

state if excessive rates of work are imposed, or if there is intercurrent infection, loss of sleep, or alcohol consumption. With reduction in thermal stress, acclimatisation is lost more slowly than it was acquired, therefore, its reinforcement could be achieved by repeated exposure as frequently as once a month.

HEAT ILLNESSES

Classification of heat illnesses

The World Health Organization (1977) lists the classifications for heat illnesses shown in Table 11.1. Heat illnesses can be grouped into minor and major disorders. The

Table 11.1 Classification of heat illnesses (WHO 1977)

992.0	Heat stroke and sunstroke	
	Heat apoplexy	Siriasis
	Heat pyrexia	Thermoplegia
	Ictus solaris	
992.1	Heat syncope	
	Heat collapse	
992.2	Heat cramps	
992.3	Heat exhaustion, anhydrotic	
	Heat prostration due to water depletion	
	Excludes: when associated with salt depletion (992.4)	
992.4	Heat exhaustion due to salt depletion	
	Heat prostration due to salt (and water) depletion	
992.5	Heat exhaustion, unspecified	
	Heat prostration NOS	
992.6	Heat fatigue, transient	
992.7	Heat oedema	
992.8	Other heat effects	
992.9	Unspecified	

minor group comprises heat oedema, heat fatigue and heat exhaustion unspecified. The three major clinical syndromes related to heat illness and resulting from exposure to hot environments, mainly heat cramps, heat exhaustion and heat stroke have been adequately covered by many investigators (Leithead & Lind, 1964; Henschel et al, 1975; Dukes-Dobos, 1981; Khogali, 1983a, 1984). Heat cramps are painful muscle spasms which occur in the extremities, back and abdomen. These would occur during or after work hours among individuals who drink large quantities of water without compensating for the salt loss during sweating.

Heat exhaustion is the most common heat illness and takes place when the blood volume is reduced by increased water or salt loss during sweating. Increased physical activity in hot environments when the ambient temperature exceeds 35°C results in an increased volume of sweat. Under such working conditions workers performing hard physical work may loose up to 5 litres of sweat in 4 hours with varying amounts of sodium chloride ranging between 0.5 and 0.75 g kg^{-1} body weight (Marriot, 1950) depending on the individual's dietary habits. Water depletion is common amongst workers engaged in road building and construction work in hot countries. It is also common amongst army recruits and soldiers during training in hot environments (Ellis, 1976). The worker with heat exhaustion may experience extreme weakness, dizziness, and fainting, loss of appetite, nausea, headaches, abdominal distress, shortness of breath and increased pulse rate (120–200 beats min^{-1} at rest) (Lind, 1983, Khogali, 1983b). Heat exhaustion may lead to trembling of the extremities and incoordination together with personality changes in which the patient becomes restless, remote and morose, or euphoric and belligerent. If unchecked or overlooked, heat exhaustion can progress to heat stroke.

Heat stroke is a complex clinical condition in which an elevated body temperature, resulting from an over-loading or failure of the normal thermoregulatory mechanism after exposure to hot environments, causes tissue damage. It is characterised by generalised anhidrosis, disturbance of the central nervous system and a rectal temperature above 40°C (104°F). The main operating factors leading to heat stroke are high body temperature, metabolic acidosis and hypoxia (Wyndham, 1973; Khogali & Mustafa, 1983). Temperatures can rise to critical levels and death may occur unless rapid, effective cooling is given. The mortality rate increases significantly when cooling is delayed. With proper management and physiological cooling, the temperature falls, but elevations of temperature occasionally recur, indicating damage to the temperature regulation centre.

Robertshaw (1983) identified dehydration, lack of heat acclimatisation and poor physical fitness as the three most important factors predisposing to heat stroke. Other factors are chronic illness such as diabetes mellitus, fatigue and obesity. The aged are also more susceptible (Veale & Cooper, 1983). In the presence of cardiovascular disease, heat stress which ordinarily would be tolerated may rapidly precipitate heat stroke.

THERMAL ENVIRONMENT

In order to make an assessment of the thermal environment, understanding of fundamental physical laws 'as expressed in the heat balance equation' governing thermal exchanges and proper use of instrumentation for measuring air temperature, mean radiant temperature, water vapour content in the air and air speed is essential.

Instruments

Air temperature

Air temperature is measured by using different measuring devices such as mercury or alcohol in glass thermometers, thermoelectronic thermometers, thermocouples and thermistors. Usually, in industry mercury in a glass thermometer with graduations marked on the stem and with an accuracy of ± 0.1°C is used to determine air temperature. The bulb of the thermometer is coated with silvery material to protect it against radiation. Air temperatures in working environments vary considerably and for this reason the range of the thermometer should be carefully selected to cover the anticipated environment. To avoid the effect of artefacts such as the effect of wall temperature, the thermometer should be hung on a stand and the reading made approximately 15 minutes from when the thermometer has been positioned. Glass thermometers are easy to use and relatively cheap. When using glass thermometers, caution is warranted because they are fragile and easy to break. The mercury spilt may cause an unnecessary risk in ill-ventilated working places. The temperature of the air is referred to as the dry bulb temperature (t°dB).

Wet bulb temperature

Relative humidity (RH) is defined as the amount of water content in the air as compared with the amount that air could contain at saturation at the same temperature (it is usually expressed as RH%). To obtain relative humidity a psychrometric chart can be used if the dry and wet bulb temperatures are known. The wet bulb and dry bulb temperatures can be measured by using sling psychrometer, motor-driven psychrometer or hair hygrometer. The sling psychrometer is the most popular one used in occupational hygiene practice (Fig. 11.3a).

Radiant temperature

The commonest method adopted to measure radiant temperature is by using a black globe thermometer, which consists of a hollow copper sphere about 15 cm in diameter, painted black, at the centre of which a glass thermometer is inserted (Fig. 11.3b). After mounting the thermometer into a stand, the globe is left for 15–20 minutes to reach equilibrium after which the temperature of the globe is measured. At equilibrium, heat loss, or gain, of the globe by convection is balanced by heat gain, or loss, by radiation. For this reason immediate determination of dry bulb temperature and air speed at the same time and in the same place is rather essential to estimate the mean radiant temperature. The mean radiant temperature is estimated either from a normogram or by using the following equation.

$$MRT = 4\sqrt{\frac{GT\,4}{100} + 2.48\,V\,(GT\text{-}AT)}, \text{ Equation (7)}$$

where: GT and AT = globe and air temperatures Kelvin; V = air speed in ms^{-1} and MRT = mean radiant temperature, Kelvin (273 + °C). It represents the temperature of a black enclosure of uniform wall temperature which would provide the same heat loss or gain as the environment measured.

Air speed

In many industrial situations wind is multi-directional or turbulent, therefore, instruments depending upon the cooling effect of a heated object e.g. thermocouples, a resistance wire and kata thermometers, provide more meaningful readings of air speed compared with directional instruments which depend on the impact pressure of the wind, e.g. a rotating or deflecting vane anaemometer. For the measurement of undirectional air speed the Kata thermometer is the instrument of choice. It is essentially an alcohol-filled thermometer with a large bulb coated with silvery material and can be obtained in three temperature ranges supplied with monographs for the rapid conversion of cooling times to air speed (Fig. 11.3c). When using the Kata thermometer, the bulb is heated in warm water until an unbroken column of alcohol rises into the upper reservoir. The bulb is then dried with a clean cloth and suspended in the air. A stop watch should be used to time the fall of the alcohol column between the two temperature marks. The cooling time of the Kata is a function of air speed and air temperature. The air speed can then be determined from the Kata factor printed on the stem, the dry bulb temperature and the cooling time, using the provided monograph.

HEAT STRESS INDICES

A heat stress index or scale is in principle an empirical relationship based on experimental trials on humans. The basic question which all heat stress indices try to answer is: which combinations of exogenous and endogenous heat loads (stress) provoke

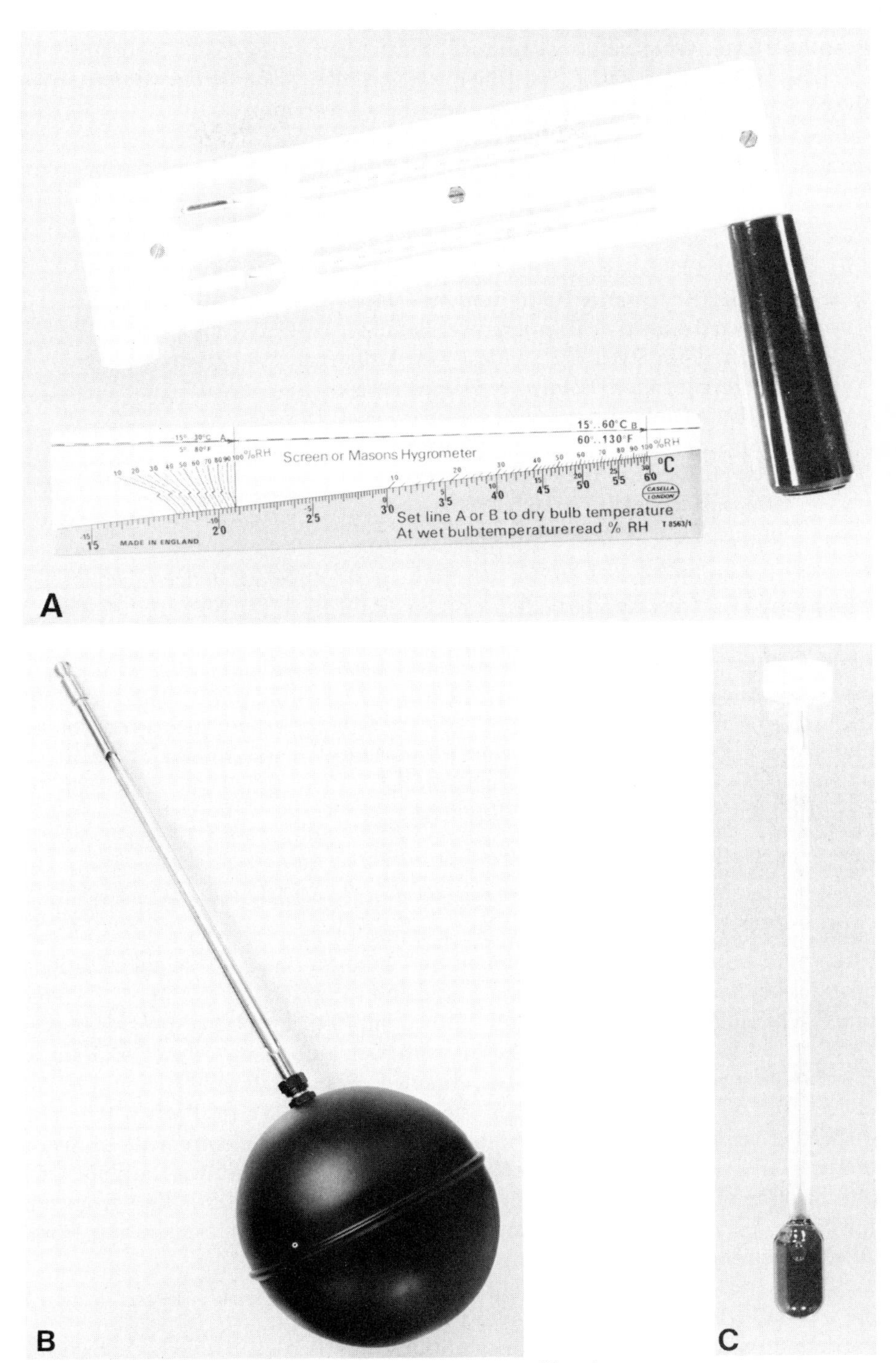

Fig. 11.3 **a** Sling psychrometer; **b** black globe thermometer; **c** Kata thermometer.

in man an equal thermal strain? Indicators of thermal strain are for example, subjective discomfort level, temperature sensation, skin temperature, core temperature, metabolic rate, pulse rate, sensible sweating onset and rate, vasodilation, vasoconstriction, shivering, time of recovery and thermal allesthesial response.

Since (Yaglou, 1927), several heat stress scale or indices have been proposed: over ninety such scales exist now (Wenzel, Personnel Communication) the best known of which are

(i) Basic Effective Temperature (Yaglou, 1927);
(ii) Index of Physiological Effect (Robinson et al, 1945);
(iii) Thermal Acceptance Ratio (Ionides et al, 1945);
(iv) Corrected Effective Temperature (Bedford, 1946);
(v) Predicted 4-h Sweat Rate Index (MacArdle et al, 1947);
(vi) Heat Stress Index (Belding & Hatch, 1955);
(vii) Wet Bulb Globe Temperature (Minard et al, 1957);
(viii) Temperature Effective Limit (Houberechts et al, 1958);
(ix) Index of Thermal Stress (Givoni, 1963);
(x) Time of Recovery (Wenzel, 1965);
(xi) Comfort Equation (Fagner, 1972).

The most commonly used nowadays in industrial hygiene practice are the effective temperature (ET) or corrected effective temperature (CET), WBGT Index, the HSI of Belding and Hatch and the predicted four-hour sweat rate P_4SR.

The P_4SR and HSI are well known indices used in industry. The HSI has many advantages, the greatest one being that it makes it possible to calculate the allowable exposure time as well as the minimum recovery time for a given heat stress condition. However, as far as P_4SR and HSI require a knowledge, or assumption, of the metabolic rate and of clothing characteristics, most probably their estimation renders the whole exercise difficult for practical and accurate use.

Effective temperature

The ET scale is based mainly on subjective assessment of the environment. The concept behind this index is to expose a group of people to various combinations of dry bulb temperature, air movement and humidity and to ask them to identify which range of climates felt the same at still and saturated environment. Two monograms were constructed, 'normal' for clothed subjects and 'basic' for those stripped to the waist. To determine ET, the dry and wet bulb temperatures and air speed should be first recorded. By using the obtained data a straight line should be drawn between the wet and dry bulbs then the values of ET are read when the value of air speed crosses with the wet-dry bulbs calibration line. The corrected effective temperature (CET) is determined using the same method by substituting the globe temperature in place of the dry bulb temperature. The ET as an index suffers from serious drawbacks, e.g. it neglects the importance of water vapour evaporation from the subject's clothing and the physical activity of the subject.

The wet bulb globe temperature index (WBGT)

The definition of hot environmental conditions proposed by NIOSH was based upon Lind's (1963) relationship between ET and rectal temperature. Essentially the WBGT index is an algebraic approximation of the ET concept (Yaglou & Minard, 1957). Although the index inherited all the limitations of the ET, it is simple, convenient to assess, needs less skill and is most suitable for industrial heat exposure and maximum permissible work conditions (Brown & Dunn, 1977). Because of its practicability, and its close correlation with other heat stress indices it was adopted by the American Conference of Governmental Industrial Hygienists (ACGIH, 1983) as the basis for its threshold limit values TLV for heat stress (Fig. 11.4). TLV-TWA is the time-

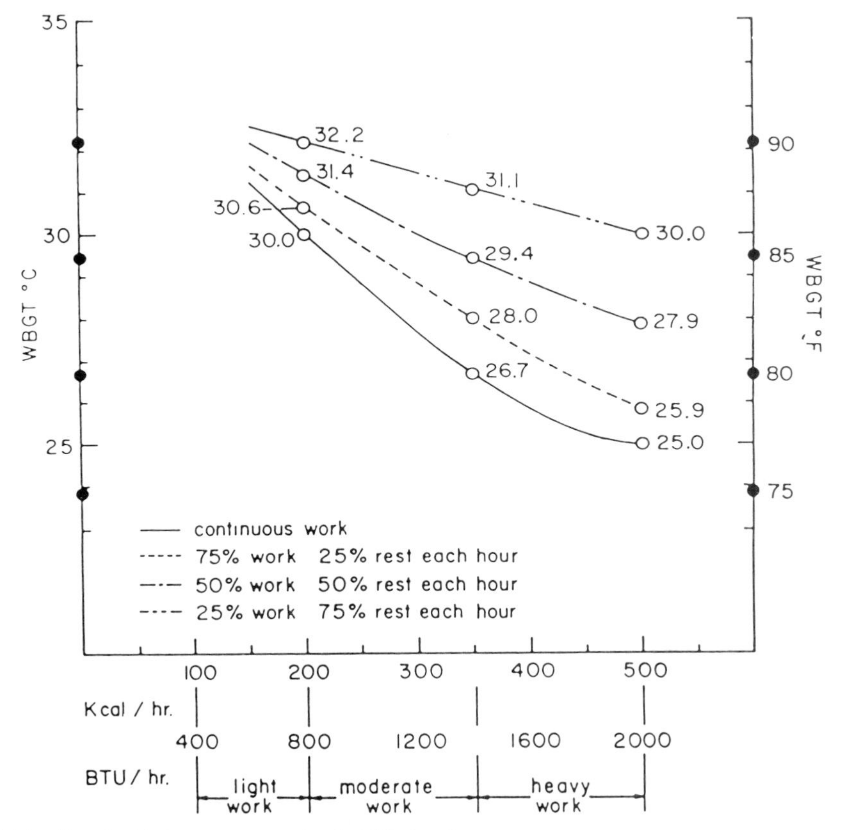

Fig. 11.4 Permissible heat exposure threshold limit value.

weighted average WBGT for a normal 8-hour workday of a 40-hour week, to which nearly all workers may be repeatedly exposed, day after day, without adverse effect. WBGT is defined as follows:

(a) Indoors or outdoors with no solar load,

$$\text{WBGT} = 0.7\,\text{NWB} + 0.3\,\text{GT}$$

(b) Outdoors when a solar load is imposed

$$\text{WBGT} = 0.7\,\text{NWB} + 0.2\,\text{GT} + 0.1\,\text{DB}$$

where, NWB = natural wet-bulb temperature, GT = globe temperature and DB = dry-bulb temperature.

The easy and simple weighing of these measurements have made it possible to produce instruments which give a direct reading in the WBGT index.

PERMISSIBLE EXPOSURE LIMITS

There is no single index of heat stress that can accurately predict physiological responses over a wide range of ambient temperature combinations and other factors such as the energy expenditure rate involved in the particular task, the clothing worn by the individual and inherent individual difference of preference.

The deep body temperature of 38°C was suggested by many investigators (Lind, 1963; Wenzel, 1968) as an upper environmental limit criterion for everyday work. Furthermore a group of World Health Organisation experts (WHO, 1969) recommended permissible exposure limits at 38°C for deep body temperature and a mean heart rate (HR) of 110 beat min^{-1} during prolonged work. They also recommended a set of environmental limits for physical work carried out for 8 hours a day, 5 days a week as follows:

Type of work	**CET** (°C)
Sedentary work (2.6 kcal/kg/h)	30
Light work (4.3 kcal/kg/h)	28
Heavy work (6.0 kcal/kg/h)	26.5

The ACGIH (1983) when adopting the WBGT index gave guidance for assessment of work load (Fig. 11.4). So far, very few studies were carried out on permissible heat exposure limits for different work-rest regimens.

CONTROL

Control of hot environments consists of two parts, one related to human factors and the other to the environment. The former can be achieved as follows.

1. Educating those exposed to heat, especially those charged with supervision. Individuals should be alerted when dangerous heat conditions exist. Heat load can be minimised by decreasing work load, increasing the frequency of rest periods and reducing the heat of the environment.
2. The loss of excessive amounts of water and salt should be replaced. The preferred method of replacement is by frequent intake of 200–300 ml of water throughout the work period.
3. Heat acclimatisation programmes for personnel should be encouraged and assessed in terms of duration and intensity of heat exposure.
4. The physical condition of the individual plays a significant role on their reaction to heat stress; the risk of heat injury is higher among obese and unfit persons compared with those who are physically fit and of normal weight. Those suffering from cardiovascular diseases are at a greater risk.
5. The work schedule should suit the climatic conditions. Work under direct sun should be discouraged and when the temperature reaches an excessively high level, physical work should be curtailed or, under some conditions, suspended. Heat injuries have been recorded at WBGT temperatures of 24°C and lower.

Control of the environment is based on several means of reducing the impact of heat upon the workers.

1. Reduction of body heat by introducing mechanisation, shared work load and increasing rest time.
2. Radiation can be reduced by increasing line-of-sight to source, insulating furnace walls and protecting the individual by using reflective screens and aprons, screens must not reduce air velocity in the working area.
3. Convection: (a) if air temperature is above skin temperature convection can be reduced by lowering air temperature and air velocity and removing clothes; (b) clothing should be loose fitting and allow air circulation under it with maximum cooling and minimum sweating. Under severe hot working conditions protective clothing is essential.
4. Increasing evaporative heat loss by increasing air velocity and decreasing humidity.
5. Work schedule shortening of exposure, frequent rest periods using air conditioned rest area for efficient cooling allow workers to self-limit their exposure on the basis of signs and symptoms of heat stroke.
6. If air temperature is below skin temperature, convection can be reduced by increasing air temperature, reducing air velocity and adjusting clothing.
7. Ventilation for heat control can be either local exhaust ventilation (which will remove heat at its source) or general, which increases air movement in the working environment.

ENERGY EXPENDITURE AND DISEASE IN HOT CLIMATE (SUDAN CASE)

In many developing countries with predominant manual labour, excessively heavy work is still common. Establishing limits for permissible physical work loads is therefore of great importance. individual variability does exist in physical working capacity, therefore, it was suggested that a work load taxing 30 to 40% of the individual's maximal aerobic power is an acceptable average upper limit for physical work carried out regularly over an 8-hour working day. For very heavy work an oxygen uptake of 1.5–2.0 l min^{-1} and a heart rate response of 130–150 beats min^{-1} were suggested for prolonged physical work for average 20-to 30-year-old individuals (Astrand & Rodahl, 1977).

Depending on occupation, there is a wide variation in the daily rates of energy expenditure ranging from 1380 up to 5000 kcal (320 to 1194 kj) (Durnin & Passmore, 1967). A man who is not engaged in heavy labour but active during leisure time daily expends about 2900 kcal (693 kj). While in the majority of professional activities e.g. light industry, house work and laboratory work the energy output is less than 5 kcal min^{-1} (less than 20 kj min^{-1}), it reaches upto 7.5 kcal min^{-1} (or 30 kj min^{-1}) in construction work, agriculture, the iron and steel industries, and the armed services. Energy output as high as 10 kcal min^{-1} (40 kj min^{-1}) were recorded among fishermen, lumberjacks, miners and dock labourers.

The energy cost of different occupational activities have been measured by several investigators (Passmore & Durnin, 1955; Durnin & Passmore, 1967; Astrand & Rodahl, 1977) but only a few studies have been made on men performing self-paced physical tasks (e.g. Davies, 1973; Davies et al, 1976; Sourr et al, 1974). In tropical and sub- tropical regions where unsupervised work in farming communities is common, the capacity to perform physical work remains an important component of productivity (Awad El Karim et al, 1981) which in turn is influenced by the presence of endemic disease (Collins, 1981). In recent years, most of the studies carried out to assess the effect of disease on energy expenditure and productivity were concerned with schistosomiasis infection. Schistosomiasis was found to affect labour output, earnings and absenteeism (Foster, 1967; Fenwick & Figenschou, 1972). During 1974–1984, a number of fundamental investigations on working capacity and schistosomiasis infection were carried out in the Sudan (Collins et al, 1976; Davies et al, 1976; Awad El Karim et al, 1980). A significant impairment of physical working capacity caused by the disease and measured under laboratory working conditions using a bicycle ergometer was estimated to amount to 18% (Awad El Karim et al, 1980). When the energy cost of different agricultural tasks of self-paced performing men was measured under natural working field conditions using the Oxylog (Humphrey & Wolff, 1977), a highly significant decrease in oxygen intake due to heavy schistosomiasis infection (1000 + eggs/g of faeces) was recorded amounting to 20–30% (Abdel Rahman, 1984). The 24-hour energy expenditure was also impaired by schistosomiasis infection. It was positively correlated with maximum oxygen intake (VO_2 max) and negatively correlated with quantitative schistosome egg excretion rate.

HEAT ILLNESSES DURING THE ANNUAL PILGRIMAGE TO MAKKAH (SAUDI ARABIA)

The annual pilgrimage to Makkah (The Hajj) is one of the five pillars of the Islamic Faith and every Muslim is expected to perform it once. The Hajj takes place on a fixed date, 9th-12th, of the 12th Lunar Month of the Arabic Calendar. It therefore occurs 11 days earlier every year on the Gregorian Calendar, and thus falls for 17 consecutive years during the hottest months May to September, when the temperature ranges between 38 and 50°C. The hot cycle started in 1981 and will continue until 1998.

Annually 2 000 000 pilgrims come to Makkah and the holy places during that period. The religious rites of the Hajj require that all pilgrims, should be in certain places in specified times. Exposed to such hot environment with added problems of overcrowding, mass movement, strained sanitary facilities and voluntary dehydration, hundreds of heat casualties are expected. During the last 5 years a total of approximately 3000 cases of true heat stroke were treated in the special centres for management of heat stroke patients. For each case of heat stroke there are usually 5–8 cases of heat exhaustion.

Heat stroke as mentioned earlier is characterised by (a) rectal temperature of above 40°C; (b) disturbance of the CNS and (c) hot dry skin. There are two types of heat strokes; exertion-induced and classical heat stroke.

The typical case of heat stroke is diagnosed easily but it may be overlooked when there is no obvious history of excessive heat load. There is a practical rule; if a patient loses consciousness under conditions of heat stress, heat stroke should be suspected, and the rectal temperature should be properly measured. The measurement in the delirious hyperactive heatstroke victim requires the application of an inlying thermistor with a scale reaching beyond the upper limit of 41°C of the standard clinical thermometer.

Management

Successful management of heat stroke cases require early detection, prompt diagnosis, and physiologically effective cooling, continuous monitoring, meticulous respiratory care and fluid replacement. The best method of cooling is evaporative cooling from a warm skin. Khogali and Weiner designed the Makkah Body Cooling Unit (MBCU) which operates on a straight forward physiological principle (Weiner & Khogali, 1980). The patient's heat is dissipated by evaporation of a spray of atomised water at 15°C and a flow of warm air at 45°C. This maintains the skin temperature at the normal level of 30–33°C. Keeping the skin warm ensures that heat is transferred efficiently via the circulation to the skin by maintaining vasodilation, which ensures a high level of blood flow, and hence heat flow, to the skin. The warm, carefully wetted skin simulates sweating and provides a high vapour pressure which facilitates the removal of water vapour. The high rate of impinging air increases the total volume of water vapour and heat removed. The cooling power of the MBCU is approximately three times that of the average person sweating at his highest rate.

While rapid and effective cooling is the corner-stone of the treatment of heat stroke, it is important to treat shock, correct fluid and electrolyte disturbances and support vital organs both during and after cooling. Treatment of the derangements brought about by hyperpyrexia, particularly convulsions, coma and respiratory failure, requires easy access to the patient. The MBCU is designed in such a way as to enable all these operations to be carried out effectively and hygienically. This method was used very successfully in the treatment of approximately 3000 cases of heat stroke during the last 5 years.

CONCLUSION

Working under thermal heat stress not only taxes an individual's physiological function but also poses a serious threat to his health status. There are many excessively hot industrial situations such as blast furnaces and coke plants in the steel industry, melting operations in the aluminium and glass industries, and boiler sections in the sugar industry which warrant careful preventive measures. Adequate water supply and salt intake, acclimatisation, first aid training, training of workers for health, safety and work practices as well as assessment and control of hot environment should be invariably implemented in all hot jobs. Special arrangements like regulation of exposure time, preplacement and periodic medical examination and protective clothes should be compulsory for jobs entailing extreme heat exposure.

Heat casualties due to direct exposure to solar energy in outdoor activities such as construction work, military training, agriculture and pilgrimages, especially during hot seasons, are significant and need special handling, precautionary measures and control. Work load, duration of work, environmental conditions and health status should be assessed for those working under direct sun. Lack of acclimatisation, cardiovascular disorders, other inter-current diseases and exposure to high intensity levels of heat are all important contributing factors to high morbidity and mortality of heat illnesses.

REFERENCES

Abdel Rahman T A 1984 Energy expenditure and schistosomiasis infection in agricultural workers of the Sudan. Ph.D. Thesis, Faculty of Medicine, University of Khartoum

American Conference of Governmental Industrial Hygienists (ACGIH) 1983 Threshold limit values of airborne contaminants, Cincinnati, Ohio

Astrand P-O, Rodahl K 1977 Textbook of work physiology, 2nd edn McGraw-Hill, New York

Awad El Karim M A, Collins K J, Brotherhood, J R, Dore C, Weiner J S, Sukkar M Y, Omer A H S, Amin M A 1980 Quantitative egg excretion and work capacity in a Gezira population infected with *Schistosoma mansoni*. American Journal of Tropical Medicine and Hygiene 29: 54–61

Awad El Karim M A, Collins K J, Sukkar M Y, Dore C 1981 The working capacity of rural, urban and service personnel in Sudan. Ergonomics 24: 945–952

Bedford T 1946 Environmental warmth and its measurements. Medical Research Council War Memo 17, London

Belding H S, Hatch T F 1955 Index for evaluating heat stress in terms of resulting physiological strains. Heating, Piping and Air Conditioning 27: 129–136

Bligh J, Johnson K G 1978 Glossary of terms of thermal physiology. Journal Applied Physiology 35: 941–961

Brown J R, Dunn G W 1977 American Industrial Hygiene Association Journal 38: 180

Burton A C 1934 The application of the theory of heat flow to the study of energy metabolism. Journal of Nutrition 7: 497

Cabanac M, 1975 Sensory pleasure. Quarterly Review of Biology 54: 1

Collins K J, Brotherhood J R, Davies C T M et al 1976 Physiological performance and work capacity of Sudanese cane cutters with schistosoma mansoni infection. American Journal of Tropical Medicine and Hygiene 23: 625–634

Collins K J 1981 Energy expenditure, productivity and endemic disease. In: Harrington G A (ed) Energy and effort, SHB Symposium, 22. Taylor & Francis, London

Crockford G W, Awad El Karim M A 1974 An assessment of a dynamically insulated heat protective clothing assembly. Am Occup Hyg 17: 111–121

Davies C T M 1973 Maximum aerobic power output in relation to productivity and absenteeism of East African sugar cane workers. British Journal of Industrial Medicine 30: 146–154

Davies C T M, Brotherhood J R, Collins K J et al 1976 Energy expenditure and physiological performance of Sudanese cane cutters. British Journal of Industrial Medicine 33: 181–186

Dukes-Dobos F N 1981 Hazards of heat exposure. Scandinavian Journal of Work and Environmental Health 7: 73–83

Durnin J V G A, Passmore R 1967 Energy work and leisure. Heinemann Educational Books, London

Ellis F P, Nelson F, Pincus L 1975 Mortality during heat waves in New York City, July 1972 and August-September 1973 Environmental Research 10: 1–13

Ellis F P, 1976 Heat illness. Transactions of the Royal Society of Tropical Medicine and Hygiene 70: 402–425

Fenwick A, Figenschou B H 1972 The effect of *Schistosoma mansoni* infection on the productivity of cane cutters on a sugar estate in Tanzania. Bulletin of the World Health Organisation 47: 567–572

Foster R 1967 Schistosomiasis on an irrigated estate in East Africa III. Effects of asymptomatic infection on health and industrial efficiency. Journal of Tropical Medicine and Hygiene 70: 185–195

Givoni B, 1963 A new method for evaluating industrial heat exposure and maximum permissible work load. Proceedings of the International Biomedical Conference

Goldsmith R 1977 Acclimatization to heat and cold in man. In: Baron D N, Compston N, Dawson A M (eds) Recent advances in medicine V 17, Churchill Livingstone, Edinburgh 299–321

Hardy J D, Soderstrom G F 1938 Heat loss from the nude body and peripheral blood flow at temperature of 22°C to 35°C. Journal of Nutrition 16: 493–499

Hardy J D 1961, Physiology of heat regulation. Physiological Reviews 41: 521–607

Henschel A, Burton L, Margolies L, Smith J E 1969 An analysis of the heat death in St Louis during July 1966. American Journal of Public Health 59: 2230–2272

Houberechts A, Lavenne F, Patigny J 1958 Le travail humain aux temperatures elevees. Maroc Med 395: 328

Humphrey S J E, Wolf H S 1977 The Oxylog. Journal of Physiology 267: 12

Ionides M, Plummer J, Siple P A 1945 The thermal acceptance ratio. Int Rep No 1 Climatology and Environmental Protection Section, Military Planning Division, Office of Quartermaster General

Kerslake D McK 1972 The stress of hot environments. Monograph of the Physiological Society, Cambridge University Press, Cambridge

Khogali M 1983a Epidemiology of heat illness during the Makkah Pilgrimages in Saudi Arabia. International Journal of Epidemiology 12: 267–273

Khogali M 1983b Heat stroke and heat exhaustion. Travel and Traffic Medicine International 1: 166–169
Khogali M, Mustafa M K Y 1983 Physiology of heat stroke: An overview. In: Hales J R S (ed) Thermal physiology. Raven Press, New York
Khogali M, 1984 The clinical management of heat stroke. Medical Forum 11: 16–19
Leithead C S, Lind A R 1964 Heat stress and heat disorders. E A Davis, Philadelphia
Lind A R 1963 A physiological criterion for setting thermal limits for everyday work. Journal of Applied Physiology 18: 51–56
Lind A R 1983 Pathophysiology of heat exhaustion and heat stroke. In: Khogali M, Hales J R S (ed) Heat stroke and temperature regulation. Academic Press, Sydney, Ch 16, 179–188
Machle W, Hatch 1947 Heat: Mans exchanges and physiological responses. Physiological Reviews 27: 200–227
Marriot H 1950 Water and salt depletion. Blackwell, England
McArdle B, Dunham W, Holling HE et al 1947 The prediction of the physiological effects of warm and hot environments. The P4SR Index Med Res Coun Roy Nav Pers Rep 47/391
Minard O, Belding H S, Kingston J R 1957 Prevention of heat casualties. Journal of American Medical Association 165: 1813–1818
Passmore R, Durnin J V G A 1955 Human energy expenditure. Physiological Reviews 35: 801–840
Robertshaw D 1983 Contributing factors to heat stroke. In: Khogali M, Hales J R S (eds) Heat stroke and temperature regulation. Academic Press, Sydney, 13–30
Robinson S, Turrell E S, Gerking S D 1945 Physiologically equivalent conditions of air temperature and humidity. American Journal of Physiology 143: 21
Spurr G B, Barrac-Neito M, Maksud M G 1974 Relationships among productivity aerobic power and energy cost in heavy physical labour. In: proceeding of XXVI Inter Cong Physiol Vol XI, 274
Veale W L, Cooper K E 1983 The elderly and their risk of heat illness. In: Khogali M, Hales J R S (ed) Heat stroke and temperature regulation. Academic Press, Sydney, 189–196
Weiner J S, Khogali M 1980 A physiological body cooling unit for treatment of heat stroke. Lancet i: 507–509
Wenzel H B 1965 Die Erholungsdauer nach hitzearbeit als mass der belastung. West deutscher Verlog Nr 1544 Koeln
Wenzel H G 1968 Pulse rate and thermal balance of man during and after work in heat as criteria of heat stress. Bulletin of the World Health Organization 38: 657–664
WHO 1969 Health factors involved in working under conditions of heat stress. Tech Rep Ser No 412. Geneva
WHO 1977 Manual of the international statistical classification of diseases, injuries and causes of deaths 9th revision 1975. Geneva
Wyndham C H 1973 The physiology of excercise under heat stress. Annual Review of Physiology 35: 193–220
Yaglou C P 1927 Temperature, humidity and air movement in industries: the effective temperature index. Journal of Industrial Hygiene 9: 297–309
Yaglou P, Minard D 1957 Control of heat casualties at military training centres. American Medical Association Archives of Industrial Health 16: 302–316

12. Assessing the hazards of whole-body and hand-arm vibration

M. J. Griffin

INTRODUCTION

The body is exposed to vibration (i.e. oscillatory motion) at work, during play and while at rest. Vibration occurs in all forms of transport, in buildings, on and around industrial machinery, and on powered tools. Whole-body vibration occurs when the body is supported on a vibrating surface, usually a seat or floor. Local vibration involves contact with the source of motion at some other point. Local vibration often involves the hand and is then called hand-arm vibration.

Whole-body and hand-arm vibration can give rise to various sensations (including pleasure, discomfort and pain), interfere with a wide a range of activities, and produce various physiological and pathological effects. Low frequency oscillation, having a frequency below about 0.5 Hz, is one of several causes of motion sickness.

Recent attention to the occupational health effects of whole-body vibration has tended to focus upon the potentially injurious effects on the back of high magnitudes of vibration experienced in, for example, off-road vehicles. Studies of the various adverse effects of the prolonged use of hand-held vibrating tool have been dominated by the investigation of vibration-induced white finger.

CHARACTERISTICS OF VIBRATION

There are four independent characteristics of vibration which must be considered in relation to any assessment of its effects: the magnitude, frequency, axis, and duration of the motion.

Magnitude

Vibration magnitudes are now most often expressed as the root-mean-square acceleration in units of metres per second per second (i.e. ms^{-2} rms). In some countries the intensity of a vibration of magnitude 'a' is expressed on a logarithmic scale of vibration level:

$$\text{vibration level (dB)} = 20 \log_{10} (a/a_{ref})$$

The reference magnitude, a_{ref}, preferred by the International Organisation for Standardisation for such a scale is $10^{-6} ms^{-2}$ (ISO 1683, 1983) but other references are in use and many workers in the field prefer the use of linear values expressed in ms^{-2}.

Root-mean-square measures of vibration are average quantities which are often useful but they underestimate the severity of exposures to some shock motions (see below). The crest factor (i.e. ratio of peak to rms of the frequency-weighted acceleration) is often used to indicate whether rms measurements will be useful (ISO 2631,

1985). The significance of a vibration magnitude depends upon the frequency, direction and point of contact with the body but, generally, magnitudes below about 0.01 ms^{-2} rms are imperceptible, those around 1.0 ms^{-2} rms are uncomfortable and those much in excess of this value may, depending on their duration, be injurious.

Frequency

Vibration frequency is expressed in terms of the number of cycles per second (i.e. hertz, Hz—ISO 266, 1975). For whole-body vibration the greatest interest is often in the frequency range from about 0.5–40 Hz. Many vehicles have their dominant motions between about 0.5 and 8 Hz and it is in this range that seats often amplify vibration (Griffin, 1978). Vibration at higher frequencies can be transmitted to the body and, in helicopters, the principal frequencies can be in the range from about 15–30 Hz. It is often possible to reduce or eliminate the effects of higher frequencies of whole-body vibration. The vibration frequencies considered in relation to effects of vibration on the hand extend from about 8 to approximately 1000 Hz.

Most existing vibration standards have been written on the false assumption that vibration exposures involve a single vibration frequency: vibration limits are defined for each frequency separately. In recent years it has been recognised that this is not a satisfactory basis for the definition of standardised vibration evaluation procedures and it is now generally preferred to report the frequency-weighted rms magnitudes of complex spectra. The shape of the frequency weighting depends upon the effect of interest. Several studies have shown that, when using an appropriate weighting, this approach is likely to be the most accurate and simple procedure for predicting the relative importance of different motions (Griffin, 1976, 1986). Even so, the determination of the spectral composition of vibration conditions is always advisable.

Axis

The direction in which vibration enters the body has a large effect on the manner in which it is transmitted. The standardised axes are defined by a right-handed coordinate system based on the body and labelled the x-, y-, and z-axis (ISO 5805, 1981). For the whole-body vibration of seated persons these represent the fore-and-aft, lateral and vertical directions respectively. Although these axes are often drawn through the human heart (for example ISO 2631, 1985) it is recommended that evaluations of vibration exposures should always involve the quantification of the motion at the interface between the body and the vibrating surface (BSI, 1987a). Vibration is therefore generally measured using accelerometers housed within suitably formed mounts placed between the surface of a seat and the occupant (SAE, 1974; Whitham & Griffin, 1977). For vibration exposures of the hand a similar set of axes is defined (labelled x_h, y_h and z_h) and the measurements are made either on the surface of the tool or with the aid of some mount or ring held by the hand (ISO 5349, 1986b; BSI, 1987b).

Duration

Exposure duration can have a complex effect on the acceptability of vibration. Since an average measure, such as the rms magnitude of vibration, does not increase with increasing exposure duration there has been a tendency to define time-dependent vibration limits (e.g. ISO 2631, 1985; ISO, 1979). This resulted in unproven, but complex, limits and the failure to define a practical method of reporting the vibration dose received during occupational exposures. Recent proposals for both whole-body and hand-arm vibration overcome these difficulties and offer other practical advantages (see below).

WHOLE-BODY VIBRATION

Causes and effects

Surveys of the health of those exposed to whole-body vibration have suffered from the difficulty of distinguishing the effects of the motion from those of many other possible causes of disorders. Many jobs involve poor sitting postures, heavy lifting and other activities which may cause, or contribute to, some of the reported problems. Furthermore, the transmission of vibration to the body and the consequences of the vibration in the body can be highly dependent on body posture. There has been little attention to determining the vibration dose received during occupational exposures or an attempt to relate vibration dose to disorders.

Disorders have been reported among various vibration-exposed groups: for example, earth moving equipment operators (Milby & Spear, 1974), tractor drivers (Kohl, 1975), road vehicle drivers (Kelsey & Hardy, 1975; Kelsey et al, 1984), coach drivers (Gruber & Ziperman, 1974), truck drivers (Gruber, 1976), helicopter pilots (Auffret et al, 1978). The most frequently reported disorders are those which might be attributed to trauma of the spine, especially degenerative changes (e.g. Sandover, 1983; Damkot et al, 1984; Seidel & Heide, 1986). Other conditions which have sometimes been attributed to vibration exposure include abdominal pain, digestive problems, urinary difficulties, prostatitis, increases in problems of balance, visual disorders, headaches, sleeplessness and similar symptoms (Griffin, 1982a).

The environments which have been associated with whole-body vibration injuries are those in which vibration is a well recognised source of discomfort: off-road vehicles (e.g. tractors, forest machines and earth-moving machinery), road vehicles (e.g. trucks, buses and cars), helicopters, and industrial machinery. Although some investigations have compared the incidence in exposed and non-exposed control groups there have been no studies in which reported disorders have been differentially related to measurements of the vibration magnitudes, frequencies, directions or durations of exposure.

Vibration evaluation

International Standard 2631 (1974, 1978) defined exposure limits for whole-body vibration. An amendment to this standard was issued in 1982 so as to clarify the procedure and reduce the ambiguities in the document. The Amendments were incorporated in a new printing of the Standard in 1985 without any deletions to the information in the previous document. ISO 2631 (1985) is therefore somewhat self-contradictory and confusing to read.

Figure 12.1 defines the ISO 2631 exposure limits for selected periods from 1 minute to 24 hours. In the z-axis the vibration limits are lowest from 4–8 Hz while in the x- and y-axes sensitivity is greatest from 1–2 Hz. The latest version of this standard does not make it completely clear whether these exposure limits are to be used with frequency-weighted vibration magnitudes as well as discrete sinusoidal components. The limits are not defined outside the frequency range 1–80 Hz of for durations below 1 minute and the time-dependency is very complex. The limits are only applicable to motions having a crest factor below about 6. No guidance is given for the assessment of motion in those environments (e.g. off-road vehicles) associated with higher crest factors. The meaning of the exposure limit is not defined in relation to any specific injury or other effect but it is stated that it is at about half the threshold of pain.

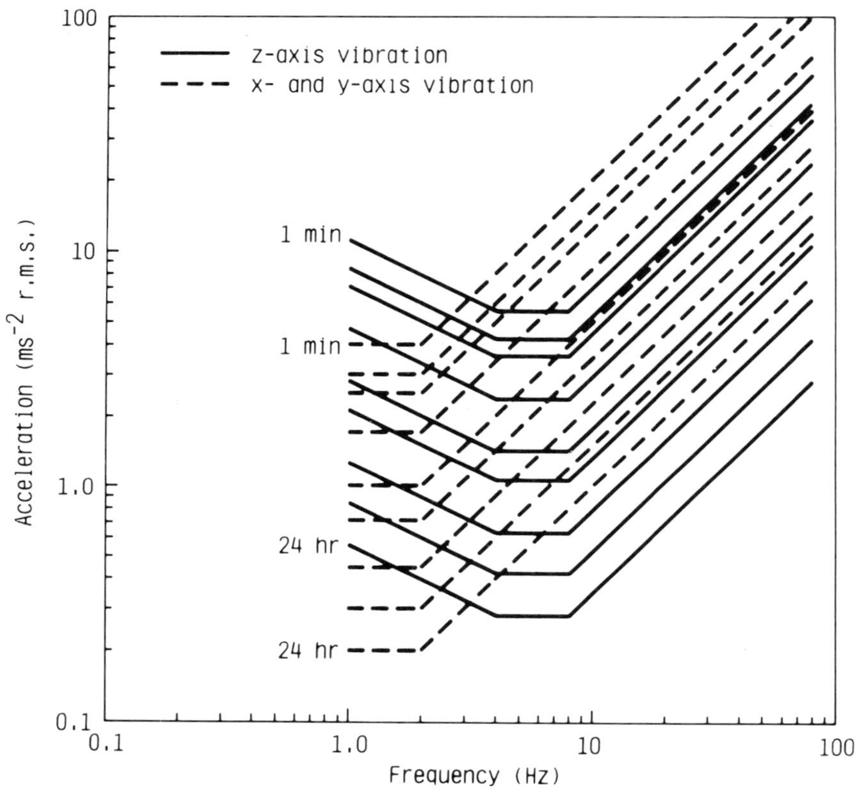

Fig. 12.1 ISO 2631 (1985) exposure limits for durations of 1, 16, 25 minutes and 1, 2.5, 4, 8, 16 and 24 hours.

International Standard 2631 has many limitations and does not fully define a means of quantifying vibration exposures. The vibration limits do not appear realistic since most persons will find some of the one-minute exposures totally intolerable while many forms of transport exceed the limits for periods in excess of about 8 hours without causing any obvious harm. The two frequency-dependencies have been widely used for obtaining environmental measures of vibration and a more complete definition of their form has been prepared (ISO, 1986a). However, there appears to be little evidence that the weighting for vertical vibration has the correct shape: most persons are subjectively more sensitive to frequencies above about 8 Hz than suggested by the standard and sensitivity to 2 Hz is usually similar to that at 1 Hz. Furthermore, the exclusion of frequencies below 1 Hz makes both the vertical and horizontal weightings unrealistic for use in those many environments containing much energy in this region.

In order to correct the errors, ambiguities and omissions in ISO 2631 a revision of the standard was commenced in the late 1970s. A major objective of the revision was to define fully a measurement procedure for all types of simple and complex vibration. Random, multiple frequency, and high crest factor motions were all to come within the scope of the revision. The concept of an internationally agreed vibration limit was abandoned in favour of a measurement procedure with, as far as possible, dose-effect information. By 1984 the fifth draft of a revision had been evolved sufficiently for it to be used experimentally. Further discussion is required to present the draft in a form which can be accepted by all members of the relevant ISO Sub-Committee by the procedure is now in use and forms the basis of several company standards (Griffin, 1986). British Standard 6841 awaits publication (BSI, 1987a).

For the assessment of the health effects of whole-body vibration the revised standard will define a new frequency-weighting and a simplified time-dependency (Griffin, 1984, 1986; Kjellberg & Wikstrom, 1985). The frequency-weighting is defined by equations such that it may be easily implemented by analogue or digital filters (BSI, 1986a; Griffin, 1986; Lewis, 1985). An asymptotic approximation to the weighting for the z-axis has greatest sensitivity in the range 5–16 Hz with sensitivity increasing at 6 dB/octave from 2–5 Hz and decreasing at 6 dB/octave above 16 Hz. Sensitivity to acceleration is constant in the frequency range between 2 and 0.5 Hz (Fig. 12.2). The frequency-weightings for the x- and y-axes are unchanged other than for an extension from 1 Hz down to 0.5 Hz. The presence of high magnitudes of fore-and-aft vibration on the backrests of seats in some vehicles has led to the definition of a weighting for this motion in which sensitivity is greatest at frequencies below 8 Hz (BSI, 1986a; Griffin, 1984, 1986).

The simplified time-dependency, which is applicable over all durations from fractions of a second to 24 hours, is given by a fourth power procedure (i.e. motions are equivalent if the fourth powers of their weighted acceleration magnitudes multiplied by their durations are similar). This time-dependency may alternatively be expressed as consisting of 1.5 dB (i.e. 1.1892 to 1) change in allowable acceleration per doubling or halving of exposure time.

In order to assess transient (including high crest factor and intermittent) vibration exposures, a means of quantifying the vibration dose is defined:

$$\text{vibration dose value } (\mathrm{ms}^{-1.75}) = [\textstyle\int a^4(t)\,dt]^{1/4}$$

where a(t) is the frequency-weighted acceleration time history and the integral is obtained over the full period when the vibration of interest may occur. The vibration dose value may be considered to be the equivalent magnitude of a 1 second vibration. This method of determining the vibration dose value is necessary when the motion is intermittent or has a high crest factor in excess of about 6.

If the crest factor is low and the motion is not intermittent (i.e. it could have been assessed by the procedures in ISO 2631) it is possible to estimate the vibration dose value from the rms value of the frequency-weighted acceleration:

$$\text{estimated vibration dose value} = [(1.4 \times \text{rms acceleration, } \mathrm{ms}^{-2})^4 \times (\text{duration, s})]^{1/4}$$

This formulation often gives a very close agreement with the preferred determination of the true vibration dose value.

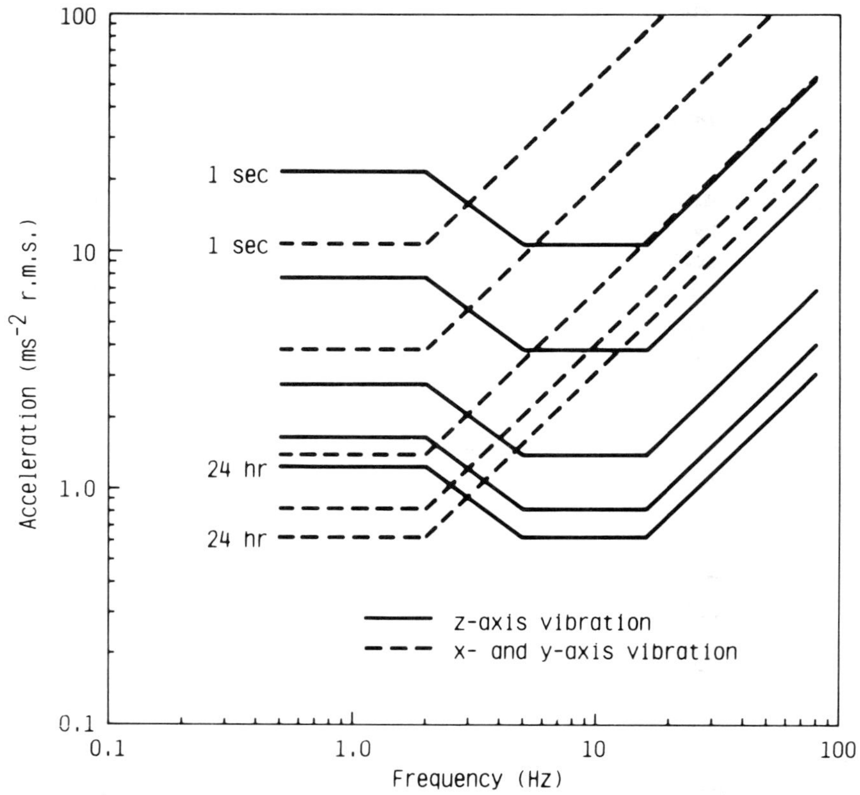

Fig. 12.2 Vibration magnitudes corresponding to an action level of 15 $ms^{-1.75}$ (estimated vibration dose value) for durations of 1 second, 1 minute, 1 hour, 8 hours and 24 hours.

There are insufficient dose-effect data to be able to offer guidance on the type or probability of any injury corresponding to any particular vibration dose value: the procedure is primarily offered as the best means currently available for quantifying the relative severity of complex vibration and shock exposures. However, it is known that high vibration dose values will cause severe discomfort, pain and injury. Vibration dose values in excess of about 15 $ms^{-1.75}$ will usually cause severe discomfort and it is reasonable to assume that increased exposure to vibration will be accompanied by increased risk of injury. Figure 12.2 shows the vibration magnitudes from 0.5–80 Hz which are equivalent to 15 $ms^{-1.75}$ for selected exposure durations from 1 second to 24 hours.

HAND-ARM VIBRATION

Causes and effects

Prolonged exposure of the fingers, hands and arms to high magnitudes of vibration is associated with a range of disorders. The term vibration syndrome may be applied to the group of symptoms but has sometimes been used to refer solely to a circulatory disorder of the fingers also called vibration-induced white fingers (VWF) or traumatic vasospastic disease (TVD). Bone and joint disorders are well-documented among users of percussive tools. Neurological effects (including decreased tactile sensitivity and increased nerve conduction times) are frequently found and there is some evidence of muscular disorders. Some researchers have reported an extensive range of other symptoms within the whole-body (Griffin, 1982a). Notwithstanding the wide range of reported symptoms, the greatest research effort has recently been orientated towards the study of the vascular disorders which, here, will be called VWF.

The most obvious symptom of VWF is the intermittent blanching of the fingers. Initially only one distal phalanx may be affected but with continued exposure to vibration the symptoms may appear in other digits and eventually extend to all phalanges on all fingers and thumbs. Some researchers believe that numbness and tingling precede symptoms of blanching and may be used as an early indication of the disease. In some persons the finger may exhibit a characteristic blueness and in extreme cases there may be skin necrosis and gangrene. An attack of white finger usually occurs when the body is cold (e.g. early in the morning) and may last up to 30 or 60 minutes. Finger sensations are decreased during an attack and there may be pain when the circulation returns. The symptoms are similar to those of Raynaud's disease and several other secondary causes of reduced peripheral circulation (Taylor & Brammer, 1982). However, the date of the first occurrence of blanching, the restriction of symptoms to those areas exposed to vibration and the absence of signs of other disorders result in the diagnosis of VWF.

The severity of symptoms of VWF may be classified by several different schemes. Various versions of the staging system proposed by Taylor et al (1974) have been used in some British literature (Table 12.1). In this system of grading, symptoms

Table 12.1 Stages of the symptoms of Raynaud's phenomenon as originally defined by Taylor et al (1974)

Stage	Condition of digits	Work and social interference
0	No blanching of digits	No complaints
0_T	Intermittent tingling	No interference with activities
0_N	Intermittent numbness	No interference with activities
1	Blanching of one or more fingertips with or without tingling and numbness	No interference with activities
2	Blanching of one or more complete fingers with numbness usually confined to Winter	Slight interference with home and social activities. No interference at work
3	Extensive blanching usually all fingers bilateral. Frequent episodes Summer as well as Winter	Definite interference at work, at home and with social activities. Restriction of hobbies
4	Extensive blanching. All fingers, frequent episodes Summer and Winter	Occupation changed to avoid further vibration exposure because of severity of signs and symptoms

of finger numbness, tingling and blanching on different areas of the hand at different times of the year are combined with reports of interference with work and leisure on a coarse scale. The inclusion of interference with activities in this staging system has made it attractive for legal purposes but it is almost entirely dependent on the individuals' reports of his symptoms and it does not record small changes which may indicate the advance of the disease. Staging is also somewhat limited for scientific purposes since it relates only to the most affected hand: neither the location nor the extent of the areas on affected fingers is defined.

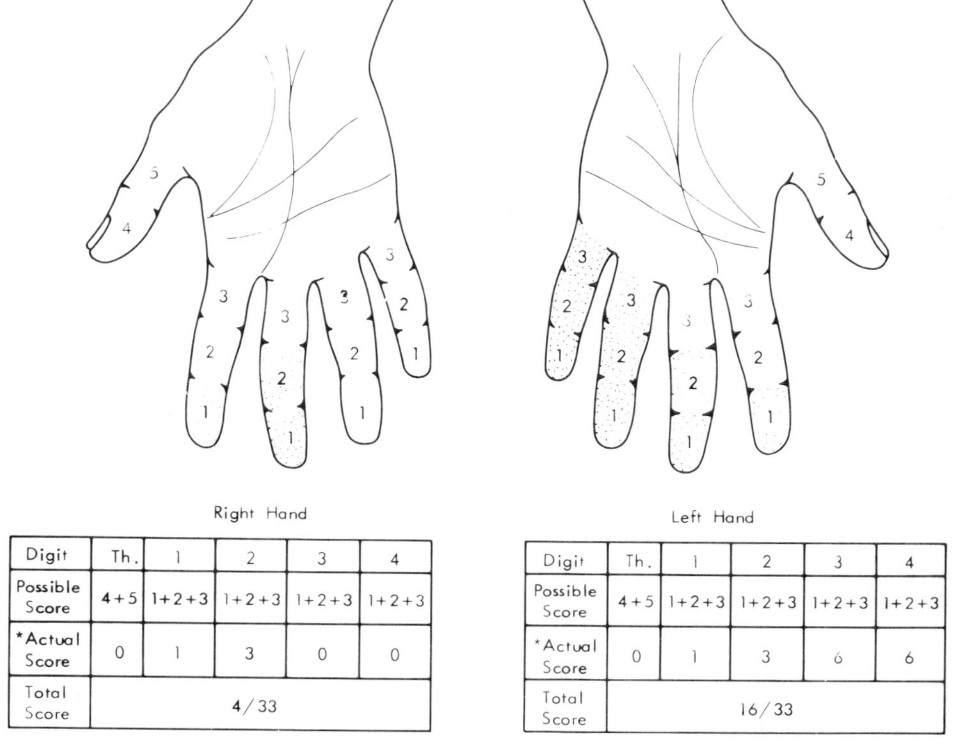

Right Hand

Digit	Th.	1	2	3	4
Possible Score	4+5	1+2+3	1+2+3	1+2+3	1+2+3
*Actual Score	0	1	3	0	0
Total Score	4/33				

Left Hand

Digit	Th.	1	2	3	4
Possible Score	4+5	1+2+3	1+2+3	1+2+3	1+2+3
*Actual Score	0	1	3	6	6
Total Score	16/33				

Fig. 12.3 A method of scoring the extent of symptoms of vibration-induced white finger symptoms (after Griffin 1982a).

In order to overcome some of the problems of staging VWF symptoms, a method of scoring the affected areas has been proposed (Griffin, 1982a, and Fig. 12.3). The scoring system isolates blanching as one identifiable symptom, numbness, tingling or other colour changes can be recorded separately using the same scheme. The scoring system may be used to document individuals' reports of blanching or independent observations of blanching. Actions (e.g. warnings, medical checks, removal from vibration work and compensation) may be related to specific scores. Scores may also be related to the areas of the hand in contact with the vibration and objective indicators of disease.

Many different objective tests of VWF have been tried but two principal methods are apparent: an indicator of blood flow or a test of neurological function. Changes in blood flow can be easily measured but peripheral circulation is modified by many factors in addition to VWF. In consequence, methods which involve direct or indirect measures of circulation require careful control of the test conditions. Although provocative cooling forms a part of some tests, simple cooling of the body or limbs is not a reliable means of provoking a visible attack of blanching. A procedure used by Welsh (1983) and based on the work of Juul & Nielson (1981) involves the warming of the body with heated water blankets at 40–45°C before and during the test — this is intended to abolish sympathetic vasoconstriction which may otherwise have a large influence on digit blood flow. The hand is then cooled in a water bath at 10°C for 3 minutes while circulation is occluded by a pressure cuff. The time taken for finger temperature to return to 21°C is then measured. Studies show that re-warming times tend to be significantly longer in those reporting symptoms of VWF. (Juul and Nielson have reported that persons with primary Raynaud's Disease tend to have normal responses in this test—possibly because their problem arises from sympathetic vasoconstriction which is abolished by the heated blankets.) Other means of monitoring changes in blood flow during, following, or between cooling or heating of the hand are in use. Lafferty et al (1983) defined a test of thermal entrainment for the diagnosis of primary Raynaud's disease which may assist differential diagnosis.

One of the most common neurological tests considered for the diagnosis of VWF has been the measurement of vibrotactile thresholds. Studies have shown that thresholds at both 31.5 and 125 Hz are significantly elevated in users of vibrating tools. (At these two frequencies the sense of vibration is normally mediated by two different end organs; the Meissner and Pacinian corpuscle respectively.) Thresholds in groups of tool users reporting symptoms of VWF are significantly greater than in groups using vibrating tools but not reporting symptoms. Also, groups of vibrating tool users without symptoms have higher thresholds than non-users (e.g. Hayward & Griffin, 1985). The value of this type of test must depend on the mechanisms which relate changes in vibrotactile thresholds to both vibration exposure and symptoms of blanching.

Neither vascular nor neurological tests provide perfect agreement with reports of blanching, some persons without symptoms have abnormal test responses and some with symptoms have responses indistinguishable from unaffected persons. However, with currently available information both types of test appear to be useful diagnostic aids.

The vibrating tools which cause VWF can be classified into several convenient groups according to their function. Prolonged exposure to moderate or high magnitudes of vibration from any source should be expected to produce symptoms of VWF but a list of tools most often involved may be useful. Table 12.2 identifies the tools

Table 12.2 Types of tool use required for compensation of vibration-induced white finger in Britain since 1 April 1985 (DHSS, 1985)

(a) The use of hand-held chain saws in forestry

(b) The use of hand-held rotary tools in grinding or in the sanding or polishing of metal, or the holding of material being ground, or metal being sanded or polished, by rotary tools

(c) The use of hand-held percussive metal-working tools, or the holding of metal being worked upon by percussive tools, in riveting, caulking, chipping, hammering, fettling or swaging

(d) The use of hand-held powered percussive drills or hand-held powered percussive hammers in mining, quarrying, demolition, or on roads or footpaths, including road construction

(e) The holding of material being worked upon by pounding machines in shoe manufacture.

or processes for which compensation for industrial injury has been prescribed in Britain since April 1985 (DHSS, 1985). (Compensation is payable if work has involved one of the named processes and at least three fingers of one hand exhibit blanching on the medial phalanx throughout the year. This is equivalent to a blanching score of at least 9.)

Vibration evaluation

Within Britain, BSI Draft for Development DD43 (1975) has provided approximate guidance on the vibration conditions which are likely to be acceptable, those which are probably unacceptable and those whose acceptability will depend on the daily vibration exposure (Griffin, 1980). British Standard 6842 is awaiting publication (BSI, 1987b) and an International Standard has just been published (ISO, 1986b). These standards will have compatible measurement procedures but the British Standard gives greater guidance on measurement while not offering the detailed predictions of the prevalence of VWF which are contained in the International Standard.

Figure 12.4 shows the acceleration frequency weighting widely used to assess the severity of hand-arm vibration exposures. This suggests that acceleration below 16 Hz is thought to have greatest effect and that the effect decreases with increasing frequency. The weighting is primarily based on subjective and biodynamic response to vibration rather than epidemiological data: VWF appears to be most often associated with vibration in the frequency range between about 20 and 400 Hz (Griffin, 1982b). It

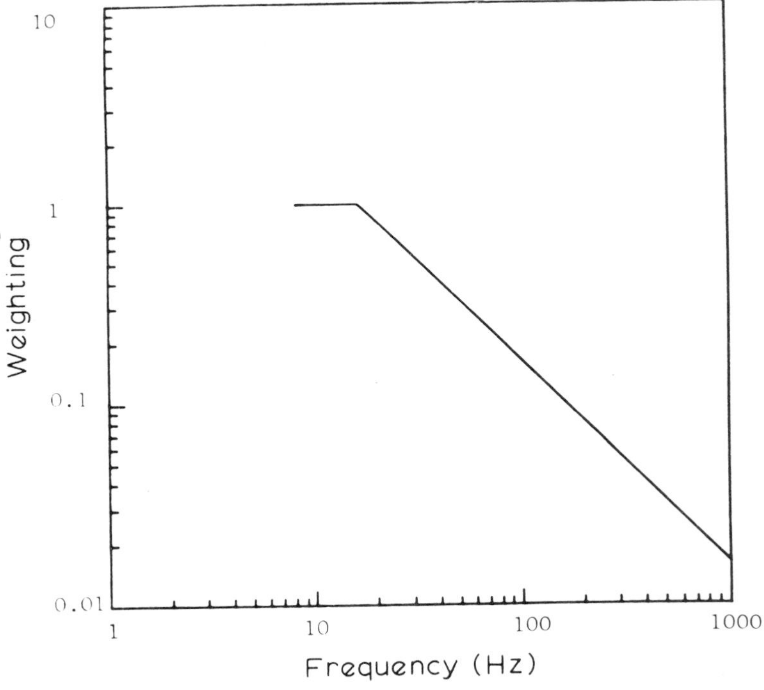

Fig. 12.4 Frequency weighting for assessing hand-arm vibration exposures.

may be argued that the weighting takes into account injuries other than VWF, but it is probably best to consider it merely as a unified procedure for measurement which is necessary for the gathering and comparison of data. In general, higher weighted values will be more likely to cause injury or disease but this may not always be the case.

Both the new British Standard and the new International Standard specify the use of the frequency weighting in Figure 12.4 and advocate that weighted rms acceleration should be reported. The characteristics of many hand-arm vibration exposures vary greatly during typical operations. For this reason a means of integrating the weighted magnitudes during a working day is defined. This is based on a squared (sometimes called energy) summation procedure in which vibration exposures are equivalent if the square of their weighted values multiplied by their exposure times are similar. This may alternatively be expressed as a 3 dB (i.e. 1.4142 to 1) change in allowable vibration magnitude for a doubling or halving of vibration duration.

The standards suggest that either the equivalent 4 or 8 hour vibration magnitude should be reported. Both the ISO and BSI Standards suggest that frequency-weighted magnitudes equivalent to 2.8 ms^{-2} rms in an 8-hour day (i.e. 4.0 ms^{-2} rms in a 4-hour period) may be expected to produce a 10% prevalence of VWF after regular daily use for about 8 years. This guidance is contained in annexes which are not parts of the standards, but, while the ISO Standard also predicts the prevalence for other exposures (Brammer, 1982) the BSI Standard concentrates upon specifying the considerable assumptions involved in any such prediction. For many purposes the 8-hour equivalent magnitude of 2.8 ms^{-2} rms may be an appropriate action level above which various preventative measures might be taken.

The reliable measurement of hand-arm vibration exposures is not always simple. On non-percussive tools which are held by handles for long periods (e.g. hand-held grinders and chain saws) it is usually possible to secure accelerometers near to the points of contact using either a jubilee clip, adhesive, or drilled and tapped screw fixing. Even with these tools the vibration magnitude can vary significantly according to the location of the measurement transducers (e.g. Eklund et al, 1984) in which case the highest magnitude in contact with the hand would normally be considered. With percussive tools (e.g. chipping hammers) there can be considerable measurement difficulties caused by the high peak acceleration. Where the hand holds the chisel of some percussive tools the acceleration can be sufficient to destroy currently available accelerometers. At other parts of such tools the peak acceleration can cause erroneous readings due to so-called dc shifts: many measurement systems will give an output which contains a large amount of spurious low frequency energy. There are several published papers which contain widely misleading values due to this phenomenon being unobserved, even though it is often apparent when inspecting the acceleration time history or frequency spectrum. (Any acceleration spectrum which rises with decreasing frequency below about 16 Hz may be treated with suspicion.) Measurements of vibration on percussive tools should always involve precautions against such problems and the use of a mechanical filter to isolate the transducer from the high frequency shocks is a solution widely adopted.

When the hand moves so as to contact various parts of the tool, or the job involves holding many pieces of work against a tool (e.g. pedestal grinding), it may be appropriate to mount the accelerometer to the body rather than on the tool. The principal of measuring the vibration at the interface with the tool should be maintained but the mounting device may be worn on the hand or finger (e.g. Rasmussen, 1982; Nelson, 1985).

Whatever method of measurement is adopted it is highly desirable to obtain several measurements so as to check repeatability. The careful inspection of the acceleration waveforms is desirable and the determination of spectra will provide useful information in addition to providing a check on the validity of the data.

DISCUSSION

Assessing the hazards of whole-body and hand-arm vibration can be complex: the quantification of vibration exposures is not always simple, the diagnosis of adverse effects can be difficult and the relation of adverse effects to the cause can be uncertain. Even so, it is now possible to identify by physical measurement those conditions which may reasonably be expected to carry some risk and therefore merit the consideration of preventative procedures. The range of individual variability in response to vibration is so great that the definition of a single limit is likely to be misleading. The action levels suggested above should provide a useful definition of those vibration conditions which require further consideration.

Many exposures to vibration may not be necessary. Exposures may sometimes be avoided or reduced with little cost or inconvenience: some consideration of the causes and suitable training on means of reducing exposure may be highly beneficial. Modifications to working methods can reduce or even eliminate the need for exposure. There are wide differences in the vibration magnitudes on different vehicles, machinery and tools so the selection of the optimum equipment may sometimes provide a beneficial reduction in exposure. However, some processes, such as those involving the holding of vibrating chisels, appear to present a fundamental problem not solved by minor redesign or simple changes in work method. Vibration reduction might be obtained with some isolation techniques but they may not always be compatible with the fine control of tool position. Alternative tools may sometimes have lower vibration magnitudes, but it may be necessary to eliminate the use of some tools or place limitations on the daily vibration exposure duration.

For some whole-body and hand-arm vibration conditions the knowledge that vibration may have adverse effects can be enough to stimulate consideration of reasonable means of reducing the vibration exposure. This may lead to a sufficient solution to the problem.

REFERENCES

Auffret R, Delahaye R P, Metges P J, Vicens 1978 Les algies vertebrales des pilotes d'helicopteres. AGARD Conference Proceedings CP 255, Paper 56

Brammer A J 1982 Threshold limit for hand-arm vibration exposure throughout the workday. In: Brammer A J, Taylor W (eds) Third International Symposium on Hand-Arm Vibration, Ottawa 1981 Vibration effects on the hand and arm in Industry. John Wiley, New York

British Standards Institution 1975 Draft for development: guide to the evaluation of exposure of the human hand-arm system to vibration. British Standard BSI DD43

British Standards Institution 1987a Guide to the evaluation of human exposure to whole-body mechanical vibration and repeated shock. British Standard 6841 (awaiting publication)

British Standards Institution 1987b Guide for the measurement and evaluation of human exposure to vibration transmitted to the hand. British Standard 6842 (awaiting publication)

Damkot D K, Pope M H, Lord J, Frymoyer J W 1984 The relationship between work history, work environment and low back pain in men. Spine 9: 395–399

Department of Health and Social Security 1985 The social security (industrial injuries) (prescribed diseases) amendment regulations 1985. Statutory Instruments 1985, 159. HMSO, London

Eklund L, Kilberg S, O'Connor D E 1984 Vibration levels along the support handle of a portable angle grinder. Arbete och Halsa, Vetenskaplig Skriftserie 1984: 42, Arbetarskyddsstyrelsen, Sweden

Griffin M J 1976 Subjective equivalence of sinusoidal and random whole-body vibration. The Journal of the Acoustical Society of America 60: 1140–1145

Griffin M J 1978 The evaluation of vehicle vibration and seats. Applied Ergonomics 9: 15–21
Griffin M J 1980 Vibration injuries of the hand and arm: their occurrence and the evolution of standards and limits. Health and Safety Executive Research Paper 9. HMSO, London
Griffin M J 1982a The effects of vibration on health. ISVR Memorandum 632, University of Southampton
Griffin M J 1982b Hand-arm vibration standards and dose-effect relationships. In: Brammer A J, Taylor W (eds) Third International Symposium on Hand-Arm Vibration, Ottawa 1981 Vibration effects on the hand and arm in industry. John Wiley, New York
Griffin M J 1984 Vibration dose values for whole-body vibration: some examples. United Kingdom informal Group Meeting on Human Response to Vibration, Heriot-Watt University, September 21–22
Griffin M J 1986 Evaluation of vibration with respect to human response. Paper presented at the Society of Automotive Engineers International Congress and Exposition, Detroit, 24 February. SAE Technical Paper Series 860047
Gruber G J 1976 Relationships between whole-body vibration morbidity patterns among interstate truck drivers. DHEW (NIOSH) Publication 77–167
Gruber G J, Ziperman H H 1974 Relationship between whole-body vibration and morbidity patterns among motor coach operators. HEW Publication (NIOSH) 751–104, Cincinnati, Ohio
Hayward R A, Griffin M J 1986 Measures of vibrotactile sensitivity in persons exposed to hand-arm vibration. Scandinavian Journal of Working Environmental Health 12: 423–427
Heide R, Seidel H 1978 Consequences of long-term occupational exposure to whole-body vibration (an abridged literature survey). Zeitschrift fur die gesamte Hygiene und ihre Grenzgebiete 24: 153–159
International Organization for Standardization 1974 Guide for the evaluation of human exposure to whole-body vibration. International Standard ISO 2631
International Organization for Standardization 1975 Acoustics — preferred frequencies for measurements. International Standard ISO 226
International Organization for Standardization 1978 Guide for the evaluation of human exposure to whole-body vibration. International Standard ISO 2631
International Organization for Standardization 1979 Principles for the measurement and the evaluation of human exposure to vibration transmitted to the hand. International Standard Draft proposal ISO/DIS 5349
International Organization for Standardization 1981 Mecahnical vibration and shock affecting man — vocabulary. International Standard ISO 5805
International Organization for Standardization 1982 Guide for the evaluation of human exposure to whole-body vibration. Amendment 1. International Standard ISO 2631-1978/A1-1982
International Organization for Standardization 1983 Acoustics — preferred reference quantities for acoustic levels. International Standard ISO 1683
International Organization for Standardization 1984 Guidelines for the measurement and the assessment of human exposure to hand-transmitted vibration. Draft International Standard ISO/DIS 5349.2
International Organization for Standardization 1985 Evaluation of human exposure to whole-body vibration — Part 1: General requirements. ISO 2631/1
International Organization for Standardization 1986a Human response vibration measuring instrumentation. ISO DIS 8041
International Organization for Standardization 1986b Guide for the measurement and the assessment of human exposure to vibration transmitted to the hand. ISO 5349
Juul C, Nielsen S L 1981 Locally induced digital vasospasm detected by delayed rewarming in Raynaud's phenomenon of occupational origin. British Journal of Industrial Medicine 38: 87–90
Kelsey J L, Githens P B, O'Connor T, Weil U, Calogera J A, Holford T R, White A A, Walter S D, Ostfeld A M, Southwick W O 1984 Acute prolapsed lumbar intervertebral disc. An epidemiologic study with special reference to driving automobiles and cigarette smoking. Spine 9: 608–613
Kelsey J L, Hardy R J 1975 Driving of motor vehicles as a risk factor for acute herniated lumbar intervertebral disc. American Journal of Epidemiology 102: 63–73
Kohl U 1975 Hazards of tractor driving — a survey in the Canton of Vaud. Arch Mal Prof 36: 145–162
Kjellberg A, Wikstrom B–O 1985 Whole-body vibration: exposure time and acute effects — a review. Ergonomics 28: 535–544
Lafferty K, Trafford J C de, Roberts V C, Cotton L T 1983 Raynaud's phenomenon and thermal entrainment: an objective test. British Medical Journal 286: 90–92
Lewis C H 1985 Frequency weighting procedures for the evaluation of human response to vibration. United Kingdom Informal Group Meeting on Human Response to Vibration, Railway Technical Centre, Derby, 17–19 September
Milby T H, Spear R C 1974 Relationship between whole-body vibration and morbidity patterns among heavy equipment operators. HEW Publication (NIOSH) 74–131, Cincinnati, Ohio
Nelson C M 1985 Evaluation of a prototype device for the triaxial mounting of accelerometers on the

fingers of users of vibrating tools. United Kingdom Informal Group Meeting on Human Response to Vibration, Railway Technical Centre, Derby, 17–19 September

Rasmussen G 1982 Measurement of vibration coupled to the hand-arm system. In: Brammer A J, Taylor W (ed) Third International Symposium on Hand-Arm Vibration, Ottawa 1981 Vibration effects on the hand and arm in industry. John Wiley, New York

Sandover J 1983 Dynamic loading as a possible source of low-back disorders. Spine 8

Seidel H, Heide R 1986 Long-term effects of whole-body vibration — a critical survey of literature (submitted for publication)

Society of Automotive Engineers 1974 Measurement of whole-body vibration of the seated operator of agricultural equipment (SAE recommended practice). SAE J10103, SAE Handbook Part II, p 1404–1407

Taylor W, Pelmear P L, Pearson J 1974 Raynaud's phenomenon in forestry chain saw operators. In: Taylor W (ed) The vibration syndrome. Academic Press, London

Taylor W, Brammer A J 1982 Vibration effects on the hand and arm in industry: introduction and review. In: Brammer A J, Taylow W (ed) Third International Symposium on Hand-Arm Vibration, Ottawa 1981 Vibration effects on the hand and arm in industry. John Wiley, New York

Welsh C L 1983 Digital rewarming timing, the assessment of VWF. United Kingdom Informal Group Meeting on Human Response to Vibration, National Institute of Agricultural Engineering, Silsoe, Bedfordshire, 14–16 September

Whitham E M, Griffin M J 1977 Measuring vibration on soft seats. Society of Automotive Engineers, SAE Paper 770253

SECTION 4

Socio-economic aspects

13. The role of the EEC in developing occupational safety and health measures

W. J. Hunter A. Berlin A. E. Bennett

INTRODUCTION

The regulation of industrial health and safety has played an important part in the work of the European Communities since the creation of the Coal and Steel Community in 1951, and the foundation of the European Economic Community and the European Atomic Energy Community in 1957. In relation to health and safety in coal mines a separate Mines, Safety and Health Commission was set up to study relevant problems (OJ C487, 1957) and, on the basis of articles 30–39 of the Euratom Treaty, standards of health and protection were established from the coming into force of that treaty (OJ C221, 1959). At first, however, the communities confined themselves to the encouragement of research, the promotion of exchanges of experience and the development of common guidelines in legislation. But the improvement of health and safety was regarded as so important by the signatories to the Treaty that it was included in the preamble of the Euratom and EEC Treaties and in Article 3 of the EEC Treaty whilst, at the same time, it was seen as an essential prerequisite to the achievement of economic integration.

The EEC Treaty, which is the principal legal basis for community action, contains two provisions which are inter alia, devoted to the promotion of health and safety, i.e. articles 117 and 118. Measures may also be taken in particular sectors such as agriculture or transport for the removal of barriers to trade on the legal bases appropriate to those sectors, i.e. articles 43 and 75. These measures may take into account health and safety considerations. In addition, because of the close connection of health and safety with the functioning of the common market the treaty's general legislative powers are used as a basis for action.

The two general bases for action in the treaty are articles 100 and 235. Article 100 refers to the ability of the Council

> 'to issue directives for the approximation of such provisions laid down by law, regulation or administrative action in member states as directly affect the establishment or functioning of the common market.'

In so doing the Council must act unanimously on a proposal from the Commission, and the Parliament and the Economic and Social Committee must be consulted in the case of directives whose implementation would, in one or more member states, involve the amendment of legislation. Two points should be noted. First, article 100 specifies the instrument to be employed, i.e. a directive, and, secondly, the powers of article 100 are available only where the lack of harmonisation of national measures directly affects the establishment or functioning of the common market. In some cases recourse may be had to article 235 which provides that

> 'if action by the Community should prove necessary to attain, in the course of the operation of the common market, one of the objectives of the Community and the EEC Treaty has not provided the necessary powers, the Council shall, acting unanimously on a proposal from the Commission and after consulting the Parliament, take the appropriate measures.'

The legal instrument used need not, in this case, be a directive and this means that other legal instruments may be adopted where the intention is to create new policy.

A new dimension has now been added to the work in health and safety, as a result of the revision of the EEC Treaty. This revision included a substantial addition to Article 118 of this Treaty so that directives in matters of safety and health at work can now be adopted by majority vote. Member states may also enforce stricter national legislation than that existing at Community level.

COMMUNITY LEGAL INSTRUMENTS

The community legislator has a wide choice of instruments which he may use. These comprise regulations, directives and decisions, and other measures such as recommendations and opinions.

Regulations

Article 189 of the EEC Treaty provides that 'a regulation shall have general application. It shall be binding in its entirety and directly applicable in all member states'. Regulations must be recognised as legal instruments and do not need national implementation — indeed it is not permissible to do so (Commission v Haly, Case 39/72 1973 ECR 101). Regulations are capable of creating individual rights which the Courts must protect (Politi v Italian Ministry of Finance, Case 43/71 1971 ECR 1039).

Directives and decisions

Article 189 of the EEC Treaty provides that

> 'A directive shall be binding, as to the result to be achieved, upon each member state to which it is addressed, but shall leave to the national authorities the choice of form and methods. A decision shall be binding in its entirety upon those to whom it is addressed.'

The Court of Justice has held that in certain circumstances, at least, directives and decisions might contain directly effective provisions (Grad v Finanzamt Transtein, Case 9/70 1970 ECR 825. Van Duyn v Home Office, Case 41/74 1974 ECR 1337. Rutili v French Minister of the Interior, Case 36/75 ECR 1219). For this to occur two conditions must be fulfilled. First there must be a clear and precise obligation and secondly it must not require the intervention of any act on the part of the institutions of the Community of the member states (EEC Directives and the Law GL Close Vol 2 Planning Law for Industry International Bar Association Cambridge 1981). Individuals may invoke before their national courts, as against the State, such provisions despite the fact that they are contained in directives or decisions rather than regulations. It is also arguable that in some circumstances, at least, obligations imposed by directives on member states can create rights in individuals against other individuals

(Case 57/76 1877 ECR 113; Case 43/75 1976 ECR 455, see Josephine Steiner, Direct applicability — a chameleon concept, Vol 98 LQR April 1982, p 229–248 for an analysis and review of direct application and direct consequences).

Other measures

Recommendations and opinions have no binding force but may be indicative of important policy orientations and member states may take such measures into account when enacting national legislation or when making administrative provisions. Similarly resolutions and declarations by the Council do not produce legal results but contain important statements of policy which serve as a guide to Community institutions and member states.

THE DEVELOPMENT OF A PROPOSED COUNCIL DIRECTIVE

It is both a complex and a time consuming task to develop a piece of Community legislation. The identification of the need to put forward such a piece of legislation can come from a variety of sources.

The scientific and technical background for this work requires considerable preparation, often by means of a specially commissioned study, aimed at establishing all the facts necessary on which to base the legislation. This study will usually include existing legislation in the member states.

The results of this study will therefore be used as a background document by the Commission staff for the preparation of a working document, which is usually the first of many documents which require translation into all the languages of the member states. This working document sets out the various issues with proposed solutions. As regards health and safety measures this working document is submitted for opinion to the Advisory Committee on Safety, Hygiene and Health Protection at Work. In developing its opinion the Committee may produce several draft opinions, leading to a final opinion.

Following the delivery of this opinion, the Commission services then develop the first draft of a proposal, using both the opinion as well as other elements from the working document. In a Directive, this will usually follow a fairly standard presentation (Table 13.1) based on texts which have been adopted to date in this field. It is important to note that wherever possible, existing texts of already adopted Directives will be used as far as possible since they are the result of long discussions often resulting in a delicate compromise at the Council of Ministers. There is thus already in existence a solid base of existing legislation with equivalent texts in the various Community languages.

This new draft is usually discussed at meetings of Government experts who are not selected by the Commission but are nominated by the Member States as a result of an invitation from the Commission which is sent to each country's Permanent Representative. The minutes of these meetings are intended to reflect the discussions and the changes which the national experts would like to see made to the Commission's text. As a result of these meetings, a modified text, as proposed by the Government experts, is prepared.

What may appear to be unnecessary changes are often the result of several hours discussion on the significance of a single word. An example of this arose recently

Table 13.1 Structure of individual Directives for the protection of workers exposed to chemical agents at work

Structure	Comments
1. Aims or objectives	General statement
2. Definitions	Terms, quantities, units
3. Scope	Coverage and exemptions
4. Specific conditions	Limitations or prohibitions of certain work practices may be necessary
5. Reporting provisions	Certain operations may require notification or authorisation
6. General control principles	Main approaches to limit exposure
7. Limitation of dose in special cases	Critical groups of workers may need to be considered
8. Assessment of exposure	Sampling and analytical strategy
9. Action levels	Ambient, and if applicable, biological values which trigger the application of certain measures
10. Exposure limits	Ambient, and if applicable, biological values which are not to be exceeded
11. Planned special exposures	For certain operations special precautions may be required
12. Exceeding the limit values	Provisions to be taken in such cases
13. Individual protection	Requirements and conditions
14. Personal hygiene	Facilities and requirements
15. Medical examinations	Frequency and guidelines
16. Records	Requirements for exposure and health records
17. Information	Information of workers regarding dangers and precautions to be taken
18. Statistics	Data for monitoring effect of Directive
19. Application	Dates when operational

as a result of the translation from English of the word 'assessment' which had been translated into French as 'evaluation'. After much discussion it became evident that evaluation involved the concept of measurement, which was not the case for the French word 'appreciation' which was finally selected. Once such discussions have been completed, it is then possible to draw up a more definitive text. At this point in time, decisions have to be made about the final form of the proposal, and to what extent it should reflect the various opinions.

Before that text can be finalised, it is necessary to obtain the opinions or approval of other interested Directorates General in the Commission. This may also result in some changes having to be made before the text is finalised, and the draft proposal can then be formally submitted to the Commission for approval.

Once the text of a Council proposal is approved by the Commission, it is sent to the Council of Ministers. If it is a health and safety proposal based on Article 100 of the EEC Treaty it is sent for opinion to the European Parliament and to the Economic and Social Committee. As a result of the opinions delivered by these two bodies, it may be necessary to draw up a modified proposal based on Article 149 of the EEC Treaty. This normally represents the final stage of this particular procedure as far as the Commission services are concerned. However, representatives of the Commission will take part in the discussions at the Council of Ministers which may well last for several years before agreement is reached on a particular proposal.

The Commission also has to ensure the correct transposition of the proposal by member states into national laws, regulations or administrative practices which is a necessary part of this procedure, and which may also require some years of work. The fact that the Commission bases its proposal on existing legislation in the member states means that on some occasions a particular member state may have to modify its legislation substantially, and on other occasions it may have to carry out only a minor modification.

The long delays involved in adopting Directives have often resulted from the requirement to obtain unanimous Council decisions on detailed technical specifications. Henceforth, in those sectors where barriers to trade are created by justified divergent national regulations concerning the health and safety of citizens, and consumer and environmental protection, the process of legislative harminisation will be confined to laying down the essential requirements, such that conformity with them will entitle a product to free movement within the Community. The task of defining the technical specifications of products so that they conform to the legislative requirements, will be entrusted to European Standards issued by the Comité Européen de la Normalisation (CEN) or by sectoral European Standards in the electrical and building sectors such as CENELEC, UEAtc or RILEM.

COMMUNITY ACTION PROGRAMMES

The Community (the three Communities were merged by the Merger or Fusion Treaty, 1 July, 1967) was annually recording by 1974, nearly 100 000 deaths and 12 000 000 injuries arising from accidents of all types for a workforce of some 104 000 000. Industrial accidents including occupational diseases, although not the major sector of risk as far as fatal accidents were concerned, represented the largest group of accidents taken as a whole and therefore constituted a priority area for Community concern. The Council of Ministers decided to strengthen the activities in health and safety at work by establishing in 1974 a committee entitled the Advisory Committee on Safety, Hygiene and Health Protection at Work, with a tripartite composition: 2 members from government, 2 from employers and 2 from workers; making a total of 6 members from each Member State. This Committee now consists of 72 members. The principal task of this Committee is to assist the Commission in the preparation and implementation of activities in the fields of safety, hygiene and health protection at work. To help it achieve this objective, the Committee has set up ad-hoc working groups on specific topics, with the aim of preparing draft opinions for the Plenary Committee. These opinions and the work of the Committee in general are regularly set out in their annual progress reports. From the moment of its creation this Committee has contributed substantially to the work of the Commission in this area.

One of its first tasks was to give consideration to initiatives that could be taken by the Commission in the field of health and safety at work. Initially the Committee limited itself to developing specific opinions, but it became evident that a more general approach was required. Thus the concept was developed of a programme of action on safety and health at work, and as a result the Commission's proposal for such a programme contained a total of six main initiatives covering the areas of safety, hygiene and health protection at work. It was conceived as a long term programme which would require many years for its achievement.

The Council of Ministers, while welcoming this proposal, considered that the programme had to be limited to certain priority actions which could be completed within a specified time period. As a result the Council in 1978 took a major step forward in furthering the protection of workers by adopting the first action programme of the European Communities on safety and health at work (OJ C165, 11 July 1978, p 1) which contained 14 actions and which was to run until the end of 1982. Of these actions six were dedicated to the protection of workers from the harmful effects of chemicals, and based on these several important initiatives have been taken at Community level. This programme thus set out the basis for a common policy aimed at increasing the protection of more than 100 million workers' health and safety throughout the member states.

In 1984 the Council adopted a second programme (OJ C67, 8 March 1984, p 2) which builds on and extends the actions provided for in the Community's first programme. It provides for a continuation of the work already in progress and for certain new actions to 'reflect the changed needs and concerns of today's society'. This programme will finish at the end of 1988, and to implement it the Commission prepares annually, after consulting the Advisory Committee and the Member States, a forward outline of work.

The programme (Annex I, see end of chapter) is divided into seven sections:

1. protection against dangerous substances;
2. ergonomic measures, protection against accidents and danterous situations;
3. organisation;
4. training and information;
5. statistics;
6. research;
7. co-operation.

The second programme is well and truly launched and the work performed in 1985 is illustrative of the progress being made.

Section I Protection against dangerous substances, actions 1–6

(a) A proposal for a Directive on the protection of workers from the risks related to exposure to benzene was submitted to Council in November 1985. Working documents were prepared for both acrylonitrile and carbon tetrachloride. These were submitted to working parties of the Advisory Committee for Safety, Hygiene and Health Protection at Work, and the opinions were adopted at its Plenary meeting on 28–29 November, 1985. This meeting also adopted an opinion on electromagnetic fields (microwaves).

(b) A meeting with Government experts was held on 5–6 March 1985 to discuss the implementation of the asbestos Directive, and the available scientific data, and an implementation meeting on the lead Directive was held on 9–10 July 1985.

(c) A study was completed on the asbestos fibre counting rules, and a review begun on the available scientific data for the different types of asbestos. A questionnaire on cadmium use and control measures was prepared for publication. In collaboration with EPA, Health and Safety Ontario, and INCO a workshop was held in January 1985 to review the available epidemiological data on nickel. A study was completed on ten of the most important organo-chlorine solvents used at the workplace.

(d) Discussions continued at Council on the proposed Directive on the protection of workers from noise. Studies began on the health effects of vibration.

(e) A report was published on human biological monitoring of industrial chemicals, i.e. acrylonitrile, aluminium, chromium, copper, styrene, xylene and zinc.

(f) The text of a proposal for a Council Directive on exposure limits was drafted.

(g) Discussions began on the theoretical and practical aspects of the measurement of exposure to dangerous agents at work, and a list was made of ISO standards already adopted or in preparation. Planning meetings were held for a workshop and a conference on training in industrial hygiene in 1986, and a workshop on passive sampling, also in 1986.

(h) A further review of chemical agents and skin absorption began.

(i) A workshop was held on 4–5 and 6 December 1985 on the chemical and biological risks in hospitals and other health care services. A report on the containment measures for pathogenic organisms was made, and a report completed on the health hazards for workers exposed to chemical and biological agents in health care establishments.

Section II Ergonomic measures, protection against accidents and dangerous situations, actions 7–11

(a) Discussions began in the Advisory Committee on safety, hygiene and health protection at work on the white Paper 'Completing the internal market' and its likely impact on safety at work.

(b) Two meeting of Chief Inspectors were held, and exchanges of inspectors were initiated between member states.

(c) A panel of experts on 'Ergonomics and Health Issues in New Technology' was established and two meetings were held to examine the guidelines which have been prepared by various organisations concerning the ergonomics of visual display units at the workplace with a view to developing a harmonised Community position. The human factors research in the field of ergonomics, and human reliability in process control operations was continued (some 21 projects participating in the Action) within the COST Programme. The validation of ISO standards in thermal stress was drawing to a close with indications of proposals for some significant amendments to the standards.

(d) An ad-hoc Committee on lumbar hazards examined the available data with a view to establishing guidelines for the prevention of back injuries. A training module on improving spinal movements and postures for agricultural workers was made available to Member States.

Section III Organisation, actions 12–14

(a) Discussions began with the Advisory Committee on safety, hygiene and health protection at work on the updating of the Commission's Recommendations of 1962 and 1966 on occupational diseases, taking into account the list agreed by ILO.

(b) A review was made of current practices of participation by workers and their representatives in matters of health and hygiene at the workplace.

(c) A report and recommendations on training in occupational medicine was made by the Advisory Committee on Medical Training.

Section IV Training and information, actions 15–16

(a) A review was made of the current practices for the provision of information to employers and workers exposed to dangerous agents at work, and an examination begun of the practical means available for the provision of health and safety information to workers.

(b) Draft information notices to workers on dangerous agents were prepared for thirteen major accident chemicals.

(c) *Extractive industries*. An International Symposium was held on Safety in Diving Operations, and an information meeting was held with representatives of workers organisations. Within the framework of the Safety and Health Commission for the Mining and Other Extractive Industries work continued on the establishment of proposals to governments for the improvement of safety and health in the oil and gas industry, and planning meetings were held for an International Symposium in 1986 on health and safety in the non-coal mining and quarrying industry.

Agriculture. A working document on the minimum safety requirements required for the use of plant protection products was submitted to the Joint Committee on Social Problems of Agricultural Workers in November 1985. A training module on the safe handling of animals was made available to member states, as was a module on the safe use of viticultural chemicals. Work on the safer design of agricultural buildings was initiated.

Fishing. Two training modules on sea fishing (small and medium scale) were completed. A proposal for a Council decision for the ratification of the Torremolinos Convention on the Safety of Fishing Vessels continued in discussion at Council.

(d) A review was made of the safety training requirements for employers, worker's representatives and safety specialists in industry. A seminar was in preparation on the special training in safety required for worker's representatives.

Section V Statistics, actions 17–18

In collaboration with the Statistical Office, a workshop was held on 2 and 3 December 1985 on the development of Community based occupational health statistics, and discussions were held with a view to harmonising the data available in member states on fatal accidents. A questionnaire was established for reviewing current practices of cancer registration.

Section VI Research, action 19

A report was in preparation in order to establish an inventory of occupational health research currently underway.

Section VII Co-operation, actions 20–21

(a) Collaboration was effected with member states at the 71st Session of the ILO on occupational health services, and safety in the use of asbestos.

(b) Meetings on neuro-toxicity, welding fumes, and hexachlorobenzene were co-organised with other organisations, including WHO, IARC and EPA. A course on occupational cancer was held in Luxembourg from 6–10 May 1985 in collaboration with IARC and the European School of Oncology. Representatives attended technical and programme meetings of WHO, IPCS, and IARC.

(c) Contact was maintained with EPA, OSHA, NIOSH, and the Swedish National Board of Occupational Safety and Health. An exchange of information took place with the Norwegian authorities.

(d) In collaboration with the Italian authorities a conference on the scientific and regulatory aspects of cancer was held in Rome on 13 and 14 June 1985.

PROTECTION AGAINST DANGEROUS AGENTS

The first programme of action placed more emphasis on protection against dangerous agents than on safety, a situation that has been redressed to some extent by the second programme. At the time of adoption of the first programme it was considered important to develop overall policies for dealing with dangerous agents as there was concern that the existing policies were considered inadequate. This was in contrast to the safety area where much work had already been performed, albeit within the context of the removal of technical barriers to trade.

That this concern about dangerous substances has not been misplaced has been confirmed by the growing awareness of the general population and more so of the workers, of the risks to their health from certain chemical, physical and biological agents. This is particularly the case for those agents which give rise to long-term effects, such as carcinogens. For the control of dangerous agents, the Commission has chosen to make in most cases Proposals for Council Directives. As has been stated earlier, once Council Directives are adopted they have to be incorporated into the laws, regulations or administrative provisions of the member states.

One of these chemicals that gave rise to concern throughout the Community in the 1970s was that of vinyl chloride monomer, and this led to the Commission's subsequent proposal. It also became evident that an overall policy was necessary for the control of dangerous agents, and this led subsequently to the development of the Framework Directive 80/1107/EEC covering all chemical, physical and biological agents at work.

Vinyl chloride monomer

The Commission acted rapidly as a result of public concern following the publication of studies regarding vinyl chloride monomer and angiosarcomas. In 1975 a first scientific and technical meeting was convened in Brussels during which technical limits were considered. During the preparation of the Community proposal for a Directive, further evidence accumulated on the carcinogenic effect of vinyl chloride monomer on man while at the same time considerable efforts were being made at the technical level to reduce worker exposure. The resulting directive which was adopted in June 1978 (OJ L197, 22 July 1978, p 12) and which has now been incorporated into member states' legislation is the first to deal with the control of worker exposure to a chemical carcinogen. It covers all workers employed in works in which vinyl chloride monomer is produced, reclaimed, discharged into containers, transported or used in any way whatsoever, or in which vinyl chloride monomer is converted into vinyl chloride polymers, and it contains the following main provisions:

establishment of atmospheric limit values and the fixing of provisions for monitoring;
technical preventive measures;
information of workers;
keeping a register of exposed workers;
guidelines regarding medical surveillance.

A technical long-term (1 year) limit value has been established of 3 parts/10^6 with equivalent limit values for shorter periods of time (Table 13.2).

Table 13.2 Limit values in parts per million of VCM in relation to reference periods

Reference period	Limit value (Parts/10^6)
1 year	3
1 month	5
1 week	6
8 hours	7
1 hour	8

In general both continuous and discontinuous methods may be used for monitoring the vinyl chloride monomer concentration in a working area, so long as the measurement systems have a sensitivity of at least 1 part/10^6. In the case of vinyl chloride polymerisation plants only continuous monitoring is acceptable. Personal protection measures have to be taken in case of abnormal increases in vinyl chloride monomer concentration levels, and also for certain operations such as the cleaning of autoclaves.

A register of exposed workers has to be kept for at least 30 years and has to contain particulars of the type and duration of work and exposure to which they have been subjected. Medical examinations are required but the frequency and type of medical examinations have to be determined by each member state. However, guidelines for the medical examinations are provided in the Directive as are the basic requirements for examinations prior to exposure.

'Framework' Directive 80/1107/EEC

An essential step in the implementation of the first programme was the adoption by Council on 27 November 1980 (OJ L327, 3 December 1980, p 8) of Directive 80/1107/EEC on the protection of workers from the risks related to exposure to chemical, physical and biological agents at work. The objectives of this directive are two fold:

elimination or limitation of exposure to chemical, physical and biological agents and the prevention of risks to workers' health and safety;
protection of workers who are likely to be exposed to these agents;

To attain these objectives it requires member states to take short and longer term measures.

The short-term measures require that appropriate information be given to workers and/or their representatives concerning the health risks due to arsenic, asbestos, cadmium, lead and mercury, and appropriate surveillance be set up of the health of workers exposed to asbestos and lead. Both of these measures have been subsumed to some extent in the individual Directives on lead and asbestos.

The longer term measures apply when a member state adopts provisions concerning an agent. In order that the exposure of workers to agents is avoided or kept at as low a level as is reasonably practicable, member states should comply with a set of requirements, but in so doing they have to determine whether and to what extent each of these requirements is applicable to the agents concerned. Some of the most important of these requirements are:

limitation of use at the place of work;
limitation of the number of workers exposed;
prevention by engineering control;
establishment of limit values and of sampling and measuring procedures, and methods for evaluating results;
collective and individual protection measures, where exposure to agents cannot be avoided by the other means, as well as hygiene measures;
emergency procedures for abnormal exposures;
information for workers;
surveillance of the workers' health.

There are additional requirements which are necessary for a total of eleven substances (Table 13.3).

Table 13.3 List of agents

Acrylonitrile
Asbestos
Arsenic and compounds
Benzene
Cadmium and compounds
Mercury and compounds
Nickel and compounds
Lead and compounds
Chlorinated hydrocarbon compounds:
chloroform
paradichlorobenzene
carbon tetrachloride

These requirements are:

provision of medical surveillance of workers by a doctor prior to exposure and thereafter at regular intervals;
access by workers and/or their representatives at the workplace to the results of ambient and biological (collective) exposure measurements;
access by each worker concerned to the results of his own biological tests indicating exposure;
information of workers and/or their representatives at the work-place of cases where the limit values are exceeded, of the causes thereof and of the measures taken to rectify the situation;
access by workers and/or their representatives to appropriate information to improve their knowledge of the dangers to which they are exposed.

Member states are also required to consult the social partners when the above requirements are being established.

This Directive has been implemented by the majority of member states, but there is some variation between countries in the way it has been implemented. Nevertheless, it represents a considerable step forward in the overall philosophy necessary for the control of workers' exposure to these agents.

Individual Directives

In the 'Framework' Directive, 11 agents (Table 13.3) are named for early action by means of individual Directives laying down limit values and other specific requirements. The Council has already adopted Commission proposals on certain of them, namely lead (OJ L247, 23 August 1982, p 12) asbestos (OJ 263, 24 September 1983, p 25) and recently on noise (OJ L137, 24 May 1986, p 28). In addition the Commission has made proposals for Council Directives on the proscription of specified agents and/or work activities (OJ C270, 10 October 1984, p 3), and also on benzene (OJ C349, 31 December 1985, p 32).

Lead

This substance is of importance because the number of workers exposed is very high (1 000 000 in the European Community) because it is used in a considerable number of industries and because of its toxic effects which are very well known. At national level, different points of view existed in 1980 on the measures to be taken, in particular regarding the protection of some categories of workers (Table 13.4).

Table 13.4 Special provisions regarding exposure of workers to lead in the member states of the European Community (1980)

	B	DK	D	F	IRL	I	LUX	NL	UK
Young workers		R		R(18)	R		R + P	R	
Young males below 16			P(16) all			P			
Young females below 18								P	
Women	R				R + P		R		
Women of child-bearing capacity									R (PbB40)
Women below 45 years			P (PbB40)						
Pregnant women	P								P

P = prohibition; R = restriction; PbB40 = Blood lead concentration of 40 μg Pb/100 ml blood.

The lead Directive was the first individual Directive to be based on the Framework Directive and it requires that all work presenting a risk of absorption of lead be assessed to determine the nature and degree of workers' exposure. Certain actions may be required as a result of the assessments made of the lead in air and the blood lead concentrations, as set out in Table 13.5).

Regular lead in air monitoring must be representative of workers' exposure and a number of technical details regarding the sampling procedure are also specified. Biological monitoring, essentially blood lead measurements, is to be performed on all workers every 6 months and clinical surveillance carried out at least once per year. There are several hygiene measures which are required, including the limitation of the risk of absorption of lead through smoking, drinking and eating; the provision of special work clothes; and, of adequate washing facilities.

There are requirements to inform workers of the dangers to health due to lead exposure, of the need for appropriate protective measures, of the monitoring results (individual lead in air values and biological group values), as well as the need to

Table 13.5 Lead in air and blood lead concentrations in relation to actions to be taken

Lead in air concentrations in μg/m³ (40 hours per week time-weighted average)	Blood lead concentrations in μgPb/100 ml blood	Actions to be taken
40	40	Minimise lead absorption of workers Provide information to workers
	40–50	Regular biological monitoring considered appropriate
75	50	All the protective measures of the Directive apply including: lead in air monitoring medical surveillance
150	70	Limit values requiring action to reduce exposure
	70–80	Limit value may be acceptable if other biological indicators below certain limits e.g. ALAU 20 μg/g creatinine

consult workers on lead in air monitoring procedures and on the measures to be taken when the lead in air limit values are exceeded.

Access to information must also be given to the doctor or authority responsible for the medical surveillance of workers in order to evaluate the exposure of workers to lead. Member states may introduce more stringent protective measures in their national legislation either for all workers or for certain categories of workers, and this enables special account to be taken of the potential risk to the embryo and to the development of the foetus which may result from a high body burden of lead.

Asbestos

The importance of the health risk due to asbestos led to the Commission's proposal for a second individual Directive based on the 'Framework' Directive. Asbestos, in addition to asbestosis, is associated with cancer of the lung, mesotheliomas and cancers of the gastro-intestinal system. Within the European Community the number of deaths from mesothelioma exceed 700 per annumn, and recent figures indicate that the numbers are going up, probably as a result of high exposures 20–30 years ago.

One of the most important means of controlling exposure is by the establishment of limit values, and these are set at 0.5 fibres/ml (5×10^5 fibres/m³) for crocidolite and 1 fibre/ml (1×10^6 fibres/m³) for non-crocidolite fibres measured or calculated over a reference period of 8 hours. A fibre is defined as having a length greater then 5 μm, with a length/breadth ratio greater than 3. The Directive includes a reference method for measuring asbestos, including a sampling and analysis strategy, and the use of the membrane filter method with counting of fibres using optical microscopy.

Other provisions to control exposure include:

the register of working sites or plants;
the prohibition of spraying of asbestos;
the precautions to be taken in working with asbestos containing materials;
the conditions under which asbestos can be handled and transported;

procedures for demolition and dismantling work, where this involves asbestos or asbestos containing materials;
information to workers so that they are properly instructed on correct handling procedures, informed of potential risks, and supplied with appropriate protective clothing and safety devices;
the cleaning and maintenance of protective clothing and equipment;
the requirements for regular medical examinations and the keeping of adequate medical records.

Member states have until 1987 to implement this Directive (until 1990 for mining activities), but there are already discussions taking place on a possible revision of the limit values.

Noise

More workers are exposed to noise and affected by it than by any other agent. It has been estimated that 10 to 15 000 000 workers in the Community are exposed to mean levels of noise which exceed 85 decibels, and of these between 6 000 000 and 8 000 000 are exposed to mean noise levels of over 90 decibels.

There are few activities without noise, which is a particular problem of many of the traditional industries. The control of noise is not only a technically difficult matter, but it is also of economic importance. The adoption by Council of a Directive aimed particularly at preventing hearing loss due to noise at work represents a considerable step forward in the control of this particular hazard. The general principle of the Directive is that the risks resulting from exposure to noise must be reduced to the lowest level compatible with technical possibilities and economic constraints. Preventive measures such as information on the risks. the provision of ear defenders, and the right of workers to have their hearing checked come into force when the mean noise level based on a normal period of eight hours exceeds 85 decibels. If this level exceeds 90 decibels these measures have to be strengthened. This Directive will be reexamined within four years with a view to reducing further the risks arising from exposure to noise.

The implementation of this Directive will not only have beneficial effects for workers by preventing their hearing loss, but it will also have beneficial effects for society as a whole since in our industrialised countries, noise induced hearing loss is one of the most important of the recognized occupational diseases.

Proposal on proscriptions

With the development of industrial society there has been increased demand and need for health protection measures in the workplace in particular as concerns carcinogens and other dangerous agents. As a result a variety of measures has been successfully undertaken at national, Community and international level. Some member states such as the UK have general or limited bans contained in their legislation. Other member states have chosen to strictly control exposure, without introducing bans, except for certain uses or work activities.

The Commission favours a ban on certain dangerous agents, and to this end has submitted to Council a Proposal for a Directive on the protection of workers by the proscription of specified agents and/or work activities [Fourth individual Directive within the meaning of Article 8 of Directive 80/1107/EEC]. It is based on the second programme of action of the European Communities on safety and health at work which contains an action to 'Develop preventive and protective measures for substances recognised as being carcinogenic and other dangerous substances and processes which may have serious harmful effects on health', and also on the 'Framework' Directive which foresees if necessary, a general or a limited ban on an agent.

Its objective is to increase the protection of workers' health by means of a general or limited ban on certain named dangerous agents and/or work processes, and to achieve this objective it requires member states to ban the agents listed in its Annex. However, the proposal does allow member states to grant certain exemptions, subject to the conditions laid down in the Annex. In such cases it requires that workers are suitably informed.

The agents that are dealt with are 2-naphthylamine and its salts, 4-aminodiphenyl and its salts, 4-nitrodiphenyl and benzidine and its salts. All these agents are dangerous and considered carcinogenic by the International Agency for Research on Cancer (IARC). Of these agents, commercially the most important is 2-naphthylamine due to its presence in 1-naphthylamine — the latter being used as a chemical intermediate in the preparation of a large number of compounds, including dyes, pesticides and antioxidants.

Other individual proposals

As been mentioned earlier, the 'Framework' Directive contains a list of agents in Annex II for the establishment of individual Directives. Of these agents, Directives on lead and asbestos have already been adopted. Of the 9 remaining agents, it was decided to give consideration to benzene first and then to other carcinogens.

As a result a proposal for an individual Directive on benzene was submitted to Council at the end of 1985. The general structure of the proposal follows that described in Table 13.1, and the specific control measures are based on those already adopted by Council for vinyl chloride monomer and for asbestos. In so doing the Commission has proposed a limit value of 16.25 mg/m^3 (5 parts/10^6) time weighted over a period of 8 hours. A proposal for an individual Directive on other carcinogens is already being drafted along the same lines, and it is anticipated that it will be sent to Council in 1987.

The Commission proposed to Council in 1986 the first modification of the 'Framework' Directive incorporating a list of 100 occupational exposure limit values.

A working document on carbon tetrachloride was discussed with the Advisory Committee on Safety Hygiene and Health protection at work, and it was decided not to draw up a proposal for an individual Directive, but to give consideration to this agent in the context of a Directive to control carcinogens

Safety signs

The Directive of Safety Signs at places of work (OJ L229, 7 September 1977, p 12) which relates to the provision of such signs, was adopted in 1977 even before the first programme of action on safety and health was adopted. It sets out a system of safety signs which has as its objective to draw attention rapidly and unambiguously to objects and situations capable of causing specific hazards; under no circumstances is this system to be considered as a substitute for the requisite protection measures.

The system may be used only to give information related to safety, and its effectiveness is dependent in particular on the provision of full and constantly repeated information to all workers likely to benefit therefrom.

Safety signs are devided into four groups, namely:

1. prohibition signs;
2. warning signs;
3. mandatory signs;
4. emergency signs.

The Directive also give details on the:

safety colours and contrasting colours;
geometrical form and meaning of safety signs;
combinations of shapes and colours and their meanings;
design of safety signs;
yellow/black danger identification.

Examples of these signs are given in Annex II (see end of chapter).

PREVENTION OF MAJOR CHEMICAL ACCIDENTS

In the 1970s a significant number of major chemical accidents occurred both within the European Community and in other countries affecting both workers and the general population. However, the potential long term consequences of accidents such as those which took place at Seveso and Manfredonia in 1976 added a new dimension and awareness to the overall problem.

The Commission's proposal on major accident hazards of certain industrial activities, aimed at preventing such accidents and reducing consequences if they occur was subsequently adopted in June 1982 by the Council (OJ L230, 5 August 1983, p 1). Due to its importance for the workers. the general population and the environment, the Directive is based on both the Action Programme on Safety and Health at Work and on the Environmental Action Programme, (OJ C112, 20 December 1973, p 1; OJ C139, 13 June 1977, p 1). A major accident is defined as a major emission, fire or explosion involving one or more dangerous substances, resulting from the uncontrolled development of an industrial activity, which could constitute a serious risk, immediate or delayed, for workers, the neighbouring population and the environment. The Directive requires that for industrial activities which involve dangerous substances:

a safety report be drawn up;
workers be informed, equipped and trained;
safety drills be organised;
the neighbouring population be informed and an emergency plan established.

In particular, for the activities subject to notification, there is an obligation on the manufacturer to examine and provide information on the substances and their behaviour under abnormal conditions, on the installations, technical processes and the number of persons exposed to risk, and on possible major-accidents. This will certainly improve the safety of the workers who are usually the first exposed to any hazard.

Table 13.6 gives examples of such substances. The notification is compulsory for 178 substances or classes of substances. For these it is envisaged to establish, in accordance with the requirements of the Action Programme on Safety and Health at Work and the 'Framework' Directive, information notices on the risks for workers. Discussions are continuing with Member States on the full implementation of this Directive, and a proposal modifying this Directive has recently been sent to Council.

Table 13.6 Examples of substances requiring notification

Substance	Quantity
2-Acetylaminofluorene	1 kg
Benzidine	1 kg
Beryllium oxide	1 kg
Arsenic (solid compounds)	500 kg
Phosgene	20 t
Chlorine	100 t
Carbon disulphide	200 t
Ammonia	1.000 t
Hydrogen	20 t
Ammonium nitrate	5.000 t
Liquid oxygen	10.000 t

CLASSIFICATION AND LABELLING OF CHEMICALS AND PREPARATIONS

A Council Directive, adopted originally in 1967, and amended several times since then deals with the classification, packaging and labelling of dangerous substances such as those that are explosive, flammable, corrosive, or irritant. It imposes an obligation on the member states of the Community to classify dangerous substances according to the nature of the hazard and to ensure that they are not put on the market unless they are packed and labelled according to the provisions of the Directive. In June 1979 the Council adopted the sixth modification of the 1967 Directive (OJ L259, 15 October 1979, p 10). This sixth modification, requires new chemical substances to be tested and notified to the competent authority in the member state, at least 45 days before being placed on the market. The Commission receives a copy of the notification dossier of a summary and keeps a list of all substances notified.

An inventory of existing chemicals in the European Community, based on the declarations made by manufacturers and importers and containing about 95 000 entries will be published in 1986. A compendium of testing methods, based on the OECD guidelines has been published in the Official Journal of the European Communities (OJ L251, 19 September 1984, p 1). The need to assist member states in providing uniform criteria for the classification and labelling of dangerous substances in terms of both risk phrases and safety advice led to the development of a Labelling Guide which was published as a Commission Directive in 1983 (Classification and labelling of dangerous substances, Doc XI/316/84).

Thus, for example, carcinogenic substances for the purpose of classification and labelling, and having regard to the current state of knowledge, are divided into categories.

Category 1
Substances known to be carcinogenic to man. There is sufficient evidence to establish a causal association between human exposure to a substance and the development of cancer.

Category 2
Substances which should be regarded as if they are carcinogenic to man. There is sufficient evidence to provide a strong presumption that human exposure to a substance may result in the development of cancer, generally on the basis of: appropriate long-term animal studies and other relevant information.

Category 3
Substances which cause concern for man owing to possible carcinogenic effects but in respect of which the available information is not adequate for making a satisfactory assessment. There is some evidence from appropriate animal studies, but this is insufficient to place the substance in category 2. For these categories the following specific risk phrases apply:

Category 1 and 2
R45 May cause cancer.

Category 3
R40 Possible risk of irreversible effects.

LIMITATIONS OF USE OF DANGEROUS SUBSTANCES AND PREPARATIONS

The first Directive in this area was adopted by Council in July 1976. It concerned polychlorinated biphenyls (PCB), polychlorinated terphenyls (PCT) and vinyl chloride monomer (VCM). Since then five additional Directives have been adopted; one of the most recent in September 1983 concerns asbestos (OJ L263, 24 September 1983, p 25). It prohibits the placing on the market and use of crocidolite and crocidolite containing products with the exception of:

asbestos-cement pipes;
acid and temperature-resisting seals, gaskets, gland packings and flexible compensators;
torque converters.

The placing on the market and the use of products containing crocidolite, chrysotile, amosite, anthophyllite, actinolite, tremolite may be permitted only if the products bear the following label.

SUMMARY

The development of measures intended to improve the health and safety of workers throughout the European Community began with the creation of the Coal and Steel Community in 1951, and with the foundation of the European Economic Community and the European Atomic Energy Community in 1957. By 1974, industrial accidents and occupational diseases represented the largest group of accidents taken as a whole, and therefore constituted a priority area for Community concern.

The Council of Ministers decided to strengthen the activities in health and safety at work by establishing in 1974 the Advisory Committee on Safety, Hygiene and Health Protection at Work, which assists the Commission in the preparation and implementation of activities in the fields of safety, hygiene and health protection at work. In 1978 a major step forward in furthering safety and health took place when Council adopted the first action programme of the European Communities on safety and health at work. This programme set out the priority actions necessary to increase the protection of workers' health and safety throughout the Member States. This programme, which ran until end 1982, has now been followed up by a second programme which runs until end 1988.

Based on these programmes, several Directives have already been adopted by Council, in particular on lead, asbestos, noise and major accident hazards. Proposals for other Directives are currently being discussed. New initiatives are being planned within the Commission's annual work programme which is discussed both with the Advisory Committee and with the member states.

ANNEX I

Actions contained in the Council Resolution of 27 February 1984 on a second programme of action of the European Communities on safety and health at work

1. Protection against dangerous substances

1.1. Continue with the establishment of Community provisions based on Council Directive 80/1107/EEC of 27 November 1980 on the protection of workers from the risks related to exposure to chemical, physical and biological agents at work.

1.2. Establish common methodologies for the assessment of the health risks of physical, chemical and biological agents present at the workplace.

1.3. Develop a standard approach for the establishment of exposure limits for toxic substances by applying the methodologies referred to in point 2. Make recommendations for the harmonisation of exposure limits for a certain number of substances, taking into account existing exposure limits.

1.4. Establish for toxic substances standard methods for measuring and evaluating workplace air concentrations and biological indicators of the workers involved, together with quality control programmes for their use.

1.5. Develop preventive and protective measures for substances recognised as being carcinogenic and other dangerous substances and processes which may have serious harmful effects on health.

1.6. Establish Community rules for limiting exposure to noise and continue work on developing a basis for Community measures on vibrations and non-ionising radiation.

2. *Ergonomic measures, protection against accidents and dangerous situations*

2.1. Work out safety proposals, particularly for certain high-risk activities, including proposals on specific measures for the prevention of accidents involving falls, manual lifting, handling and dangerous machinery.

2.2. Examine the major-accident hazards of certain industrial activities covered by Directive 82/501/EEC.

2.3. Work out ergonomic measures and accident prevention principles with the aim of determining the limits of the constraints imposed on various groups of the working population by the design of equipment, the tasks required and the working environment so that they are not prejudicial to their health and safety.

2.4. Work out proposals on lighting at the workplace.

2.5. Organise exchanges of experience with a view to establishing more clearly the principles and methods of organisation and training of the departments responsible for inspection in the fields of safety, health and hygiene at work.

3. *Organisation*

3.1. Make recommendations on the organisational and advisory role of the departments responsible for dealing with health and safety problems in small and medium-sized undertakings, by defining, in particular, the role of specialists in occupational medicine, hygiene and safety.

3.2. Draw up the principles and criteria for monitoring workers whose health and safety are likely to be seriously at risk, such as certain maintenance and repair workers, certain migrant workers and certain workers in subcontracting undertakings.

3.3. Draw up the principles for participation by workers and their representatives in the improvement of health and safety measures at the workplace.

4. *Training and information*

4.1. Ensure that employers and workers who are liable to be exposed to chemicals and other substances at their workplace have adequate information on those substances. Prepare information notices and manuals on the handling of certain dangerous substances, particularly those covered by Community Directives. If need be, and taking account of existing Community regulations on the matter, draw up proposals on the establishment of systems and codes for the identification of dangerous substances at the workplace.

4.2. (a) Draw up programmes aimed at better training as regards occupational hazards and measures for the safety of those at work (safety training) and

(b) Training schemes intended for specific groups:

young workers;

groups who have special need of up-to-date information, i.e. workers doing a job to which they are not accustomed or who have difficulty in acquiring information through the usual channels, or to whom it is difficult to communicate information;

workers in key positions, i.e. people who lay down working conditions, communicate information, etc.

5. *Statistics*
   5.1. Establish comparable data on mortality and occupational diseases and collect data from existing sources on the frequency, gravity and causes of accidents at work and occupational diseases, including, as far as possible, data on vulnerable groups of workers and absenteeism due to illness.
   5.2. Compile an inventory of existing cancer registers at local, regional and national level in order to assess the comparability of the data contained in them and to ensure better coordination at Community level.

6. *Research*
   6.1. Identify and coordinate topics for applied research in the field of safety and health at work which can be the subject of future Community action.

7. *Cooperation*
   7.1. Within the framework of existing procedures, continue to cooperate with international organisations such as the World Health Organisation and the International Labour Office and with national organisations and institutes outside the Community.
   7.2. Continue to cooperate on other actions by the Community and by member states, where it proves worthwhile.

ANNEX II

SÆRLIG SIKKERHEDSSKILTNING — BESONDERE SICHERHEITSKENNZEICHNUNG — SPECIAL SYSTEM OF SAFETY SIGNS — SIGNALISATION PARTICULIÈRE DE SÉCURITÉ — SEGNALETICA PARTICOLARE DI SICUREZZA — BIJZONDERE VEILIGHEIDSSIGNALERING

1. Forbudstavler — Verbotszeichen — Prohibition signs — Signaux d'interdiction — Segnali di divieto — Verbodssignalen

a) 

Rygning forbudt
Rauchen verboten
No smoking
Défense de fumer
Vietato fumare
Verboden te roken

b) 

Rygning og åben ild forbudt
Feuer, offenes Licht und Rauchen verboten
Smoking and naked flames forbidden
Flamme nue interdite et défense de fumer
Vietato fumare o usare fiamme libere
Vuur, open vlam en roken verboden

c) 

Ingen adgang for fodgængere
Für Fußgänger verboten
Pedestrians forbidden
Interdit aux piétons
Vietato ai pedoni
Verboden voor voetgangers

d) 

Sluk ikke med vand
Verbot, mit Wasser zu löschen
Do not extinguish with water
Défense d'éteindre avec de l'eau
Divieto di spegnere con acqua
Verboden met water te blussen

e) 

Ikke drikkevand
Kein Trinkwasser
Not drinkable
Eau non potable
Acqua non potabile
Geen drinkwater

Fig. 13.1

2. Advarselstavler — Warnzeichen — Warning signs — Signaux d'avertissement — Segnali di avvertimento — Waarschuwingssignalen

a)

Brandfarlige stoffer
Warnung vor feuergefährlichen Stoffen
Flammable matter
Matières inflammables
Materiale infiammabile
Ontvlambare stoffen

b)

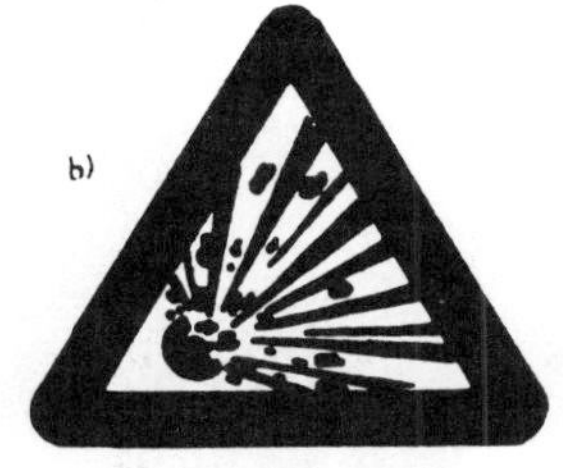

Eksplosionsfarlige stoffer
Warnung vor explosionsgefährlichen Stoffen
Explosive matter
Matières explosives
Materiale esplosivo
Explosieve stoffen

c)

Giftige stoffer
Warnung vor giftigen Stoffen
Toxic matter
Matières toxiques
Sostanze velenose
Giftige stoffen

d)

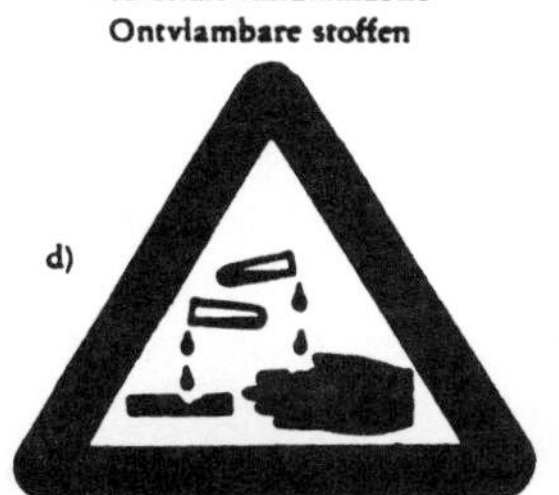

Ætsende stoffer
Warnung vor ätzenden Stoffen
Corrosive matter
Matières corrosives
Sostanze corrosive
Bijtende stoffen

e)

Ioniserende stråling
Radioaktivitet/Røntgenstråling
Warnung vor radioaktiven Stoffen oder ionisierenden Strahlen
Radioactive matter
Matières radioactives
Radiazioni pericolose
Radioactieve stoffen

f)

Kran i arbejde
Warnung vor schwebender Last
Beware, overhead load
Charges suspendues
Attenzione ai carichi sospesi
Hangende lasten

g)

Pas på kørende transport
Warnung vor Flurförderzeugen
Beware, industrial trucks
Chariots de manutention
Carrelli di movimentazione
Transportvoertuigen

h)

Farlig elektrisk spænding
Warnung vor gefährlicher elektrischer Spannung
Danger: electricity
Danger électrique
Tensione elettrica pericolosa
Gevaar voor elektrische spanning

i)

Giv agt
Warnung vor einer Gefahrenstelle
General danger
Danger général
Pericolo generico
Gevaar

Fig. 13.2

3. Påbudstavler — Gebotszeichen — Mandatory signs — Signaux d'obligation — Segnali di prescrizione — Gebodssignalen

a) 

Øjenværn påbudt
Augenschutz tragen
Eye protection must be worn
Protection obligatoire de la vue
Protezione degli occhi
Oogbescherming verplicht

b) 

Hovedværn påbudt
Schutzhelm tragen
Safety helmet must be worn
Protection obligatoire de la tête
Casco di protezione
Veiligheidshelm verplicht

c) 

Høreværn påbudt
Gehörschutz tragen
Ear protection must be worn
Protection obligatoire de l'ouïe
Protezione dell'udito
Gehoorbescherming verplicht

d) 

Åndedrætsværn påbudt
Atemschutz tragen
Respiratory equipment must be used
Protection obligatoire des voies respiratoires
Protezione vie respiratorie
Adembescherming verplicht

e) 

Fodværn påbudt
Schutzschuhe tragen
Safety boots must be worn
Protection obligatoire des pieds
Calzature di sicurezza
Veiligheidsschoenen verplicht

f) 

Beskyttelseshandsker påbudt
Schutzhandschuhe tragen
Safety gloves must be worn
Protection obligatoire des mains
Guanti di protezione
Veiligheidshandschoenen verplicht

Fig. 13.3

4. Redningstavler — Rettungszeichen — Emergency signs — Signaux de sauvetage — Segnali di salvataggio — Reddingssignalen

a) 

b) 

Førstehjælp
Hinweis auf „Erste Hilfe"
First aid post
Poste premiers secours
Pronto soccorso
Eerste hulp-post

c) 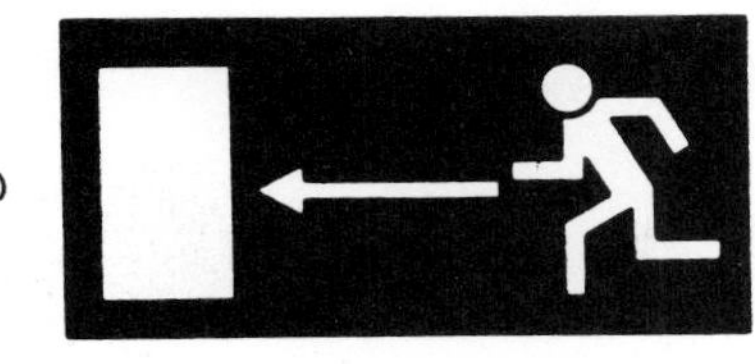

eller/oder/or/ou/o/of

d) 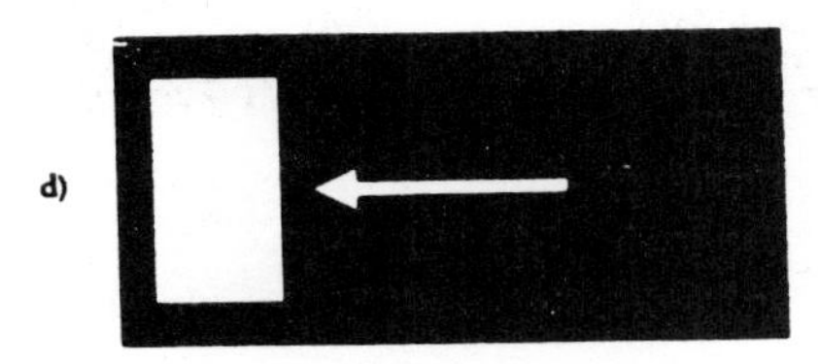

Retningsangivelse til nødudgang
Fluchtweg (Richtungsangabe für Fluchtweg)
Emergency exit to the left
Issue de secours vers la gauche
Uscita d'emergenza a sinistra
Nooduitgang naar links

e) 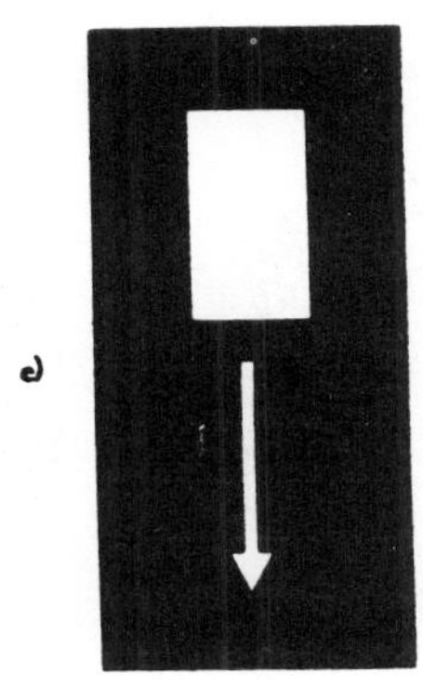

Nødudgang
(anbringes over udgangen)
Fluchtweg
(über dem Fluchtausgang anzubringen)
Emergency exit
(to be placed above the exit)
Sortie de secours
(à placer au-dessus de la sortie)
Uscita d'emergenza
(da collocare sopra l'uscita)
Nooduitgang
(te plaatsen boven de uitgang)

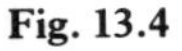
Fig. 13.4

14. Some recent developments in the common law

G. Applebey

INTRODUCTION

The phrase 'recent advances' has a pleasing scientific ring about it and, by implication, presupposes a belief in the march of progress. Whatever law is, and lawyers themselves have an in-built tendency to disagree about this, it certainly cannot be counted as one of the physical or natural sciences.[1] What counts as progress in the law is therefore usually a matter for debate. For this reason it is better to talk of changes or developments in legal rules or policy. It is clear that over the last decade this rate of change hs not been uniform. The second half of the 1970s saw a period of activity in social legislation with the Health and Safety at Work Act 1974 and Employment Protection Acts (1975–78) as two major examples of this process. Since then there has been less new legislation directly applicable to occupational health, indeed there has even been discussion of de-regulation, as a part of Government policy, over and above the official emphasis since the Robens Committee in 1972 on 'self regulation' and less law. In actual fact, despite this, the process of producing important new health and safety regulations, such as those dealing with the Control of Lead at Work (1980), Notifications of New Substances (1982), and the Classification, Packaging and Labelling of Dangerous Substances (1984), has continued apace, and the proposed new Control of Substances Hazardous to Health Regulations, if they become law, would be potentially far reaching in their impact.

The subject of this short chapter is not legislative or regulatory reform, however, but the slower moving stream of the common law with its sometimes imperceptible but constant capacity for change. By common law we mean judge made law,[2] deciding individual cases and providing by their judgment, or within that part of it known to lawyers as the ratio decidendi, a precedent or rule of law to be followed by judges in future cases. It is within this context that I should like to discuss recent developments under both case law and statute, in employers liability for ill-health at work. It has to be noted that the law dealt with is that of England and Wales and would not of course apply elsewhere, except in Scotland where the principles of negligence are similar. However it might be added, that throughout the British Commonwealth

[1] Even this statement would be questioned by those adherents of the theory of natural law. Based largely on the writings of Aristotle and St Thomas Aquinas (1224–1274) law is said to be based on divine reason, and therefore its workings are as capable of understanding as any other aspect of the natural order. The fallacies of this argument were demonstrated by David Hume and others in the 18th Century.

[2] The expression 'common law' can also mean, the English legal system and those derived from it i.e. USA, Canada, Australia, anglophonic Africa, India etc., as opposed to the 'civil law' systems of Western Europe, derived from ancient Roman Law. Scots law is a hybrid of these two. A third meaning applies to those cases such as contract, tort, criminal law etc. which were within the jurisdiction of the common law courts, as opposed to 'equity' which was administered in the Court of Chancery. This distinction was abolished in 1873 but the terminology survives.

the principles of negligence will be largely based upon English law and English cases can be cited, as persuasive if not binding authority.

LIABILITY FOR LOSS OF HEARING

It is worth while starting with deafness for two reasons. First, because it has produced the largest number of claims in recent years, and secondly because of what the case law reveals about the ever expanding tort of negligence.

Apparently it has been known for at least 150 years that persons working in conditions of excessive noise are liable to suffer some form of deafness. As long ago as 1886 Thomas Barr delivered a paper to the Glasgow Philosophical Society entitled 'Inquiry into the effects of loud sound upon the hearing of boilermakers and others working in noisy surroundings'. He described many of the most important features of what is now called noise induced hearing loss. In particular shipyard workers, riveters, caulkers, platers, and 'holders-on' were liable to the loss of the higher frequencies, onset of the condition was rapid, and there was a need to use protective devices particularly in the early stages of employment.

Yet nothing was done about it. Noise was treated as an inescapable fact of shipyard and industrial life. The Factories Acts never specifically referred to noise. Only two sets of regulations, the Woodworking Machines Regulations 1974 and the Agriculture (Tractor Cabs) Regulations 1974 (both of course comparatively recently) required protection for the ears of workers and only in 1975 did occupational deafness become a prescribed disease. The significance of this absence of a clear statutory duty was important, as it denied to potential plaintiffs the action known as Breach of Statutory Duty, much used by persons claiming injury at work. As far as civil liability was concerned this left only the possibility of claiming damages for common law negligence, and it is along this avenue that the law has developed.

If one may be excused a brief excursion back into history it is instructive to consider again the first reported case on deafness, Berry v Stone Manganese Marine (1971). The plaintiff had worked for the defendants since 1957 on a job which involved the use of a pneumatic hammer. He was one of several doing similar work in the same area, so there was a great deal of noise, amounting to between 115 and 120 decibels. The defendants did not seek advice about the noise, but did provide ear plugs in two out of seven possible sizes, which in fact gave insufficient protection, and left the choice of ear plugs to the plaintiff himself without any further supervision. The court decided that from 1956 a type of ear muff had been available commercially and if correctly worn would have protected his hearing fully. Ear muffs were finally provided in 1966 but they took no steps to encourage or persuade their wearing by workmen. By 1960 the plaintiff knew that the noise was causing him loss of hearing but only commenced legal action in 1970. The defendants denied liability and argued that his claim was statute-barred since the cause of the action had occured more than 3 years before the commencement of the action.

The High Court held, giving judgement for the plaintiff, that the defendants were negligent from the time that the plaintiff began to work for them in that (1) they had not sought advice and (2) not supplied more than two sizes of ear plugs and not provided supervision over their selection. The defendants were also (3) negligent in not supplying ear muffs for, with adequate encouragement, the plaintiff would probably have worn these from the outset. However the defendants were not negligent in failing to arrange for the plaintiffs hearing to be tested by audiometry for their duty as employers did not involve taking steps to find out if an employees hearing was being affected. Finally the plaintiff's claim in respect of matters arising more than 3 years before the writ was issued was statute-barred because the plaintiff's long history of increasing deafness would have led any reasonable man to seek advice before 1967 and the plaintiff did not do so. Since the defendants negligence in not taking steps to encourage or persuade the wearing of ear muffs continued after 1967 while the plaintiffs hearing became worse as a result he was entitled to compensation for the increase in loss of hearing since 1967, determined on the evidence of the audiograms and his own evidence, giving sufficient weight to the fact that making a man already deaf still deafer was to increase his handicap considerably. Mr Berry was awarded £1250 at the conclusion of the trial in 1971.

This case created a precedent and should have been followed by a flood of cases. In fact this turned out not to be so. In practice it seems some sort of agreed scale was worked out between the trades unions and insurers, though this eventually collapsed. For a variety of reasons the insurance companies of the employers decided to contest the majority of claims. The obvious answer is not hard to find. By 1983 over 20 000 claims had been made by employees particularly those who had worked for years in the shipyards. With an estimated 100 000 people affected by noise induced hearing loss, the cost to insurers could have been enormous. Meanwhile Northern Ireland with its own legal system, and significantly, civil juries, managed to decide questions of liability and settle, it is said, over 8000 claims during the period.

Eventually in September 1983 the Queens Bench Division of the High Court sitting at Newcastle-upon-Tyne heard two consolidated actions involving three groups of workers, each group being regarded as a test case. The case is reported under the name Thompson v Smiths Ship Repairers (North Shields) Ltd (1984). The plaintiffs were all employed as labourers or fitters in shipbuilding and ship repairing yards over a long period, in most cases beginning in the 1940s (or earlier) until the 1970s, and all had been exposed to excessive noise which had impaired their hearing. The employers at all material times knew that noise levels in the yards exposed their workers to the risk of hearing loss but they failed to provide the workforce with any protection against noise until the early 1970s. The Court decided that since 1963 when the Ministry of Labour issued a pamphlet on industrial noise and thereafter, the defendants should have done something about the problem in the form of taking expert advice and using adequate and reasonable protective devices which were by then available to them.

The plaintiffs had begun their actions in negligence in 1980 and 1981. Although it was impossible to quantify precisely what particular degree of damage had been caused to the plaintiffs, the evidence clearly established that a substantial part of the damage had occurred before 1963 and that after 1963 the plaintiffs' exposure to noise had merely aggravated the existing damage to their hearing. The employers aruged that their failure to provide proper hearing protection amounted to actionable negligence only after 1963 and that they were only liable to pay damages for the plaintiffs' loss of hearing since then.

Mr Justice Mustill held that since all of the plaintiffs had been employed before 1963 only the proportion of their deafness from that date would be taken into account in assessing the amount of compensation, bearing in mind also that most of the plaintiffs would have suffered some loss of hearing anyway with the ageing process. However, the Court also stated that in assessing the plaintiffs' loss and the damages to which they were entitled the Court would make allowances in the plaintiffs favour to take account of the uncertainties involved in making such an apportionment.

The result of this finding as applied to the plaintiffs themselves was that two of those who were suffering 'a moderate hearing loss' and had to use hearing aids were respectively awarded £1350 and £1250 (the latter received £295 as special damages for the cost of his hearing aid). The others, less seriously damaged received from £600–£900 by way of damages. One's reaction to these sums must be how small they seem. Mustill J. himself refers to this at the close of his judgement.

> 'it may be said that these are small awards, even for men who are not very deaf. So they are, considering that the impairment of the plaintiffs' hearing comes from a working lifetime of exposure to excessive noise. It is, however, essential to remember what these actions are about. The court is not here to set up and administer a scheme which would provide these men with a complete indemnity for their loss of hearing, whoever their employers might have been, and irrespective of whether and to what degree those employers might have been, and irrespective of whether and to what degree those employers were at fault. If it is felt that the existing provisions for industrial injury benefits in the field of deafness are too ungenerous, the remedy must lie elsewhere. Here the court is faced with six actions claiming damages for negligence. The plaintiffs have chosen to allege that their employers were in breach of their duty at common law to take reasonable care on their behalf, and that their employers' wrongful failure to take such care has caused them to suffer damage. Having made these allegations they must prove them. They have succeeded only in part. In an action at law, justice looks to the interest of both sides, and although it may seem harsh that these men should have no ampler remedy, it would be unjust to the defendants to hold them liable for more than the proven consequences of their default'.

It is interesting to compare these damages with an earlier, unfortunately only briefly reported, case in 1982. In Tripp v Ministry of Defence (1982) the plaintiff was a retired dockyard labourer who as a result of excessive noise at work over 20 years, had developed moderately severe deafness, accompanied by a degree of tinnitus. Audiometric tests showed loss of hearing capacity of 63 decibels in his right ear, and 55 decibels in the left one. His handicap had greatly affected his social life, particulary as a member of the club of which he was a committee member as he was unable to hold a normal conversation, or hear the telephone. Nor could he watch television, since in order to hear it he had to turn the volume up to maximum, which meant everyone else left the room. For these and other hardships he was awarded damages of £7500, a very large sum compared with the Thompson case.

One does expect compensation to be relatively low for deafness mainly because the plaintiff is usually unable to argue loss of income and the main heads of damages will be loss of amenity and enjoyment of life. It should be noted also that the amount of damages awarded do not, strictly speaking, create a precedent as each plaintiff is treated as an individual. Recently two insurers representing the iron trades and two trades unions have negotiated a scale of compensation for different degrees of hearing loss with damages ranging from £270 up to a maximum of £6000.

The scale of compensation which has emerged since Thompson is not binding on other plaintiffs in the legal sense and they would be entitled to take their case to trial for a higher sum but in practice almost all claimants will be advised by their solicitors and trades unions to take what is on offer. As one would expect there have been a lot more claims since the judgement by former shipyard and engineering workers for an assured if modest sum as compensation for their disability. For their part the insurance companies must also be highly relieved. Whether the outcome really does satisfy the requirements of justice as well as efficiency, requires further research. In the meantime what was said to be the United Kingdom's most widespread occupationally caused disability has, for the time being, been legally resolved.[1]

Other recent cases

What these deafness cases show is the protean nature of the common law action for negligence expanding into new areas of liability and capable of dealing with everything from a snail in a bottle of ginger beer to being hit by a straying cricket ball. At the heart of common law negligence is the question of foreseeability. If the defendant is under a duty of care and the type of injury which occurs is foreseeable (it is not necessary to prove he could foresee the precise injury) then the defendant is liable. This is illustrated by a well known case from a few years ago called Tremain v Pike (1969). A farm worker had contracted Weil's disease as a result of working in premises which had been infected with rats. It was recognised that the cause was contact with rats' urine, but since the annual incidence of the disease had been falling to only a few cases a year among sewer workers, coal miners and fish workers, it was not considered reasonably foreseeable that the farm's employees would be at risk and the employer was held not to be liable. However the courts frequently show a more liberal attitude towards plaintiffs and the requirement of what is foreseeable is clearly not a static one, taking into account recent medical and technological advances.

TENOSYNOVITIS

As an example of this the High Court in Burgess v Thorn Consumer Electronics (1983) awarded £11 230 to a woman assembly line worker suffering tenosynovitis caused by working 8 hours a day fitting components, some of them into tiny holes in plastic printed circuit boards. Bristow J. noted that the complaint was a significant occupational hazard for people working on production lines in the electronics industry. The defendants were liable because they had failed to recognise that tenosynovitis was an occupational hazard for its workers and no steps of any sort had been taken for their safety. The defendants were in breach of their duty of care in failing to give the plaintiff warning which might have prevented her condition from deteriorating to the extent of requiring surgery. The judge considered what steps should have been taken if the hazard had been recognised as it should have been. In fact nothing could have been done to prevent assembly line workers from contracting the condition

[1]Punch, 24 July 1985 carried the following item in its 'Country Life' column. 'A telephone engineer who suffers from constant ringing in his ears following an accident at work was awarded £500 damages today in the High Court'. Tinnitus Telecomitis perhaps?

except for some job rotation which would have reduced the risk. However the company should have warned workers that if they started having pain in their wrist or arm they should report it at once and consult their doctor because it might indicate the presence of a condition which would be innocuous if dealt with promptly but might otherwise become serious. Having failed to do any of these things the company was held to be negligent.

VIBRATION INDUCED WHITE FINGER

Another commonly occurring industrial condition, vibration induced white finger or VWF has been the subject of quite a lot of litigation in recent years, and there has been a tendency for plaintiffs to be unsuccessful, which was in fact the outcome of the latest of these, White v Holbrook Precision Castings Ltd (1985). The plaintiff was offered a job as a grinder. It was common knowledge among other employees that grinding work caused a condition involving numbness of the fingers. After 3 years he had numbness in the second joint of a finger. Raynauds phenomenon was diagnosed, and he was taken off grinding work and offered a less well paid job as a castings inspector. He accepted that job but made a claim for damages against his employers. The Court of Appeal dismissed the employees appeal, deciding that there was no duty on the company when offering Mr White the job as a grinder to warn him that if he took the job there was a likelihood that he would develop VWF. Lawson L. J. had this to say:

> 'Generally speaking if a job has risks to health and safety which are not common knowledge but of which an employer knows or ought to know and against which he cannot guard by taking precautions, then he should tell anyone to whom he is offering a job what those risks are if, on the information then available to him, knowledge of those risks would be likely to affect the decision of a sensible, level headed prospective employee about accepting the offer'.

This is a statement of general principle which can be extrapolated and used in future cases, but applying it to the facts of the case we are discussing, the Court of Appeal held that the present case did not fall within this category. The judge at first instance (lawyers language for when the case was first heard) had found that the plaintiff accepted the job knowing that other employees in the grinding shop were suffering from the condition involving numbness of the fingers.

> 'The information about Raynaud's phenomenon which the respondents knew or ought to have known when they offered the appellant the job as a grinder was that it caused minor discomfort which did not affect capacity for work at all in the vast majority of cases, and minor and trivial inconvenience in the carrying out of outside activities. On that knowledge they would have inevitably concluded that the appellant would have regarded Raynaud's phenomonon as of no consequence when deciding whether or not to take the job'.

One is left to draw ones own conclusions from this statement. In the meantime it is believed that unions and employers are now trying to negotiate a scale of compensation in order to settle cases out of court without litigation similar to the one now operating in deafness claims.

BYSSINOSIS

By way of contrast a recent High Court decision (Brooks v J. and P. Coates (UK) Ltd, 1984) shows how far the courts are on occasion prepared to go in finding negligence, almost to the point of imposing strict liability for the plaintiff's ill-health. In this case the plaintiff worked in the defendants cotton spinning mill from 1935 until 1965. He had been exposed over the years to substantial quantities of very fine cotton dust given off in the course of spinning. After years of suffering a bronchitic condition, in 1979 the plaintiff was diagnosed as suffering from byssinosis. In 1980 the plaintiff brought action against his former employers claiming negligence and breaches of s. 4 and s. 63 of the Factories Act 1961. Boreham J. held that the defendants were in breach of their statutory duty for the following reasons:

1. they had failed to carry out their duty under section 4 to render harmless as far 'as practicable' such dust as might be injurious to health. Although the defendants knew that cotton dust could cause byssinosis and thereafter ought to have known that it might be dangerous to the plaintiff's health, they had failed to provide adequate ventilation of the room where the plaintiff worked until 1963 although it was practicable to do so much earlier;
2. although it was not established that the defendants knew or ought to have known (there was no medical evidence proving this) that cotton dust in a fine cotton mill was likely to be injurious to their employees, the dust given off was in sufficient quantities to be offensive to the plaintiff and was present in substantial quantities. On the facts, the defendants had failed to take all practicable measures to protect the plaintiff from inhaling the dust or to prevent the dust accumulating, or to fit an exhaust appliance, and were therefore in breach of s. 63 of the 1961 Act.

The second of these headings is the most noteworthy. The defendants were liable even although there was no clear medical evidence which they could have known about warning them of this risk involved at the relevant time. The plaintiffs damages were however reduced by a half on the basis that his condition was probably 50% attributable to his cigarette smoking.

ASBESTOS

The extent to which courts are now aware of the risks associated with contact with asbestos dust is shown clearly in two recent judgements. In the first of these Church v Ministry of Defence (1984) the plaintiff claimed damages for injury to his lungs from asbestos dust while working on vessels in the Royal Naval Dockyard at Chatham. The plaintiff had pleural plaques on the lung but otherwise his injuries were symptomless. The defendants admitted liability but contended that the plaintiff had suffered no damage, and was therefore *de minimis* ('de minimis non curat lex'). Peter Pain J. held that in the light of what was known about asbestosis it would be wrong for the law to disregard the plaques as insufficient to constitute damage in law. Upon discovery of the plaques the plaintiff had naturally been extremely worried that he was likely to contract asbestosis. The anxiety had flowed from the defendants negligence and was in itself actionable damage. It was now clear that he was unlikely to suffer further incapacity but there remained a risk, which was not very substantial, that he might and he should therefore be awarded damages. In the second case Sykes v Ministry of Defence (1984) the plaintiff also had calcified pleural plaques resulting from irritation of the pleura by asbestos fibres. These amounted in this case also to actionable damage even though they were symptomless and there was no possibility that symptoms could arise from them. (In the Church case the plaintiff was said to have symptomless insipient fibrosis of the lung which could possibly develop into asbestosis in addition to pleural plaques). According to Otton J.

> 'Looking at the evidence as a whole those definite structural changes having been caused by exposure to asbestos did amount to significant and therefore actionable damage which would entitle the plaintiff to recover damages. . . . Once the plaintiff has established some actionable physical damage which has been caused by the defendants negligence in permitting the exposure to asbestos he was also entitled to be compensated for the albeit slight risk of other independent complaints arising from the same cause. The appropriate figure for general damages to compensate him for the plaques for the slight risk of developing lung complaints caused by asbestos, but not parasitic upon the plaques, and for anxiety, was £1500.'

Both these cases would, it is suggested, be suitable for inclusion under the new provisional damages rules, of which we shall have more to say later.

BACK INJURIES

Finally, it may be worth looking at a case involving a back injury, a common accident at work problem. The case is Bailey v Rolls Royce (1984) in which the plaintiff had sprained his back while loading a heavy circular component weighing 192 lb and trying to pull it into position on a turntable. The employers knew he had injured his back on previous occasions but had not given any specific instructions to him regarding his load lifting capabilities. The judge after hearing conflicting medical evidence found that the plaintiff had a weak back, that the defendants knew of that condition, and that they had been negligent and in breach of section 72(1) of the Factories Act in failing to ensure that the plaintiff was not required to lift a weight exceeding more than about 56 lb.

The Court of Appeal thought otherwise and reversed this decision holding that,

> 'once there was a finding that the plaintiff had a weak back, the defendants would be liable in negligence if they had knowledge that the likelihood and consequences of injury from the type of work undertaken by the plaintiff was such that they ought reasonably to have taken action to protect the plaintiff from the risk'.

Since there were no grounds for preferring one of the conflicting views held by equally qualified medical practitioners on whether there had been a need before the plaintiff's injury to take measures to protect him from the risk of such injury, the allegation that the defendants had been negligent had not been made out.

Furthermore, the defendants would only be liable for breach of their duty under section 72(1) of the Factories Act 1961 if the plaintiff had been employed to lift the particular load and, in so doing, it was likely that he personally would suffer injury; that 'likely' in the subsection was to be construed as 'probable' or 'more probable than not'; and that, since there was no evidence that the plaintiff in lifting the particular object and placing in on the turntable would probably suffer injury, the defendants were not in breach of their statutory duty under the section.

This case illustrates, if illustration is needed just what a burden it can be (no pun intended) to prove negligence even for a relatively straight forward injury particularly where the expert medical opinion is divided. Indeed the case is of interest largely because of the discussion of the question of conflicting medical evidence. The Court of Appeal approved the words of Lord Scarman in the House of Lords in a recent case involving medical negligence, Maynard v West Midlands Regional Health Authority (1984). Lord Scarman said this (at page 638) in a passage which could apply, *mutatis mutandis*, to employers liability as well as medical negligence.

> 'A case which is based on an allegation that a fully considered decision of two consultants in the field of their special skill was negligent clearly presents certain difficulties of proof. It is not enough to show that there is a body of competent professional opinion which considers that theirs was a wrong decision, if there also exists a body of professional opinion which supports the decision as reasonable in the circumstances. It is not enough to show that subsequent events show that the operation need never have been performed, if at the time the decision to operate was taken it was reasonable in the sense that a responsible body of medical opinion would have accepted it as proper. I do not think that the words of Lord President Clyde in Hunter v Hanley, 1955 SLT 213, 217 can be bettered: "In the realm of diagnosis and treatment there is ample scope for genuine difference of opinion and one man clearly is not negligent merely because his conclusion differs from that of other professional men . . . The true test for establishing negligence in diagnosis or treatment on the part of a doctor is whether he has been proved to be guilty of such failure as no doctor of ordinary skill would be guilty of if acting with ordinary care . . ." I would only add that a doctor who professes to exercise a special skill must exercise the ordinary skill of his speciality. Differences of opinion and practice exist, and will always exist, in the medical as in other professions. There is seldom any one answer exclusive of all others to problems of professional judgement. A court may prefer one body of opinion to the other; but that is no basis for a conclusion of negligence.'

LIMITATION ACTS

A common feature of most of the cases in this chapter and many other occupational health cases generally is the fact that the plaintiff is suffering a condition, illness or disability which took a long time to develop and of which the date of onset is imprecise. The plaintiff may also have had the condition or been aware he was ill for some time before he began legal proceedings. The law has rules about this, known as statute of limitations, to prevent actions arising where the cause of action or facts of the case are so old or out-dated that the quality of the evidence is in doubt. Limitation periods also serve to encourage potential plaintiffs to act quickly, but of course conversely they penalise those who fail to do so (what lawyers used to call 'sleeping on your rights'). The interval for starting an action was reduced to 3 years in personal injuries cases in 1954 and this could often work an injustice, and deny the plaintiff his remedy. The classic case illustrating this is Cartledge v Jopling (1963) in which a miner contracted pneumoconiosis but was not aware of this for several years thereafter and failed to bring his action on time and therefore had no legal remedy. Since then, Parliament has made several attempts to find a reasonable solution to the time-limits problem. The latest of these is the consolidating Limitation Act 1980, and this statute probably does constitute a 'recent advance' in the law, although it is so full of difficulties that this has to be immediately questioned. Essentially the act allows the court a discretion (first introduced in 1975) to override the time-limits if it considers it equitable to do so. The worse excesses of the old law should now be avoided.

It is worth looking at how the law now stands. Section 11 of the 1980 Act provides that in personal injury cases the limitation period is 3 years from (a) the date on which the cause of action accrued; or (b) the date of knowledge, if later, of the person injured. Personal injury includes any disease or any impairment of a person's physical or mental condition. The 'date of knowledge' provision is designed for those such as Mr Cartledge who had suffered damage from disease, but was unaware of this fact until many years later. 'Date of knowledge' is defined in section 14 of the Act as the date on which the person concerned first knew:

(a) that the injury in question was significant;
(b) that the injury was attributable in whole or in part to the act or omission which is alleged to constitute negligence;
(c) the identity of the defendant;
(d) where it is alleged that the act or omission is that of some person other than the defendant, the identity of that person and the additional facts supporting the bringing of an action against the defendant.

This last ground might include the question of the employers vicarious liability for example and whether the employee was acting in the course of his employment. Knowledge that the acts or omissions constituted negligence in a legal sense is irrelevant. Knowledge also includes constructive knowledge i.e. what might reasonably be expected to be acquired from facts observable or ascertainable by the plaintiff or obtainable with medical or other expert help which it would be reasonable to expect the plaintiff to seek. In Simpson v Norwest Holst Southern (1980) the plaintiff had no actual knowledge of the identity of the defendant employer where the company was one of a group, and had given the plaintiff an incorrect name on payslips. The Court of Appeal held that he could not reasonably have been expected to acquire knowledge of the identity of his employer. In Leadbitter v Hodge Finance Ltd (1982) it was held that knowledge of a fact ascertainable only with the help of expert advice is not attributed to a person so long as he has taken all reasonable steps to obtain and to act on the advice.

The most important part of the 1980 Act is contained in Section 33. The courts are now empowered to override the time limits if it appears to the court that it would be equitable to do so having regard to the degree to which: (a) the provisions of section 11 and 12 of the Act prejudice the plaintiff and (b) any decision of the court under the subsection would prejudice the defendant.

In deciding whether to exercise this discretion the court has to have regard to all the circumstances of the case and in particular to:

(a) the length of and reasons for the delay on the part of the plaintiff;
(b) the extent of loss of cogency of the evidence because of the delay;
(c) the conduct of the defendant after the cause of action arose;

(d) the duration of any disability of the plaintiff after the cause of action arose;
(e) the extent to which the plaintiff promptly and reasonably ... once he knew he might have an action for damages;
(f) the steps, if any, taken by the plaintiff to obtain medical, legal or other expert advice and the nature of any such advice he may have received.

There has recently been a number of cases on the question of whether this discretion is available in all cases, or only in exceptional ones. In Firman v Ellis (1978) the Court of Appeal held that the court had an unfettered discretion to extend the three years period in any case in which it considers it equitable to do so, and this view was also expressed by the House of Lords in Thompson v Brown Construction (1981). However in another case, Walkley v Precision Forgings Ltd (1979) the House of Lords also made it clear that the discretion was excluded where the plaintiff had commenced a first action within the normal limitation period.

Although many cases will still be caught by the three year rule, the courts' present attitude is probably best epitomised by the treatment of delay in a case we have already looked at (Brooks v J. and P. Coates). In that case the judge held that since 1965 the plaintiff knew or ought reasonably to have had knowledge that his condition was due to the 'acts or omissions on which he relied to bring his action' (i.e. negligence) and therefore since his writ had only been issued in 1980, his action was prima facie barred by virtue of s. 11 of the 1980 Act. However despite the long delay the plaintiff was not blameworthy and although there was real prejudice to the defendants by reason of the delay, nevertheless, applying the test of where would the greater prejudice fall, the prejudice suffered by the plaintiff if he was denied the right to litigate his claim substantially outweighed the prejudice which would be suffered by the defendants if the action was allowed to continue. Accordingly the court exercised its discretion under Section 33 and allowed the action to proceed.

Obviously some actions involving occupational illnesses are still going to be caught by the Limitations Acts but it is hoped this recent liberalising of the rules will allow more cases to go on to trial and to be heard on their merits, rather than be trapped because of what may appear to non-lawyers to be the technicality of a procedural bar.

DAMAGES

Reference has already been made at several places in this chapter to the question of damages—it will be obvious, even to the outsider, that the quantification of loss is a problematic exercise. Damages are usually classified as either 'special' i.e. actual amounts of lost income, both past and future, and 'general' for items such as pain and suffering, loss of amenity (loss of enjoyment of some activity, including pastimes, from which plaintiff is now precluded). This latter category is of course fairly subjective and varies from person to person. In England, damages are meant to be compensatory and unlike the United States, punitive damages are not awarded in personal injury actions.

The heads of damages and the sort of sums awarded for serious illness are illustrated by Hobbs v British Rail Engineering decided in January 1985. The plaintiff was a 59-year-old man at the date of the hearing. A former gas fitter and plumber he had been in considerable contact with asbestos between 1942 and 1962. He was now suffering from asbestosis, the symptoms of which appeared in early 1984, with breathlessness and chest pains. His condition deteriorated and by March, 1984 he was compelled to give up his employment. By August, 1984 he was completely bedridden and totally dependent on his wife. His life expectancy was about 2 months, though until recently he had been a very active man. Damages awarded were as follows:

(1) General damages:	
Past loss of earnings	£2785
Loss of future earnings	£21 424
Loss of pension rights	£3000
Value of DIY services (based on multiplier of 6)	£2100
Wife's care and attendance	£1400
Cost of future care	£500
(2) Special damages:	
Pain and suffering and loss of amenities (bearing in mind short life expectancy)	£18 277
Loss of expectation of life	£1500
Total awarded	£50 986

As noted earlier the amount of each award does not of itself create a precedent as each plaintiff is treated as an individual, though courts do try to achieve rough parity for the same injury. Damages are generally still lower in England than other common law countries. There have been several awards of over one million dollars to asbestosis victims in the United States.

'THE LOST YEARS'

In 1979 the House of Lords overruled a long line of leading cases in Pickett v British Rail Engineering. Mr Pickett had worked for the defendants between 1949 and 1974. In 1974 he developed symptoms of mesothelioma of the lung because of his exposure to asbestos dust during his employment. Damages totalling £14 947 were awarded to him at the trial in 1976. At that time the life expectancy of Mr Pickett who was 51, was 1 year. Before his illness the plaintiff was very fit, and a first rate cyclist. The Plaintiff appealed against the award of damages but died before his appeal could be heard. When the appeal was heard the House of Lords held on the point of law involved that where a person's life expectancy is reduced because of an accident or industrial disease he is entitled to be compensated for loss of earnings during the period of which he has been deprived, the so called 'lost years'. The House of Lords regarded as unjust the reasoning behind the old law which reduced Mr Picketts claim for lost earnings to only the one year he was expected to live. Plaintiffs should be able to claim for the years that they otherwise would have lived. Their Lordships foresaw clear problems with their ruling in Picketts case and these have turned out to be well founded. In particular the estates of many deceased plaintiffs were being over-compensated, particularly since the dependents of a deceased victim could now claim under the wrongful death statutes, the Fatal Accidents Act 1976 and the Law Reform (Miscellaneous Provisions) Act 1934 and receive additional damages more than once, clearly an anomalous situation.

The result has been a statutory amendment to deal with this loophole in the law. Under Section 2(3) of the Administration of Justice Act 1982, claims for 'lost years' may only be brought by the person himself while still alive. After that persons death, in an action brought by his dependants (which is in fact a different legal action based on the Acts mentioned above) the damages under one of these, the 1934 Act, shall not now include any damages for loss of income in respect of any period after that persons death. Since, sadly, many people still die from industrial diseases, changes in this area of law can have considerable impact on the families financial security after a victims death.

PROVISIONAL DAMAGES

The same Administration of Justice Act 1982 also provides for a novel development in English law, the award of 'provisional damages'. Section 6 allows such an award to be made in personal injury cases where there is proved or admitted to be a chance that at some definite or indefinite time in the future the victim will develop some serious disease or suffer some serious deterioration in his physical or mental condition caused by the original tort. In such cases the court will assess damages on the assumption that the deterioration will not occur, while recognising that further damages could be awarded at a later date should the assumption prove to be false. The detailed rules on provisional damages came into force on 1st July 1985.[1] Potentially this new power to award more damages at a later date could have a very important impact on many personal injuries actions arising from occupation hazards.

CONCLUSIONS: OBSERVATIONS ON CIVIL LIABILITY GENERALLY

As we have seen the common law has developed in a number of ways over the last few years and of course will continue to do so. At the present time major structural reform of the compensation system in the United Kingdom seems unlikely though calls are still often heard that this is long overdue. The opportunity afforded by the Pearson Commission (1973–1978) to produce a 'no fault' system, such as exists in New Zealand and elsewhere, was not taken up. There will always be gaps in coverage as new areas of illness and disability are taken on board by the law. For example, there is a need for litigation on radiation linked diseases.

According to a recent article by Barrett (1981) the three main legal difficulties in establishing liability for ill-health remain:

(a) showing that the ill-health was caused by conditions for which the defendant is liable;
(b) establishing that the defendant was negligent in failing to take adequate steps to protect the plaintiff;
(c) circumventing the Limitation Acts.

We have looked at all three of these in this chapter. However, in the opinion of the author, the major problems in relation to occupational illness and disability claims are those of the legal system as a whole. First, delay in getting cases to trial which can be 2–3 years in many cases, with the possibility of long delays in appeals afterwards. Second, the staggering cost of civil litigation, which is a cause for concern—(it should perhaps be noted that in Britain unlike many other countries, legal aid is available

[1]See Practice Note 1985 2 All ER 895

in civil cases, but contingent fees for lawyers are not allowed). Third, the absence of a class action in English law, suitable for dealing with personal injuries. Until changes are made in the legal system as a whole, and civil procedure in particular, the lot of the victim of occupational disease and injury seeking compensation is likely to be in Ison's phrase 'a forensic lottery'. The reward of course is that if he is successful he gets full compensation which can be a substantial sum of money. In 1978, according to Pearson only 10.5% of victims achieved this [in the valuable Oxford study, Harris et al (1984) gave the figure as 12% of their sample], and this is probably a lot lower in relation to occupational illnesses. It is safe to say that no other system devised can offer such large financial recoveries, even bearing in mind that damages in England are low by North American standards. A 'no fault' system would certainly mean more people would recover damages but in smaller amounts (as is illustrated by the diminishing returns from workers compensation in the USA). State schemes such as the British Industrial Injuries Scheme, which remains narrow in its treatment of industrial disease offer even less usually, and the British one gives little more then income maintenance at subsistence level, with a tariff of very small amounts for disabling injury. [On the problems of this system and recent developments the article by Wilson (1982) is useful.]

The other major development in Western Countries, that of Products Liability (used in the sense of a scheme of strict liability for defective products) has also failed to materialise in the UK, where actions against manufacturers still have to be taken in negligence. In America, on the other hand, the gradual acceptance by the courts of strict liability against manufacturers has led to an explosion in personal injury suits. Starting with Borel v Fibreboard Paper Products Corp (1973) strict liability of producers enabled thousands of asbestosis victims to recover substantial damages for their inuries. There have been four initiatives to introduce Products Liability into the UK in the 1970s. Taken in chronological order these were: (1) the EEC Draft Directive (1976); (2) European Convention on Products Liability in regard to personal injury and death (Council of Europe—'Strasbourg Convention' 1977); (3) The joint reports of the Scottish and English Law Commissions in 1977; (4) Pearson Commission Report (1978). In 1985, an EEC Directive was finally issued. This will have to be implemented in the UK in order to comply with community law. Meanwhile the length to which the courts have gone in holding manufacturers liable in negligence means that in many cases, liability close to strict liability already exists.

CASES REFERRED TO IN TEXT

Bailey v Rolls Royce 1984, ICR 688, CA
Berry v Stone Manganese Marine Ltd 1982, 1 Lloyd's Rep, 182
Borel v Fibreboard Paper Products Corp 493 F 2d 1076 (5th Cir 1973)
Brooks v J and P Coates (UK) Ltd 1984, 1 All ER 702
Burgess v Thorn Consumer Electronics 1983, The Times, 16 May 1983
Cartledge v Jopling and Sons Ltd 1963, 1 All ER 341 1963 AC 758
Church v Ministry of Defence 1984, The Times, 7 March 1984 QBD
Firman v Ellis 1978, 2 All ER 851
Leadbitter v Hodge Finance Ltd 1982, 2 All ER 167
Pickett v British Rail Engineering 1979, 1 All ER 774
Simpson v Norwest Holst Southern Ltd 1980, 2 All ER 471
Sykes v Ministry of Defence 1984, The Times, 23 March 1984 QBD

Thompson and others v Smiths Shiprepairers (North Shields) Ltd 1984, 1 All ER 881
Thompson v Brown Construction 1981, 2 All ER 296
Tremain v Pike 1969, 3 All ER 1303
Tripp v Ministry of Defence 1983, CLY 1017
Walkley v Precision Forgings Ltd 1979 1 WLR 606
White v Holbrook Precision Castings Ltd 1985, 1 RLR 215

REFERENCES

Barr T 1986 Proceedings of the Philosophical Society of Glasgow Vol XVIII
Barrett B 1981 'Employers liability for work related ill-health'. Industrial Law Journal 101–112
Harris et al 1984 Compensation and support for illness and injury (Oxford Socio-Legal Studies). Clarendon Press, Oxford
Ison 1967 The forensic lottery. Staples Press, London
Pearson Royal Commission on Civil Liability and Compensation for Personal Injury Cmnd 7054 (1978)
Wilson S R 1982 'Occupational disease — the problems of a comprehensive system of coverage'. Industrial Law Journal, 141–155

15. Ethnic factors in health and disease

D. G. Beevers J. K. Cruickshank

The practice of medicine in the inner cities of Europe has been transformed over the last 30 years by the health problems of migrants from Third World countries. Both the Caribbean, Asian (Indian sub-continent) and Chinese sub-cultures have become permanent variants of the dominant white culture in Western countries. The special medical and health-care problems often arising from social and cultural differences in these minority groups, mean that all members of the health-care professions need special training in previously neglected areas, including some tropical medicine as well as medical anthropology and sociology.

This chapter considers the cultural and social influences on the ethnic factors related to health and disease in people of employment age and then discusses some of the common medical problems encountered in blacks and Asians, particularly in England, as well as Western countries.

DEMOGRAPHY

Most countries of the developed world now have ethnic minorities in their midst. Whilst the black population of the USA is long established, there remain major differences in mortality and morbidity between blacks and whites, and serious inequalities of health care. There are several other settled ethnic minorities in many countries. For example there are large Asian communities in South and Central Africa, as well as well established ethnically Chinese groups in Malaysia, Britain, the USA and the West Indies. It causes no surprise to refer to Chinese Americans, Black Americans and Hispanic Americans, as well as White Americans. Similar terms such as Black Frenchmen, or Asian Britains or Vietnamese Swedish are not yet in common use in Europe possibly because of the newness of the problem.

The history of migration in the black population of the Caribbean is unusual because the black majority in Jamaica and other islands is itself a settled migrant group, having been transported there as slaves from West Africa. This community once established was later joined by significant numbers of Indian Asians and Chinese. More recently there has been a migration of many black people back across the Atlantic to Britain. The simultaneous arrival of other groups, principally Asians from India and Pakistan, to Britain has produced another ethnic minority which differs as much from the Afro-Caribbeans as from the whites. In addition, there are other ethnic sub-populations including Chinese, Cypriots, Vietnamese, Yemenis and Italians (Black, 1985a,b,c and d).

In France substantial numbers of Black migrants came from West Africa, in the Netherlands there are Indonesians, South Molluccans and Surinamese. There are many blacks in Belgium and even West Germany has large numbers of Turkish guest workers.

THE REASONS FOR MIGRATION

The reasons for population migration are complex, but are mainly economic (Layton-Henry, 1985). There is evidence that migrants differ from their peers who remain at home; they have usually been fit enough to migrate, but many have migrated because of financial or social disadvantage in their country of origin. Many migrants travelled in search of jobs, better pay and living conditions. This led to the arrival in Europe and America of large numbers of relatively unskilled men with little formal education. Many were prepared to take on unpopular and unpleasant jobs, for low wages, during the post-war economic boom. After this economic migration there followed a smaller social migration of wives, fiancees and parents many of whom were also of employment age.

Involuntary migration, that is the resettling of refugees from political oppression or war, has tended to produce a different picture. The Asian refugees who came to Britain from Kenya and Uganda are clearly different from the economic migrants who came directly from the Punjab or Bangladesh. Most East African Asians in Britain have settled in with relative ease and have restarted businesses and resumed the professions they had to vacate in Africa.

It is not possible, therefore, to generalise about the ethnic minorities in the West nor about their health problems. There are major differences in the disease pattern seen in blacks, Asians, whites and the other ethnic minorities. It is important to note, however, that these ethnic differences are almost completely due to differences in environmental factors, including dietary habits, poverty and disadvantage. There are no diseases which are exclusively related to racial factors, and very few which have any genetic basis (see Chapter 9); even sickle cell disease is encountered in Mediterranean people. The biggest influence on the health of the ethnic minorities in the West is better described as cultural and social as it is related to the enormous differences in life-style of blacks, whites and Asians (Johnson, 1984).

CULTURAL EFFECTS ON DISEASE

When migrants arrive in their host country they may choose to live in relatively close communities and continue to exhibit the same social and medical patterns of their original country. Later generations may take on some of the health characteristics of the host population but despite this, important differences remain. The voluntary or involuntary isolation and at times ostracism of immigrants means that they retain many of their own sub-cultural attitudes to health and disease. Differences in choice of life-style, for instance the tightly knit nature of Asian families, and other factors like the retention of Indian style food mean that second and third generation offspring of migrants remain socially and medically very different from the host population.

In the future major changes in the ethnic patterns of disease may be expected. An example of this is the change in the incidence of coronary heart disease in Japanese Americans. Whilst this group has retained its ethnic identity in the West coast of the USA, their heart disease incidence is becoming increasingly similar to that of white Americans and their stroke incidence is becoming lower than that seen in Japan (Winkelstein et al, 1975).

Now that major migrations of Third World populations have substantially diminished, the practicalities of health care provision are likely to change a great deal.

One obvious problem, that of language, will become much less common over the next 30 years. At the moment many Asians in Britain cannot speak English, and this must increase their isolation from the white community and also limit their access to medical care. It is important to note that many non-English speaking Indians and Pakistanis are also illiterate in their own language, so the provision of multi-lingual direction signs in hospitals or places of work, or in health care booklets may not be adequate (Leatherdale et al, 1978). At the moment all major institutions serving Asian immigrants must employ interpreters; otherwise the provision of care to its users will be insensitive and inefficient (Cox, 1977). This problem will be less prevalent as Asian children pass through schools in England, but the major cultural differences will remain.

At present many of the migrants from Asia and the Caribbean have concepts of human anatomy, physiology and disease which are unrelated to that which is taken for granted by whites. Whilst most white children are aware of the concept of bacteria, viruses, immunisation and infections, these are new to many migrants particularly those from rural areas in the Third World. Many Asians are unaware that the limbs, the thorax and the abdomen are structurally and functionally different units. Their body image differs from that of whites (Bakhshi, 1978). Their symptoms, therefore, reflect this lack of knowledge, with pains that cross tissue planes and extend from the lower limbs to the head in a manner not reported by whites or even blacks. Concepts of nerves, arteries and veins may be limited and instead their own culturally based attitudes remain, often based on superstition, and a distinct lack of information. All too often an Asian patient tries to resort to a literal translation of his symptoms: too much pain. Even the most careful and sympathetic history taking with the aid of an interpreter may baffle the Western trained doctor and provide little diagnostically useful information.

THE PROVISION OF HEALTH CARE

Specialists in all branches of medicine, and other health care professionals like nurses and receptionists need to be aware of the ethnic factor in health and health care. Certain ethnic groups have particular potential medical problems which need to be appreciated if patients are to receive a high quality of medical care with accurate diagnosis and effective treatment. In both preventive medicine, primary health care and in hospitals there is a need for special training in the common medical problems of all population sub-groups. Just as nobody doubts the problems of chronic chest disease in unskilled workers in industrial cities, so nobody should fail to be aware of the problems of tuberculosis, malaria and heart disease in Asians, and sickle cell disease and strokes in Afro-Caribbeans.

The practicalities of providing excellent medical care must be conditioned by local requirements, and sadly also locally available resources. Whilst there is some use of unqualified local hakims by the Asian community, there is evidence that most Asian patients seek only for good quality medical care (Jain et al, 1985). Some patients even say that after retirement, they wish to remain in Britain because of the ready availability of both primary and secondary health care. However, there is also evidence that Asian patients seek out general practitioners who originate from their own country, mainly because this obviates language problems (Johnson et al, 1983). There is surprisingly very little evidence that Asian patients in any way provide a burden to the National Health Service (Blakemore, 1983). There may be some dissatisfaction with the health care they receive, and in Afro-Caribbeans this may explain a greater reliance on second opinions from private doctors. It may be that these private doctors are more sympathetic and spend more time with their patients, but there must be some suspicion that vulnerable patients are being exploited.

Attention should be paid to the special requirements of some ethnic minority patients. The need for interpreters has already been mentioned. Hospitals must consider the provision of special diet kitchens, with dieticians trained in the use of Indian types of food, as well as vegetarian diets. Both doctors and nurses should be aware of the mass grief in Asian families when a relative dies. European trained doctors and nurses should make efforts to know more about the structure of Asian societies, the system of familial and first names and attitudes of their patients to hospitals, doctors and disease. Most Asian women are keen not to expose their bodies so that at times clinical examination can be difficult. This is less of a problem if the doctor is female, and a distinct preference for female obstetricians and gynaecologists is evident.

Some Afro-Caribbeans may place relatively more trust in herbal or alternative medicine than whites and Asians. One other striking phenomenon is the tiny numbers of black West Indian medical students and doctors, but the visibly large numbers of black nurses. The cultural reasons for this are complex.

It is not the intention of this review to cover relatively rare life-threatening diseases, but rather to discuss conditions which are likely to be seen in relatively fit people of employment age. They can be divided into four categories: the purely genetic diseases, the imported tropical diseases, the ethnic differences in common diseases seen in all races, and nutritional deficiencies.

GENETIC DISEASE

Sickle cell disease

This condition occurs almost exclusively in Afro-Caribbean black people. Until relatively recently most homozygous SS patients died before reaching adult life. Now with improved medical care (Thomas et al, 1982) significant numbers of young people are reaching employment age. Most are chronically unwell and not capable of heavy work but some are employed in light work or white-collar trades. Employment is likely to be difficult as sickle cell disease sufferers are anaemic and prone to sickle cell crises with pulmonary, splenic or bone infarctions in response to minor infections, anoxia or even cold. They also have an increased susceptibility to renal failure, strokes and septicaemia and an acute chest syndrome which is associated with a high mortality (Davies et al, 1984). The rarer sickle cell genotypes sickle SC and sickle thalassaemia tend to produce a milder disease, and sufferers can work productively, although it is important that employers are aware of the diagnosis and are supportive.

The sickle cell trait (AS) is present in about 10% of otherwise fit black people. Very rarely, and only in response to extreme hypoxia, some individuals may develop sickle cell crises. People with sickle cell trait are completely healthy and capable of any degree of rigorous work. However, carriers of the sickle cell trait should be aware that if they have children by a spouse who also has sickle trait, there is a 25% chance of the offspring having sickle cell disease.

Screening and genetic counselling facilities are now being developed in Britain and it is to be hoped that sickle cell disease will gradually become less common with the appropriate genetic counselling which is necessary and overdue in Europe.

Glucose 6 phosphate dehydrogenase deficiency

This condition occurs mainly in men (being X linked) of African, Chinese, Mediterranean and Indian stock. In childhood it presents with jaundice and anaemia. Most affected men are fit and in adult life the main problem is that some drugs can precipitate sudden haemolysis. Ideally all Asian men with malaria should be screened for G6PD deficiency, before being prescribed primaquine, although the haemolysis is rarely severe. Chloroquine in the acute malarial illness is quite safe. Haemolysis can also be provoked by some of the sulphonamide drugs including cotrimoxazole as well as nitrofurantoin (Winthrobe et al, 1981).

Beta thalassaemia

This is an inherited autosomal recessive disease leading to chronic anaemia in Asian and Mediterranean people, and is unlikely to be encountered in occupational health. Thalassaemia minor is of no clinical importance but its detection is important in preventing thalassaemia major. It is much rarer than sickle cell trait, and, with effective genetic counselling should become even less prevalent (Weatherall & Clegg, 1981).

IMPORTED DISEASE

There is an enormous number of tropical diseases that are seen occasionally in European countries. These diseases, which are usually acquired in the Third World, affect all ethnic groups and should be considered in all people who have recently travelled abroad. It is not the purpose of this review to cover all tropical diseases, although employment medical officers must be aware of the possibility in all recent travellers, and if there is any doubt, local tropical and infectious disease experts should be consulted. There are some conditions however, which may occasionally concern employers, and do merit consideration here.

Malaria

Plasmodium vivax malaria is now a common condition; it is easy to diagnose and treat and patients do not usually need hospital admission. Those at particular risk include recent travellers to the Indian sub-continent who have omitted to take malaria prophylaxis or have failed to persist with it (Warwick et al, 1979). The clinical presentation may be either insidious, or relatively acute, up to 12 months after the patient has left the endemic area. The main symptoms are tiredness, fever and occasionally rigors. When taking a history, the clinician can be misled when he asks his patient 'when did you come to Britain?' The patient may then tell his doctor when he first arrived, as much as 20 years previously. A supplementary question 'have you been abroad recently?' may then reveal recent foreign travel.

The diagnosis of *Plasmodium vivax* is simple, requiring examination of a thick blood film for malaria parasites. Treatment with chloroquine should be commenced immediately for 3 to 4 days and after this primaquine should be taken for 2 weeks. It is best to exclude G6PD deficiency before prescribing primaquine to men, as in susceptible cases a mild haemolytic anaemia may be precipitated. However, the delay and trouble of screening for this may not be justified if it means that treatment is withheld. Normally a patient with *Plasmodium vivax* malaria only need take about 1 week off work.

Falciparum malaria is a severe and potentially fatal disease which is contracted by travellers from Africa, South America and the middle and far East. Chloroquine resistance is now a major problem and all patients require expert medical care urgently.

Yaws

This disease is discussed here, not because it presents any problems of illness in Europe, but because it causes diagnostic difficulties. Yaws is a treponemal disease, endemic in Africa, and formerly in the Caribbean, which causes inflammatory swellings on the legs. After treatment or after the natural progression of the disease, thin circular depigmented scars remain on the lower legs; there are no other long term sequelae.

The clinical problem faced in Britain is because it is not possible to differentiate bacteriologically between yaws and syphilis. Both diseases leave permanently positive serological tests for syphilis (Cardiolipin Wasserman Reaction and Venereal Disease Reference Laboratory tests) as well as positive treponemal tests. People who had childhood yaws are therefore in danger of being falsely labelled of having syphilis and may appear never to have received treatment. It is important to take a careful history from black patients with positive tests for syphilis. If discrete enquiry reveals no past history of sexually transmitted disease, but there is a past history of sores on the legs then it is probable that yaws and not syphilis explains the abnormal blood tests.

Hepatitis B

In white people, hepatitis B is relatively uncommon, except in drug addicts or homosexuals. The commonest agent causing hepatitis in whites is normally sporadic hepatitis A. In the Indian sub-continent, south-east Asia and the Pacific islands however, hepatitis B is common, and almost endemic, but is not associated with homosexuality or contaminated intravenous injections (Weatherall et al 1981). Jaundice in Asian people needs careful investigation. Hepatitis B may become chronic and progress to hepatic cirrhosis and hepatoma. Positive hepatitis B blood carries some biological hazard to staff taking specimens as well as to laboratory workers, who must therefore be alerted accordingly.

ETHNIC DIFFERENCES IN COMMON DISEASES

It is important to note that there are relatively few medical conditions which are specific to ethnic minorities. The main patterns which are discussed next are differences in the incidence of common diseases in the various ethnic groups rather than any true racial effect. These differences are almost all related primarily to environmental factors including diet, social conditions and overcrowding. Whilst there may be some genetically determined tendency for some people to be prone to particular disease processes, there is no evidence that any particular genotype is related primarily to racial origin.

Tuberculosis

Whilst the incidence of tuberculosis has steadily declined in all western countries over the last 20 years (Medical Research Council, 1980), there is an increasing proportion of cases who are of near, middle or far eastern origin. This high incidence is seen in first and second generation migrants, so whilst some tuberculosis is imported, a great deal is acquired in the West. There is a striking difference between migrants from Asia compared with Afro-Caribbeans (Jackson et al, 1981) as there is only a modest excess of tuberculosis in blacks compared with whites and this is commonly only in close knit groups including Rastafarians (Packe et al, 1985). This difference between Asians and blacks may be related to family structure, as Asians usually tend to live in closer and larger family units even though they share the same sort of relatively small city-centre housing units as blacks.

The clinical manifestations of tuberculosis also differ in Asians and whites. Whilst bovine pulmonary tuberculosis is commonest, Asians have a marked excess of extrapulmonary infection including lymph node involvement, and spinal, bony and abdominal tuberculosis. Furthermore streptomycin resistant tuberculosis is commoner in Asians. Death from tuberculosis is now very uncommon, but is seen particularly in elderly Asians who frequently first present with advanced disease or even clinically moribund (Innes, 1981).

The diagnosis of tuberculosis should be considered in all Asian people and all patients complaining of fevers, weight loss and bone or abdominal pain. These symptoms may be insidious so may present to occupational health departments. Extrapulmonary tuberculosis can be difficult to diagnose but pulmonary disease can usually be detected on plain chest X-ray. Whilst a routine pre-employment chest X-ray is probably useless in white people, a case can be made for this screening test in Asians and Middle Eastern people especially if employment is in the catering or allied industries or in the health services. Similarly in occupational medicine, Heaf testing can be justified in selected employments in high risk groups. A negative test indicates that the person does not have tuberculosis and also is not immune to future infection. A weakly positive Heaf test suggests either previous tuberculous infection or acquired immunity. A very strongly positive test makes the diagnosis of active tuberculosis probable, and a clinical history, general examination and chest X-ray should be conducted immediately.

Liaison between occupational physicians and local chest and infectious disease physicians is mandatory. Modern regimes for the treatment of tuberculosis mean that patients are not usually admitted to hospital and the patient should be fit enough to resume work within 2 to 3 months. Family members and work mates who have been in close contact with the patient should be checked for tuberculosis by chest X-ray and Heaf testing.

Other respiratory diseases

There is good evidence that other chronic chest conditions are commoner in Asians compared with whites in Britain. Both asthma and bronchitis are common causes of illness in Asian children, adolescents and adults, and it is likely that the long term sequelae of these diseases, with emphysema, respiratory failure and cor pulmonale will follow as the years go by (Jackson et al, 1981). There is also an excess of asthma in young blacks. It is interesting that many Asians who were born in the Indian sub-continent date the onset of their respiratory disease to their arrival in Britain. However, one recent investigation has demonstrated asthma as a common reason for admission to hospital in India and is also common in people of Asian origin born in Britain (Greaves 1986). Allergy to house dust mites, and food allergy is commoner in Asians and black asthmatics whilst allergy to pollens and animals is commoner in whites (Morrison-Smith & Cooper, 1981; Wilson, 1985). These variations to allergens are almost certainly related to differences in life style. Asians rarely keep household pets and are unlikely to expose themselves to pollens, but the relative overcrowding in large Asian families may explain excess house dust allergy.

People with reversible airways obstruction can usually continue to work if they are adequately treated. The use of aerosol inhalers containing beta agonists and corticosteroids can greatly improve the clinical condition but adequate training of the patient on the correct use of these inhalers is important. Time spent with Asian asthmatics, if necessary with an interpreter, and using a dummy or empty inhaler for training purposes may greatly improve a patient's capacity to work.

Sarcoid

This condition is mentioned briefly here because its presentation can mimic tuberculosis. Sarcoid appears particularly common in black people and is rare in Asians in comparison with whites. It often presents with extra-thoracic manifestations with enlarged lymph nodes, renal and hepatic lesions and erythema nodosum. Although classical pulmonary sarcoid is usually a mild and self-limiting disease, the extra-thoracic disease in blacks is more severe and may necessitate corticosteroid therapy (McNicol & Luce, 1985).

Coronary heart disease

There are striking ethnic differences in coronary heart disease (CHD) in England with a low incidence in Afro-Caribbeans and a high incidence in Asians (Adelstein, 1978; Cruickshank et al, 1980). However, in the USA, CHD has now become a leading cause of death in black Americans, indicating the powerful environmental and dietary factors of this modern epidemic. Heart attacks are relatively uncommon in black people in England although there is evidence from both Jamaica, Africa and North America that the incidence is rising (Cooper, 1986). The reasons for this relative protection are uncertain but black populations have lower plasma total cholesterol and higher HDL cholesterol levels and lower circulating levels of fibrinogen and factor VII than whites (Cassel, 1971). Blacks in Britain may also smoke less. It is very probable that coronary heart disease rates will rise in black people in Britain. However, major efforts in health promotion, with advice on the avoidance of cigarette smoking and care over healthy diets may have some impact.

The picture in the Asian community in Britain is very different as there is evidence from several studies of an excess of coronary heart disease compared with whites (Marmot et al, 1984). Whilst the national incidence of heart attack may be falling in Britain, it remains high in Asians of both Indian and Pakistani origin. This trend is clinically evident in coronary care units where one frequently sees Asian patients but rarely sees blacks. The high incidence of coronary disease in Asians is hard to explain. In India, heart attack is not a common reason for admission to hospital, and it seems possible that factors associated with migration or possibly the stresses of migration are also important. There is evidence that in Britain, Asian people in general smoke less than whites or blacks so it is unlikely that cigarette smoking habits explain the ethnic differences (Jackson et al, 1981). High dietary and plasma levels of cholesterol may be a major factor, and many Asian foods do contain large quantities of milk based products. Furthermore non-insulin dependent diabetes mellitus may be a factor, as it appears common in Asians in Britain (Mather & Keen, 1985). The ethnic differences in blood pressure discussed next do not explain the observed differences in coronary heart disease. Strategies for prevention of coronary heart disease are the responsibility of all medical and health personnel. The implications of the ethnic differences in heart disease discussed above are, therefore, very relevant to occupational health services in city centres.

Strokes and hypertension

Any clinician working in large cities in Britain will have noticed that hemiplegic strokes are common in blacks (Cruickshank et al, 1980). The commonest underlying cause of stroke is high blood pressure and it is readily apparent that hypertension is common in blacks. Population studies both in Britain and the USA have been able to document a higher prevalence of hypertension in the black population compared with whites. There remains some doubt whether the average blood pressure of black people is also higher than whites (Cruickshank & Alleyne, 1981). As blood pressure is related to body mass index and to low social class, it is possible that if these factors are taken into account, there remain no real ethnic differences in blood pressure. Obesity is very common in black women in Britain and both hypertension and strokes are major causes of illness.

An important factor related to high blood pressure is high salt intake, but there really is no convincing evidence that black people in Britain or the USA eat more salt than whites, although their potassium intake may be lower (Langford, 1981).

It is possible that a genetically determined tendency to raised blood pressure in response to a high salt diet is commoner in blacks than whites but no one has been able to demonstrate this effect. Alcohol excess, probably the commonest identifiable underlying cause of hypertension, is not unduly common in blacks.

Whatever the reasons for the ethnic differences in blood pressure, hypertension is the commonest chronic disease in western countries and it is commonest of all in blacks. Occupational health screening programme need only examine a few hundred people in each ethnic group before this trend becomes apparent. Hypertension, being eminently treatable is the most important parameter justifying screening of healthy people and many occupational health teams have made positive attempts to identify cases. With efficient care in collaboration with general practitioners and hospital doctors hypertensive patients really should not need to take time off work although some evidence from North America suggests that absenteeism may rise in response to case detection programmes (Haynes et al, 1978).

Diabetes mellitus

There appears to be a genuine excess of maturity onset (type II) non-insulin dependent diabetes in Asians and probably blacks in Britain (Mather & Keen, 1985; Odugbesau & Barnett, 1985). This tendency may be explained by the high incidence of obesity in both black and Asian women, but obesity does not explain the excess in men.

It is possible that selective migration of people without insulin dependent (Type 1) diabetes may explain why this condition is relatively rare in blacks and Asians. Control of blood glucose is often poor possibly due to poor compliance or inadequate understanding of dietary and drug regimes. As a consequence of this, many patients are inappropriately treated with insulin. The vascular and ophthalmic consequences of diabetes are common in both ethnic minorities with obvious implications for the continued employment of such people (Cruickshank & Alleyne, 1981). Help given by the occupational health team, with compliance with diet and medication would certainly be helpful. In particular, encouragement with calorie restriction, with the provision of healthier foods in work-place canteens could benefit the health of all employees.

Diabetic coma is a major problem in black people. In city centre hospitals, the majority of admissions for hyperosmolar non-ketotic diabetic coma are in black people. The prognosis of this type of coma is poor even with expert treatment (Nikolaides et al, 1984). The onset is usually insidious and early diagnosis in patients at their place of work could prevent progression.

Other diseases

There are many other differences in the incidence and nature of common diseases in the three main ethnic groups in Britain, as well as disease patterns which are more characteristic of people of Chinese or Middle Eastern origin. It is not proposed to discuss them in detail here. Leukaemias, peptic ulcers and thyroid disease are all common in Asians whereas the common cancers seen in white people (breast, stomach, colon and lung) are relatively uncommon (Potter et al, 1984; Marmot et al, 1984). There is an excess of gynaecological cancers in blacks and Asians have a high incidence of rheumatic heart disease. The reasons for these many and diverse differences are open to speculation but it is the opinion of the authors that they are predominately related to environmental factors, social class and living conditions and so should be remediable.

Pregnancy

Female employees may become pregnant so it is important that occupational physicians are aware of ethnic factors in obstetrics. The perinatal mortality rate is unacceptably high in Asians and the reasons for this are uncertain (Terry et al, 1980). Whilst antenatal care in the west is generally good, Asian women may present for the first time in late pregnancy often with obstetric crises. There is an excess of congenital malformations, but only in Muslim (mainly Pakistani and Bangladeshi) women and not in Indians or blacks. This may be due to the high frequency of consanguinous (first cousin) marriages. Occupational health workers should insist that Asian women in their employment who become pregnant should obtain early expert obstetrical care.

NUTRITIONAL DISEASES

Diseases related to nutrition are common, particularly in the Asian population. It is necessary that the various dietary habits of migrants from different parts of the world are understood. In general all migrant groups tend to continue to consume the same sorts of foods as they did in their country of origin, and furthermore second generation migrants also retain their ethnic dietary patterns.

Chinese migrants in Britain tend to continue to eat Chinese style food, with its high salt content. There is little evidence on this point but the very high salt diet in China and Japan has been blamed for the high incidence of stroke in these countries. West Indians in Britain still tend to consume imported yams, sweet potatoes and green bananas and the high calorie content of these foods may explain the high incidence of obesity, type II diabetes and hypertension (Cruickshank et al, 1985).

The diets of Asian people similarly may influence their disease patterns. The high cholesterol content of Asian sweets which are made from concentrated milk products may explain high blood cholesterol levels and coronary heart disease. Many Asian people, both Hindus and Sikhs are strict vegetarians and folic acid deficient macrocytic anaemias or iron deficient hypochromic anaemias are common and often undetected until low levels of haemoglobin are reached (Chanarin et al, 1985).

Finally there is the problem of adult Asian osteomalacia. This disease, which manifests itself with many multiple symptoms, including bone pain, pseudofractures and a distinctive waddling gait, is common particularly in Asian women and which cannot be explained, by intestinal malabsorption of vitamin D or chronic renal failure. Deficient vitamin D intake, together with lack of exposure to sunlight are probable explanations. Abnormalities of vitamin D metabolism have also been implicated as well as the high phytate content of chapati flour. This disease can be very disabling but once diagnosed it responds rapidly to dietary supplements of calcium and vitamin D (Heath, 1983). Elevated serum alkaline phosphatase levels on biochemical profiling, make a simple if relatively unselective screening test applicable at the place of work.

PSYCHIATRIC DISORDERS

A detailed discussion of this topic is relevant to occupational health but not feasible here. Enormous cultural and social pressures are experienced by minority immigrant groups. The nature of African and Asian societies is hugely different from Caucasian Western culture; thus many apparent abnormalities of behaviour may be explained by culture rather than illness. It is uncertain, for example, whether the excess of schizophrenia in blacks is really an excess of an illness, or rather a form of value judgement by white doctors of behaviour which would not be considered abnormal in other countries. The response of elderly West Indian men and women to the very obvious disaffection and bitterness of young black youths and the lack of sympathy or understanding of the dominantly white police, social workers and doctors can hardly fail to lead to depression. Unemployment, a blight which affects all age groups is highest in immigrants and the psychosocial consequences of this are all too apparent to doctors particularly in inner cities. Occupational physicians need to know the pressures experienced by ethnic minority families, and should influence employing authorities to be sensitive to the special problems of such people.

The psychiatric morbidity in the Asian community is itself different from other groups. In older Asian women, the lack of knowledge of the English language only compounds loneliness and at times bewilderment. Somatic symptoms for which no physical diagnosis can be found may well be primarily related to the stresses of being a visibly different non-English speaking minority (Griggs, 1986). Second generation Asian women, brought up with western orientated schooling and media now face the problems of rejection of traditional Asian patterns of arranged marriages, male domination of the family and society and closely confined large and at times claustrophobic family units. Why is alcoholism so common in Asians, when alcohol in any amount is proscribed in both the Muslim and Sikh faiths? It is not possible in this brief review to speculate further but an understanding of some of these pressures is important for occupational health teams in most cities in Europe and North America.

CONCLUSIONS

The differences in health and disease in ethnic minorities in western countries are complex. The problems encountered by new migrants are very different from second or third generation Asians or blacks. There is no scientific reason to suppose that there are any truly racial factors in disease; they are better defined as being of ethnic origin. Furthermore, with a few exceptions, the differences that are seen are dominantly of environmental rather than genetic origin. Major population migrations have brought new and challenging medical and social problems, and no doctors in any branch of medicine can ignore them.

REFERENCES

Adelstein A M 1978 Current vital statistics — methods and interpretation. British Medical Journal 2: 983–5
Bakhshi S 1978 Communication with Asian diabetics. British Medical Journal 2: 1500
Black J 1985a Asian families I: Cultures. British Medical Journal 290: 762–764
Black J 1985b Families from the Mediterranean and Aegean. British Medical Journal 290: 923–925
Black J 1985c Afro-Caribbean and African families. British Medical Journal 290: 984–988
Blakemore K 1983 Ethnicity, self-reported illness and the use of medical services by the elderly. Postgraduate Medical Journal 59: 668–670
Britt R P, Hollis Y, Keill J E, Weinrich M, Keil B W 1983 Anaemia in Asians in London. Postgraduate Medical Journal 59: 645–547
Cassel J C 1971 Summary of the major findings of the Evans County Cardiovascular studies. Archives of Internal Medicine 128: 887–889
Chanavin I, Malkowska V, O'Hea A-M, Riusler M G, Price A B 1985 Myeloblastic anaemia in a vegetarian Hindu community. Lancet ii: 1168–1172
Cooper R 1986 Race, disease and health. In: Rathwell T, Phillips D (eds) Health race and ethnicity. London, Croom Helm
Cox J L 1977 Aspects of transcultural psychiatry. British Journal of Psychiatry 30: 211–221
Cruickshank J K, Beevers D G, Osbourne V L, Haynes R A, Corlett J C R, Selby S 1984 Heart attack, stroke, diabetes and hypertension in West Indians, Asians and whites in Birmingham, England. British Medical Journal 281: 118–1189
Cruickshank J K, Alleyne S A 1981 Vascular disease in West Indian and white diabetics in Britain and Jamaica. Postgraduate Medical Journal 57: 766–768
Cruickshank J K, Jackson S H D, Beevers D G, Bannan L T, Beevers M, Stewart V L 1985 Similarity of blood pressure in blacks, whites and asians in England: The Birmingham Factory Study. Journal of Hypertension 3: 365–371
Davies S C, Luce P J, Win A A, Riordan J F, Brozovic M 1984 Acute chest syndrome in sickle cell disease. Lancet i: 36–38
Fleming A F (ed) 1982 Sickle cell disease. Churchill Livingstone, Edinburgh
Greaves I 1986 A comparison of acute hospital admission in Britain and India. Unpublished observations
Griggs J 1986 Ethnic status and mental illness in urban areas. In: Rathwell T, Phillips D (eds) Health race and ethnicity. Croom Helm, London
Haynes R B, Sackett D L, Taylor D W, Gibson E J, Johnson A L 1978 Increased absenteeism from work after detection and labelling of hypertensive patients. New England Journal of Medicine 229: 741–4
Heath D A 1983 Thoughts on the aetiology of vitamin D deficiency in Asians. Postgraduate Medical Journal 59: 649–651
Innes J A 1981 Tuberculosis in Asians. Postgraduate Medical Journal 57: 779–780
Jackson S H D, Bannan L T, Beevers D G 1981 Ethnic differences in respiratory disease. Postgraduate Medical Journal 57: 777–778

Jain C, Narayan N, Narayan K, Pike L A, Clarkson M E, Cox I G, Chatterjee J 1985 Attitudes of Asian patients in Birmingham to general practitioner services. Journal of the Royal College of General Practitioners 35: 416–418

Johnson M R D, Cross M, Cardew S A 1983 Inner city residents, ethnic minorities and primary health care. Postgraduate Medical Journal 59: 664–667

Johnson M R D 1984 Ethnic minorities and health. Journal of the Royal College of Physicians of London 18: 228–230

Langford H 1981 Is blood pressure different in black people? Postgraduate Medical Journal 57: 749–754

Layton-Henry Z 1985 The new commonwealth migrants 1954–62. History Today 35: 27–32

Marmot M G, Adelstein A M, Bulusu L 1984 Lessons from the study of immigrant mortality. Lancet ii: 1455–1457

Mather H M, Keen H 1985 The Southall Diabetes Survey: Prevalence of known diabetes in Asians and Europeans. British Medical Journal 291: 1081–1084

McNicol M W, Luce P J 1985 Sarcoidosis in a racially mixed community. Journal of the Royal College of Physicians 19: 179–183

Medical Research Council 1980 National survey of tuberculosis notifications in England and Wales. British Medical Journal 281: 895–898

Morrison-Smith J, Cooper S M 1981 Asthma and atopic disease in immigrants from Asia and the West Indies. Postgraduate Medical Journal 57: 774–776

Nickolaides K, Barnett A H, Spiliopovlos A J, Watkins P J 1981 West Indian diabetic population of a large inner city diabetic clinic. British Medical Journal 283: 1374–1375

Odugbesan O, Barnett A H 1985 Asian patients attending a diabetic clinic. British Medical Journal 290: 1051–1052

Packe G E, Patchett P A, Innes J A 1985 Tuberculosis outbreak among Rastafarians in Birmingham. Lancet i: 627–688

Potter J F, Dawkins D M, Pandha H S, Beevers D G 1984 Cancer in blacks, whites and Asians in a British hospital. Journal of the Royal College of Physicians of London 18: 231–235

Terry P B, Condie R G, Settatree R S 1980 Analysis of ethnic differences in perinatal statistics. British Medical Journal 281: 1307–1308

Thomas A N, Pattison C, Sergeant G R 1982 Causes of death in sickle-cell disease in Jamaica. British Medical Journal 285: 633–635

Warwick R, Swimer G, Britt R 1979 Increasing incidence of malaria in Britain. Lancet i: 1242–1243

Weatherall D J, Clegg J B (eds) 1981 Thalassaemia syndromes, 3rd edn. Blackwell Scientific Publications, London

Wilson N M 1985 Food related asthma: a difference between two ethnic groups. Archives of Disease in Childhood 60: 861–865

Winkelstein W, Kagan A, Kato H, Sachs S 1975 Epidemiological studies of coronary disease and stroke in Japanese men living in Japan, Hawaii and California: Blood pressure distributions. American Journal of Epidemiology 102: 502–510

Winthrobe M M, Lee G R, Boggs D R (eds) 1981 Clinical haematology, 8th edn. Lea and Febiger, Philadelphia

16. Shiftwork

P. Knauth J. Rutenfranz

INTRODUCTION

There is a substantial volume of literature on shiftwork. In a bibliography of shiftwork research covering the years 1950 to 1982 Schroeder & Goulden (1983) have cited altogether 1328 articles and books. An excellent critical review of the literature on shiftwork and health has been written by Harrington (1978). In the following article the publications from 1975 to 1985 considered by the authors to be typical or important will be reviewed.

However, first the term 'shiftwork' has to be defined. The definition 'shiftwork is work either permanently or frequently at unusual times or at changing times' covers all types of shiftwork including irregular working hours.

The two main categories of shift systems are: permanent shift systems and rotating shift systems. From the medical point of view it is important to differentiate between shift systems with and without nightwork. Further subgroups are discontinuous and continuous shift systems. Whereas discontinuous shift systems operate Monday to Friday or Monday to Saturday, continuous shift systems include weekend working as a feature of the shift system. Other features characterising shiftwork systems are: the number of consecutive nightshifts, the time at which shifts start and finish, the length of each shift, the order of shift rotation, the length of the shift cycle, and the regularity of the shift system.

As there are probably a lot of different shift systems in use in the world which have different negative and positive effects on man, it is important always to describe the main characteristics of the shift systems under study. However, even persons working in the same shift system vary widely in their tolerance of shiftwork.

This phenomenon is taken into consideration in the model of stress and strain of shiftworkers (Colquhoun & Rutenfranz, 1980). The objective stress, i.e. the phase-shifting of working and sleeping hours in relation to the phase of the circadian system as well as the combination of shiftwork and unfavourable working, organisational, and environmental conditions, is the same for all workers at a particular working place.

This stress induces a state of subjective strain in the shiftworker which may affect his well-being, working efficiency, family and social life or even his health. However, the magnitude of the effects observed in different persons at the same working place varies very much depending on 'intervening' variables. These 'intervening variables' include:

individual characteristics such as age, personality, circadian phase position (morningness/eveningness) and physiological adaptability;
job-related factors such as coping with the physical or mental load of the job, with the environmental conditions at the working place and with the type of shift system or;
domestic and social circumstances such as housing conditions, marital state, age of children, attitude of the family of the shiftworkers towards shiftwork in general or towards particular negative aspects of the shift system.

STRESS OF SHIFTWORKERS

The disruption of the daily variation of physiological functions caused by having to be awake and working at unusual hours, and to sleep during the daytime is one of the major stresses in connection with shiftwork (in particular nightwork).

Re-entrainment of physiological functions to nightwork

The time-structure of human biological rhythms covers periodicities from 10^{-3} seconds up to many years. Those rhythms with a period of approximate 24 hours are called 'circadian rhythms' (from the Latin 'circa diem', i.e. about a day).

There are many studies dealing with the question whether a complete adjustment of circadian rhythms of physiological functions to nightwork may be achieved or not. However, there is a lack of systematic longitudinal studies of shiftwork systems. The problem has mainly been investigated in field studies and experimental shiftwork studies covering only some weeks. To judge the alterations of circadian rhythms when changing from daywork to nightwork or within longer periods of night shifts it seems to be necessary to collect data, which cover the whole 24-hour period, i.e. during waking and sleeping. Many researchers could not get data from 0.00 to 24.00 h (Meers, 1975; Vokac & Rodahl, 1975; Patkai et al, 1977; Akerstedt et al, 1977; Lille et al, 1981; Dahlgren, 1981a). If only acrophases (determined by a cosinor analysis) are published it is not possible to judge changes of the shape of circadian rhythms (Reinberg et al, 1975).

However, even if the data are complete the interpretation is always difficult, because the observable circadian rhythms reflect not only the endogenous oscillatory drive but also the influence of exogenous influences. For instance, activity and sleep may mask the overt rhythm (Aschoff, 1978; Kiesswetter et al, 1981; Vokac & Vokac, 1985). To minimise such 'masking effects', some research groups try to study the entrainment of the endogenous clock under sensibly constant conditions ('constant routine'), e.g. taking an identical small meal at regular intervals in combination with a light sedentary activity for 24 h (Mills et al, 1978) or with bedrest for 24 h (Moog & Hildebrandt, 1982, 1986).

Discussing re-entrainment to nightwork it is necessary to distinguish between short-term adjustment which is defined as changes in the circadian rhythm over a period of successive night shifts, and long-term adjustment. Long-term adjustment is expected to occur only after considerable experience of shiftwork and to result in a facilitation of short-term adjustment (Folkard et al, 1978).

Most authors who have studied short-term adjustment to night work came to the conclusion that re-entrainment always remains incomplete because important synchronisers (Zeitgebers) are conflicting (Akerstedt et al, 1977; Folkard et al, 1978; Knauth et al, 1978, 1981, Chaumont et al, 1979; Vieux et al, 1979; Smith, 1979; Dahlgren, 1981a,b; Vokac et al, 1981; Minors & Waterhouse, 1985). Whereas some authors consider social and cognitive Zeitgebers, i.e. social contacts and awareness of time of day, as most important for humans (Aschoff, 1978; Dahlgren, 1981b; Knauth et al, 1981) other authors emphasise the importance of light as a synchroniser for human circadian rhythms (e.g. Czeisler et al, 1981).

As Zeitgeber effects are different for different physiological functions, an internal dissociation of these functions has to be expected in connection with nightwork. Different velocities of adjustment to night work have been observed by all authors who have registered paralled several physiological functions (Reinberg et al, 1975; Akerstedt et al, 1977; Patkai et al, 1977; Sen & Kar, 1978; Vokac et al, 1981; Kiesswetter et al, 1981, Knauth, 1983).

The correct methods to study long-term adjustment to nightwork aparently are longitudinal studies. However, there is only one study comparing circadian rhythm of body temperature of six shiftworkers 1 month, 1 year and 3 years after the introduction of a new rotating shiftwork schedule (Dahlgren, 1981a,b). The long-term adjustment during the awake periods was characterised by a general flattening of the curve within the nightshift week. Even after 3 years of experience there was still a short-term adjustment of the curve within the nightshift week.

The importance of life-style for long-term adjustment has been discussed by Folkard et al (1978) who showed that part-time nurses showed less evidence of adjustment from the first to the second night shift than full-time nurses. The part-time nurses were scheduling their lives to a predominantly day-oriented activity pattern whereas the full-timers were making much more of a commitment to a nocturnal way of life. These observations support the idea of the stress-strain model that strain is caused by the combination of stress factors and intervening variables. However, before discussing intervening variables other stresses besides the phase-shifting of work and sleep have to be mentioned.

Combined effects of shiftwork and unfavourable working, organisational, and environmental conditions

Some recent studies have shown that shiftwork often occurs in combination with unfavourable work load and with unfavourable environmental and organisational factors (Werner et al, 1980; Knauth, 1983).

Although the need of research concerning these combined effects is evident, only a few studies have dealt with this problem:

shiftwork and noxious agents (Guillerm et al, 1975; Reinberg et al, 1986)
shiftwork and noise (Cesana et al, 1982; Irion et al, 1983; Seibt et al, 1983; Seibt, 1986; Smith & Miles, 1986)
shiftwork and heat (Rutenfranz et al, 1986; Ottmann et al, 1986; Pokorski et al, 1986)
shiftwork in combination with paced or unpaced work (Pokorski et al, 1986).

INTERVENING VARIABLES

The fact that persons vary widely in their tolerance of shiftwork may be explained by the possible influence of many intervening variables. The intervening variables which have been studied in the last years are first individual factors such as age, personality and circadian rhythm characteristics and second situational factors.

According to Akerstedt & Torsvall (1981b) age plays an important role when regarding individual differences in adjustment to shiftwork. However, the importance varies depending on which shift is considered. Difficulties in connection with nightshift in particular related to sleep seem to increase at around 40–45 years of age (Akerstedt, 1976; Akerstedt & Torsvall, 1981b; Koller et al, 1981; Wynne et al, 1985). However, the cause-effect relation is yet unclear. Haider et al (1981) assumed that sensitisation reactions may develop in older shiftworkers.

The following aspects of personality have been studied in shiftworkers:

extraversion/introversion (e.g. Akerstedt, 1976; Colquhoun & Folkard, 1978; Colquhoun & Condon, 1981; Akerstedt & Torsvall, 1981b; Wynne et al, 1986);
neuroticism (e.g. Nachreiner, 1975; Akerstedt, 1976; Colquhoun & Folkard, 1978; Akerstedt & Torsvall, 1981b);
rigidity of sleeping habits (e.g. Folkard et al, 1979; Frese & Rieger, 1981; Wynne et al, 1986).

The factor 'rigidity of sleeping habits' seems to be a more powerful predictor of problems of shiftworkers than extraversion/introversion and neuroticism (Wynne et al, 1986).

Concerning the circadian rhythm characteristics some studies have dealt with the phase position (e.g. Hildebrandt et al, 1977; Breithaupt et al, 1978; Foret et al, 1982; Ishihara et al, 1983; Moog & Hildebrandt, 1985) or the dominant periodicity of body temperature rhythms in morning and evening types. Evening types seem to adjust quicker to nightwork than morning types. The diurnal type, however, is not a stable dimension. 'Morningness' increases with age (Torsvall et al, 1981).

Based on several studies Andlauer & Reinberg have published the hypothesis that shiftworkers with a high amplitude of circadian rhythms tolerate shiftwork better than persons with a small amplitude (e.g. Andlauer et al, 1977, 1979; Reinberg et al, 1978, 1980). However the results of other studies did not support this hypothesis (Costa & Gaffuri, 1983; Härmä et al, 1985). As an alternative hypothesis Minors & Waterhouse (1983, p 239) have speculatively proposed that a high amplitude

> 'might imply that the group is more regular in its habits of sleep and wakefulness and that the high amplitudes are a consequence of the associated masking effect'.

Finally the internal desynchronisation of the oral temperature circadian rhythm has been studied by Reinberg et al (1984). Desynchronisation of rhythm was found in most subjects with poor tolerance.

Besides individual characteristics situational factors seem to be of importance for the amount of problems reported by shiftworkers. Küpper et al (1980) and Knauth (1983) found that shiftworkers with daysleep often or always disturbed by noise complained more frequently about neurovegetative and gastrointestinal symptoms than shiftworkers with undisturbed or rarely disturbed sleep. However, longitudinal studies are needed to decide whether some shiftworkers become more sensitive and therefore complain more or if their health really is impaired by the accumulation of sleep loss.

STRAIN OF SHIFTWORKERS

The combination of shiftwork and intervening variables may have effects on well-being, health, performance efficiency, social and family life.

Well-being of shiftworkers

Sleep is a major concern in the life of shiftworkers. Reviewing the literature (up to 1981) on sleep disturbances of shiftworkers Rutenfranz et al (1980a) and Knauth (1983) conclude, that the highest frequency of complaints (up to 95%) was related to shift systems which include nightshifts whereas shift systems without nightshifts (up to 30% complaints) or daywork (up to 40% complaints) seemed to be less problematic. Complaints about sleep disturbances are related to a reduced duration of sleep, an insufficient quality of sleep or both.

In general the shortest sleep duration is found in connection with nightshifts as has shown a review of studies up to 1979 (Knauth & Rutenfranz, 1981) or a recent study by Frese & Harwich (1984). For example (based on 14 000 diary records of 1230 shiftworkers) the average duration of daysleep between two night shifts was 6.1 hours and the duration of night sleep before morning shifts 7.0 hours (Knauth et al, 1980). However, the duration of sleep before morning shifts is depending on the start time of the morning shift and the commuting time. The earlier the morning shift begins and the longer the travel from home to work the shorter will be the duration of sleep before morning shifts (e.g. Hak & Kampman, 1981).

Comparisons of sleep duration for workers on permanent night shift and rotating shifts (Tasto et al, 1978; Lortie et al, 1979; Tepas et al, 1981; Dahlgren, 1981c; Smith et al, 1982) or comparisons of quickly rotating shift systems with more slowly rotating ones (Melton et al, 1973; Reinberg et al, 1975; Smith, 1979; Knauth et al, 1983a) produce no clear results. The main methodological problem is the comparability of the groups in different shift systems.

There seems to be only one longitudinal study of sleep length of six rotating shiftworkers 1 month, 1 year and 3 years after a change of the shift system (increase of nightwork). However, as the number of subjects and type of data did not allow a statistical analysis, only tendencies were reported: with longer experience sleep length increased and sleep difficulties decreased. Cross-sectional studies, however, comparing different age classes found a significant decrease in sleep length after nightshifts with age (Akerstedt & Torsvall, 1981a).

The number of EEG studies in the home of the shiftworkers (Tilley et al, 1981; Torsvall et al, 1981) near to the work place (Matsumoto, 1978; Foret & Benoit, 1980) or in special sleep laboratories (e.g. Goncharenko, 1979; Walsh et al, 1981; Dahlgren, 1981c; Webb, 1983; Kiesswetter et al, 1985) is not very big. However, all these studies show that day sleep is not only a phase-shifted nightsleep but it also has a different sleep structure. In most studies a shorter REM-latency, a shorter amount of REM- and stage-2-sleep as well as another distribution of REM-periods in day sleep compared with night sleep were found.

More studies are needed to investigate the sleep quality during successive days with nightshifts. There are some EEG-data supporting the hypothesis of an accumulation of sleep deficits but other data show a tendency towards recovery which, however, depend strongly on individuals (Foret & Benoit, 1980; Kiesswetter et al, 1985). Meijman (1981) who studied the 4 days following a night shift period of 7 nights concluded from data of subjective judgement of sleep quality and fatigue that a recovery period of 2 or 3 days is too short. In other studies comparing slowly rotating and rapidly rotating shift systems less sleep problems or less fatigue effects were observed in the quickly rotating shift systems with maximal two or three nightshifts in succession (Knauth & Schmidt, 1986; Williamson & Sanderson, 1985).

Health effects

Since the very careful epidemiologic study of Taylor & Pocock (1972) on mortality of shiftworkers no comparable study has been performed. In a retrospective survey on invalidity and mortality over 18 years Hannunkari et al (1978) compared drivers of locomotives and their assistants with two reference groups of trainmen and railroad clerks. The observed number of deaths/person-years was 5.3 o/oo in the group of engineers, 4.6 in the group of railroad clerks and 4.1 in the group of trainmen. The number of observed cases of death or disability due to diseases of the circulatory system was higher than the expected number of cases in the group of engineers. The inverse was true for the two reference groups. The stress of irregular working hours including nightshifts seems to be combined with other stresses as noise vibration, draft and poor seats as has shown an additional questionnaire study in a group of locomotive engineers. However, there was no information about the working hours and working conditions of the two reference groups.

In another study unfavourable environmental effects also seemed to be of importance. Teiger (1984) analysed data from a professional retirement fund of four occupational groups in newspaper companies. The working conditions of rotary printers (permanent night workers) and plate makers (permanent evening workers) were characterised by physical load, time pressure, noise, toxic exposure and heat whereas compositors (afternoon and evening workers) and especially correctors (afternoon and evening workers) had better environmental conditions. At age of 70 years there was a significant difference in the percentage of living population between the rotary printers (63.5% survivors) compared with the other groups combined (plate makers 66.9%, compositors 70.5% and correctors 74.3%); 43.0% of the rotary printers retired early compared with 10.8% of the correctors. The reasons for most of the early retirements was medically attested working incapacity.

All studies analysing absence from work attributed to sickness of groups in different shift systems have the methodological problem of comparability of groups. Therefore it is not astonishing that results are contradictory, i.e. some are in favour of permanent shift systems, others are in favour of rotating systems (e.g. De Groote, 1975; Colligan et al, 1979; Fischer, 1986). With regard to rotating shift systems, however it is recommended to differentiate between slowly and rapidly rotating shift systems.

In a study of medically certified absence in a quickly rotating 2-2-2 shift system, no significant effects of time of the week, type of shift or position within the cycle were observed (Nicholson et al, 1978). To take the problem of comparability of groups into consideration Koller (1983) used a procedure to match shiftworkers to dayworkers with regard to age and years on work. She did not find differences in sickness absence between shift and dayworkers, but there was a significantly excess rate in drop-outs. An overall 'health score' was computed from data concerning absence attributed to sickness, morbidity, distribution and severity of diseases and subjective complaints. This health score showed a steep decrease during the first years of shiftwork, followed by a continued slight decrease in middle age and again a pronounced decrease from the age of 41 years onwards. In dayworkers a stabilisation in scores was observed up to middle age with a distinct decrease from about 40 to 50 years. In an analysis of specific effects of shiftwork on health Koller (1983) found that the shift-working population significantly exceeded dayworkers in diseases of the digestive system and the circulatory system.

Many older studies were dealing with gastrointestinal complaints and diseases of shiftworkers. In a review of the literature (up to 1979) on ulcer incidence in a total of 34 047 persons with day- or shiftwork Rutenfranz et al (1980a) found that:

about 0.3–7% of persons with daywork;
about 5% of persons with morning and afternoon shifts;
about 2.5–15% of persons with rotating shift systems including nightwork;
about 10–30% of persons who had given up shiftwork; had ulcers.

The incidences of ulcer in these groups are widely overlapping, i.e. the type of working time, organisation cannot be the single cause for ulcers. Wolf et al (1979) report of 17 factors which increase the probability of duodenal ulcer not including nightwork.

In the retrospective cohort study of Angersbach et al (1980) the indicence of ulcers in former shiftworkers was much higher than in other shiftworkers and dayworkers. The process of self selection seemed to be still incomplete at the end of the study period of 11 years because the relative risk of suffering from a gastrointestinal disease for the first time was higher in the last years of the investigation.

In the last few years the interest of researchers seems to have focussed more on cardiovascular diseases of shiftworkers. In studies of Koller et al (1978) and of Angersbach et al (1980) the drop-outs had a higher incidence of cardiovascular diseases compared with reference groups, i.e. other shiftworkers or dayworkers.

A register study of a representative sample of the Swedish population showed a small but statistically significant excess risk of ischaemic heart disease in shiftworkers (Alfredsson et al, 1982). However, this kind of study cannot control for selection effects. A continuous increase in risk of coronary heart disease with increasing shift exposure was found in a 15-year follow-up study (Orth-Gomer et al, 1985). The increase became significant only after 11 years or more of shiftwork. However, in the exposure class of 21 or more years of shiftwork there was no excess risk of ischaemic heart disease. It might be speculated that this phenomenon is caused by selection.

Cesana et al (1983) investigated stress and risk of coronary heart diseases in shiftworkers. They observed a progressive increase of adaptation and selection process with the aging of the working population. There is not enough evidence up to now to conclude that cardiovascular diseases are more prevalent in shiftworkers.

Reviewing studies related to neurotic reactions and psychosomatic disorders Koller et al (1981) state that no clear conclusions about higher frequencies of psychiatric disorders in shift or dayworkers can be drawn.

As a new aspect of shiftwork research the effects of shiftwork on the human immune system were studied by Nakano et al (1982) and Curti et al (1982). A statistically significant depression of lymphocyte DNA synthesis has been observed in three-shift workers compared with dayworkers (Curti et al, 1982). In three-shift workers studied by Nakano et al (1982) the circadian variation of mitogen-induced blastogenesis of lymphocytes was lost whereas in permanent nightworkers only the response to concanavalin A was depressed.

Effects on social life

From older studies it is known that depending on the type of shift and type of shift system shiftworkers may have social problems because their working times and leisure times are shifted compared with the 'normal' social environment. These problems may be related to the:

time for marital-related activities;
time for children;
frequency of visiting friends and relatives;
frequency of leisure activities;
possibility to be an active member of voluntary organisations;
time for continuing education.

More recent studies have supported the older findings or added more detailed information (e.g. Nachreiner et al, 1975; Tasto et al, 1978; Ulich & Baitsch, 1979; Rutenfranz et al, 1980b; Gordon et al, 1981; Baer et al, 1981; Balck & Vajen, 1982; Jansen et al, 1983). Some authors try to quantify effects of different types of shifts and shift systems on leisure time (e.g. Rutenfranz et al, 1980b; Knauth et al, 1983a) by using time budget studies. Clark (1984) used this method to get information on the social network and social contacts of the shiftworkers. Furthermore, utility scaling of time off work shows a differential attribution of utilisation potential for leisure time activities for different times of the day and different days of the week (Wedderburn, 1981; Baer et al, 1985).

The hypothesis of Mott et al (1965) that the workers' psychosocial reaction to shiftwork is depending to a great extent on the attitude of his wife and her willingness and ability to adjust to his working hours has not yet been studied in detail. Clark (1984) has investigated the range of social ties, the joint conjugal role relationship and the domestic division of labour, however, without analysing the attitudes of the shiftworkers to shiftwork.

Performance and accidents

Circadian rhythms in human performance exist for a wide range of tasks studied under laboratory and field conditions (reviews: Rutenfranz & Colquhoun, 1979; Folkard & Monk, 1979). Studies of the activities of operators in a control room of the chemical industry showed that the human operator perceives different amounts of information and uses different working procedures at different times of the day (Queinnec et al, 1984). Although in most cases performance is best during daytime and worst during nighttime there is at least one exemption; performance of memory-loaded cognitive tasks is best during nighttime (Folkard et al, 1976; Monk et al, 1978; Monk & Embrey, 1981).

In a review of 24 studies up to 1983 there were almost as many studies with a higher frequency of accidents at night as studies with a higher frequency of accidents at daytime (Knauth, 1983). However, this result may not be attributed to the classification in memory-loaded tasks and other tasks. The main methodological problem in this kind of accident research is the fact that working, environmental, and organisational conditions at night and at daytime are not comparable (Colquhoun, 1976; Carter & Corlett, 1984).

COMPENSATION FOR THE NEGATIVE EFFECTS OF SHIFTWORK

Nightwork may be regarded as a risk factor for health and often results in negative effects on the well-being and social contacts of shiftworkers. The best way to solve this problem would be the elimination of nightwork. As this approach is unrealistic another approach (the reduction of nightwork) is recommendable. As an example Knauth et al (1983b) in cooperation with the management and the occupational health practitioner reduced the number of persons needed during night-time (2–6 am) for loading and unloading aircraft in a first step from 104 to 66 and in a second step down to 38 persons by redesigning the shift system.

If it is not possible to reduce the amount of nightwork or the amount of shiftwork Thierry et al (1975) recommend specific measures ('countervalues') for the compensation of specific deficits of the shiftworkers. For example an adequate compensation for sleep deficits caused by noise during daysleep after nightshifts is not money but rather a sound-attenuated sleeping room. If many shiftworkers complain because they cannot participate in regular evening courses for continuing education, courses should also be offered in the morning or afternoon.

Rutenfranz (1982) has proposed the following special health provisions for night workers:

selection of workers;
regular health checks;
preventive health care;
additional free days;
providing proper meals at night;
sleeping allowances during nightshifts;
constructing shift schedules based on objective physiological, psychological and social criteria.

Selection of workers

The following persons should be excluded from nightwork (further details see Rutenfranz, 1982): persons with a history of digestive tract disorders, diabetics, people with thyrotoxicosis, epileptics (e.g. primary generalised epilepsies), persons suffering from chronic sleep disturbances, patients with heart diseases exhibiting a significant reduction of physical performance capacity, active and extensive tuberculosis patients, alcoholics, other drug addicts and persons with visual impairment that is too severe for effective correction to be possible.

Regular health checks

A committee of occupational health practitioners and scientific experts has proposed the following procedure (Herrmann, 1982; Rutenfranz, 1982) for regular health checks:

all persons working at least 6 hours on a regular shift system or at least 5 hours on an irregular shift system during the hours from 10 pm to 6 am should have regular health checks;
the recruiting medical examination should exclude all persons from nightwork using the above mentioned negative health criteria;
there should be a second health check at the latest 12 months after starting nightwork;
regular health checks depending on the age of the worker at least at the following intervals:

aged under 25 years:	24 months
25–50 years:	60 months
50–60 years:	24–36 months
over 60 years:	12–24 months

Preventive health care and additional free days

The discrepency between the time structure of the shiftworkers and the time structure of other persons, e.g. the family, friends, relatives and neighbours causes most problems of shiftworkers. Therefore it seems reasonable to give the shiftworker the chance to live under 'normal' conditions with respect to circadian rhythms as often as possible. Some large German plants and communities have begun to offer regular treatment for shiftworkers over 50 years old in specialised rehabilitation clinics (Kurkliniken) for 2 to 3 weeks at 2- or 3- year intervals. By following a regular sleep-wake-routine with proper meal-times the shiftworker may normalise his circadian rhythms. Furthermore physiotherapeutic measures and a general health check are offered.

The recommendation of additional free days for persons in continuous shift systems also tries to increase the number of 'normal' days for shiftworkers. However, there are no epidemiological longitudinal studies about the efficiency of extra rest days for health.

Providing proper meals at night

In some studies shiftworkers complained about disturbances of appetite, in particular in connection with nightshifts (e.g. Shift Work Committee, 1979; Werner et al, 1980; Rutenfranz et al, 1980b). Therefore the French industry is obliged by law to provide the nightworkers with facilities at least to warm-up meals during nighttime. Reinberg et al (1979) report that carbohydrate intake appeared to be lower in 1974 than in another study in 1963 in the same oil refinery. This change may be caused by the introduction of a cafeteria. The shiftworkers preferred the regular meals at the cafeteria to home made sandwiches.

Some studies of the eating behaviour of shiftworkers show that it might be desirable to improve this behaviour by information and 'training' (Masuda & Takagi, 1975; Matsumoto et al, 1978; Reinberg et al, 1979; Cervinka et al, 1984; Croft, 1984).

Sleeping allowances during night shifts

Minors & Waterhouse (1981) conclude from their laboratory studies that a short 'anchor sleep' taken at the same time each night could stabilise the circadian rhythms. In Japan about half of three-shift workers took naps in the form of short anchorsleep-periods during midnight or early morning hours (Kogi, 1981). Many Japanese plants seem to have such more or less legalised sleeping allowances during night shifts and many of the shiftworkers even had the possibility to use a bed in the factory.

Although it is known that shiftworkers in other countries also take naps during nightshifts, most employers are not ready to allow this. The advantages of such naps are prevention of fatigue and improvement of efficiency and safety (Kogi & Sakai, 1985; Andlauer et al, 1982). However, problems may arise if the shiftworker has difficulties in falling asleep or if he has to react correctly immediately after having been woken up, e.g. by an alarm bell.

The daytime sleep which follows a night-time nap seems to be shorter (Torsvall et al, 1986) and seems to have less stage four sleep (Matsumoto et al, 1982). In general it seems to be recommended to help shiftworkers to develop an individual sleep strategy to cope as well as possible with their shift system (Kogi, 1982; Tepas, 1982; Croft, 1984).

Construction of shift schedules based on objective criteria

It may be assumed that there are thousands of different shift schedules in use all over the world. Many shiftworkers who do not know other shift systems think that the schedule they are used to is the best one. It seems, however, more reasonable to base the judgement of a shift system on objective criteria as the adjustment of physiological functions to nightwork, performance and accidents, well-being (e.g. sleep, fatigue, appetite), health, personal and social problems.

Based on these criteria Knauth & Rutenfranz (1982) concluded from a review of the literature that the following recommendations should be kept in mind when constructing or judging a shift system:

(a) a shift system should have few nightshifts in succession;
(b) the morning shift should not begin too early;
(c) the shift change times should allow individuals some flexibility;
(d) the length of the shift should depend on the physical and mental load of the task, and the night shift could be shorter than the morning and afternoon shifts;
(e) short intervals of time between two shifts should be avoided;
(f) continuous shift systems should include some free weekends with at least two successive full days off;
(g) in continuous shift systems a forward rotation should be preferred;
(h) the duration of the shift cycle should not be too long;
(i) shift rotas should be regular.

The recommendations are in good agreement with conclusions of the Japanese 'Shift Work Committee', (1979). In this connection the book of Queinnec et al, (1985) is also of interest because they have discussed in detail the positive and negative consequences of changing different characteristics of a shift system.

Two of the recommendations of Knauth & Rutenfranz (1982) should be explained in more detail. The first and most important recommendation is based on the following reasons (Knauth, 1983): Even after many nightshifts in succession reentrainment of physiological functions to nightwork seems to remain incomplete. In connection with few consecutive nightshifts least disturbance of circadian physiological functions is observed.

An accumulation of fatigue and in some studies of sleep, deficits has been observed in the course of longer periods of consecutive nightshifts. Furthermore there are some hints in the literature that the probability of accidents may increase after the second nightshift. In two studies (published after 1975) in which shiftworkers had changed from slowly to rapidly rotating shift systems the majority reported an overall preference for the new shift system or had a higher job satisfaction (Knauth & Schmidt, 1985; Williamson & Sanderson, 1986). It would, however, be desirable to have more controlled eperimental field studies.

The last aspect to be mentioned is the length of shift. There are several studies of 12-hour shifts (e.g. Nachreiner et al, 1975; Bajnok, 1975; Crump & Newson, 1975; Ganong et al, 1976; Gardner & Dagnall, 1977; Northrup et al, 1979; Kelly & Schneider, 1982; Saito, 1982; Sakai et al, 1982). Notable advantages include less time necessary for commuting, greater blocks of time off and even with an average working time of 40 (42) hours per week (in 4 weeks) it is possible to meet the recommendations cited above.

However, some older workers have problems to adjust to the long shifts (Northrup et al, 1979). Furthermore in a compressed workweek (e.g. three consecutive 12-hour shifts) fatigue seems to be a problem (Saito, 1982). Gardner & Dagnall (1977) studied absence attributed to sickness of shiftworkers who had experience of 'at least a year'. They conclude that there is no evidence that the change to a 12-hour shift had any significant effect on this kind of absence. However, the study period is very short and longitudinal studies are needed. In general the mental and physical load of the task should be one of the main criterias when discussing the introduction of twelve-hour shifts. Another problem is the (not yet existing) tolerable limits of exposure to toxic agents for 12-hour shifts.

CONCLUSION AND NEED FOR FUTURE STUDIES

Although the studies of the last 10 years have increased our knowledge about possible effects of shiftwork on health, working efficiency, family and social life, more epidemiological studies are needed. Cause-effect relationships may only be investigated by carefully controlled longitudinal studies. It is of course very difficult to perform a controlled study in a complex field situation with a combination of shiftwork and unfavourable environmental and organisational conditions and it is not always possible to find adequate control groups.

It is also very important to study the long-term effects of changes of the shift system or of the accompaning unfavourable stresses within groups of shiftworkers. No shiftworker can really judge the effects of a shift system which he does not know from experience. Therefore experimental changes of shift systems should be supported for a trial period of at least half a year.

Furthermore as there is not one best way to organise sleep, meals and leisure time to cope with different kinds of shiftwork, every shiftworker should be encouraged to try several strategies to find out which one best fits his needs.

REFERENCES

Akerstedt T 1976 Interindividual differences in adjustment to shift work. In: International Ergonomics Association, IEA (ed) Proceedings. 6th Congress of the International Ergonomics Association: 'Old world, new world, one world', 11–16 July 1976, University of Maryland, College Park, Maryland, p 510–514
Akerstedt T, Patkai P, Dahlgren K 1977 Field studies of shiftwork: II. Temporal patterns in psychophysiological activation in workers alternating between night and day work. Ergonomics 20: 621–631
Akerstedt T, Torsvall L 1981a Age, sleep and adjustment to shiftwork. In: Koella W P (ed) Sleep 1980. Karger, Basel p 190–195
Akerstedt T, Torsvall L 1981b Shift work. Shift-dependent well-being and individual differences. Ergonomics 24: 265–273
Alfredsson L, Karasek R, Theorell T 1982 Myocardial infarction risk and psychosocial work environment: an analysis of the male Swedish working force. Social Science and Medicine 16: 463–467
Andlauer P, Carpentier J, Cazamian P 1977 Ergonomie du travail de nuit et des horaires alternants. Editions Cujas, Paris
Andlauer P, Reinberg A, Fourre L, Battle W, Duverneuil G 1979 Amplitude of the oral temperature circadian rhythm and the tolerance to shift-work. Journal de Physiologie (Paris) 75: 507–512
Andlauer P, Rutenfranz J, Kogi K, Thierry H, Vieux N, Duverneuil G 1982 Organization of night shifts in industries where public safety is at stake. International Archives of Occupational and Environmental Health 49: 353–355
Angersbach D, Knauth P, Loskant H, Karvonen M J, Undeutsch K, Rutenfranz J 1980 A retrospective cohort study comparing complaints and diseases in day and shift workers. International Archives of Occupational and Environmental Health 45: 127–140
Aschoff J 1978 Features of circadian rhythms relevant for the design of shift schedules. Ergonomics 21: 739–754
Baer K, Ernst G, Nachreiner F, Schay T 1981 Psychologische Ansätze zur Analyse verschiedener Arbeitszeitsysteme. Zeitschrift für Arbeitswissenschaft 35: (7NF) 136–141
Baer K, Ernst G, Nachreiner F, Volger A 1985 Subjektiv bewertete Nutzbarkeit von Zeit als Hilfsmittel zur Bewertung von Schichtplänen. Zeitschrift für Arbeitswissenschaft 39: (11NF), 169–173
Bajnok I 1975 'It's good for nurses, but is it good for the patients?' Hospital Administration in Canada 17: 25–26
Balck F B, Vajen H 1982 Auswirkungen der Schichtarbeit auf die Gesundheit und die sozialen Beziehungen. Medizin, Mensch, Gesellschaft 7: 7–11
Breithaupt, H, Hildebrandt G, Döhre D, Josch R, Sieber U, Werner M 1978 Tolerance to shift of sleep, as related to the individual's circadian phase position. Ergonomics 21: 767–774
Carter F A, Corlett E N 1984 Shiftwork and accidents. In: Wedderburn A, Smith P (eds) Psychological approaches to night and shift work. International research papers, seminar paper 3.1–3.11 Heriot-Watt University, Edinburgh, Scotland
Cervinka R, Kundi M, Koller M, Haider M, Arnhof J 1984 Shift related nutrition problems. In: Wedderburn A, Smith P (eds) Psychological approaches to night and shift work. International research papers, seminar paper 14.1–14.18 Heriot-Watt University, Edinburgh, Scotland
Cesana G C, Ferrario M, Curti R, Zanettini R, Grieco A, Sega R, Palermo A, Mara G, Libretti A, Algeri S 1982 Work-stress and urinary catecholamines excretion in shift workers exposed to noise. Epinephrine I (E) and nor-epinephrine (NE). Medicina del Lavoro 73: 99–109
Cesana G C, Zanettini R, Galli M, Ferrario M, Panza G, Grieco A 1983 Stress and coronary risk in shiftworkers. Medicina del Lavoro 74: 351–360
Chaumont A-J, Laporte A, Nicolai A, Reinberg A 1979 Adjustment of shift workers to a weekly rotation (study 1). Chronobiologia 6: Suppl 1: 27–34
Clark D Y 1984 Types of household adjustment to rotating shiftwork: employees in manufacturing industry. In: Wedderburn A, Smith P (eds) Psychological approaches to night and shift work. International research papers, seminar paper 10.1–10.16 Heriot-Watt University, Edinburgh, Scotland
Colligan M J, Frockt I J, Tasto D L 1979 Frequency of sickness absence and worksite clinic visits among nurses as a function of shift. Journal of Environmental Pathology and Toxicology 2: 135–148
Colquhoun W P 1976 Accidents, injuries and shift work. In: Rentos P G, Shephard R D (eds) Shift work and health. A symposium, p 160–175 US Department of Health, Education, and Welfare, National Institute for Occupational Safety and Health, Cincinnati, Ohio
Colquhoun W P, Condon R 1981 Introversion-extraversion and the adjustment of the body-temperature rhythm to night work. In: Reinberg A, Vieux N, Andlauer P (eds): Night and shift work. Biological and social aspects, p 449–455 Pergamon Press, Oxford (Advances in the biosciences, vol 30)
Colquhoun W P, Folkard S 1978 Personality differences in body-temperature rhythm, and their relation to its adjustment to night work. Ergonomics 21: 811–817

Colquhoun W P, Rutenfranz J 1980 (eds) Studies of shiftwork. Taylor & Francis, London
Costa G, Gaffuri E 1983 Circadian rhythms, behaviour characteristics and tolerance to shiftwork. Chronobiologia 10: 395
Croft G 1984 Shiftwork: how to cope. A three part series. Claremont, New Hampshire, Community Health Network
Crump C K, Newson E F P 1975 Implementing the 12-hour shift: a case history. Results of a study conducted at the University Hospital in London, Ontario. Hospital Administration in Canada 17: 20–24
Curti R, Radice L, Cesana G C, Zanettini R, Grieco A 1982 Work stress and immune system: lymphocyte reactions during rotating shift work. Preliminary results. Medicina del Lavoro 6: 564–569
Czeisler C A, Richardson G S, Zimmerman J C, Moore-Ede M C, Weitzman E D 1981 Entrainment of human circadian rhythms by light-dark cycles: a reassessment. Photochemistry and Photobiology 34: 239–247
Dahlgren K 1981a Temporal patterns in psychophysiological activation in rotating shift workers — A follow-up field study one year after an increase in nighttime work. Scandinavian Journal of Work, Environment and Health 7: 131–140
Dahlgren K 1981b Long-term adjustment of circadian rhythms to a rotating shiftwork schedule. Scandinavian Journal of Work, Environment and Health 7: 141–151
Dahlgren K 1981c Adjustment of circadian rhythms and EEG sleep functions to day and night sleep among permanent nightworkers and rotating shiftworkers. Psychophysiology 18: 381–391
De Groote C 1975 Ploegenarbeid en absenteisme. Extern 4: 407–413
Fischer F M 1986 Absenteeism among shiftworkers of the automotive plants. In: Haider M, Koller M, Cervinka R (eds) Night and shiftwork: longterm effects and their prevention. Lang, Frankfurt am Main, Bern (Studies in industrial and organizational psychology, vol 3, p 189–196)
Folkard S, Knauth P, Monk T H, Rutenfranz J 1976 The effect of memory load on the circadian variation in performance efficiency under a rapidly rotating shift system. Ergonomics 19: 479–488
Folkard S, Monk T H 1979 Shiftwork and performance. Human Factors 21: 483–492
Folkard S, Monk T H, Lobban M C 1978 Short and long-term adjustment of circadian rhythms in 'permanent' night nurses. Ergonomics 21: 785–799
Folkard S, Monk T H, Lobban M C 1979 Towards a predictive test of adjustment to shift work. Ergonomics 22: 79–91
Foret J, Benoit O 1980 Predictable effects on individual sleep patterns during a rapidly rotating shift system. International Archives of Occupational and Environment Health 45: 49–56
Foret J, Benoit O, Royant-Parola S 1982 Sleep schedules and peak times of oral temperature and alertness in morning and evening 'types'. Ergonomics 25: 821–827
Frese M, Harwich C 1984 Shiftwork and the length and quality of sleep. Journal of Occupational Medicine 26: 561–566
Frese M, Rieger A 1981 Beschreibung und Kritik einer Skala zur Prädiktion von psychophysischem Befinden bei Schichtarbeitern. Zeitschrift für Arbeitswissenschaft 35: (7 NF), 95–100
Ganong W L, Ganong J M, Harrison E T 1976 The 12-hour shift: better quality, lower cost. Journal of Nursing Administration 6: 17–29
Gardner A W, Dagnall B D 1977 The effect of twelve-hour shift working on absence attributed to sickness. British Journal of Industrial Medicine 34: 148–150
Goncharenko A M 1979 Electrophysiological investigation of sleep in shift workers exposed to emotional stress due to work. Human Physiology 5: 468–474
Gordon G C, McGill W L, Maltese J W 1981 Home and community life of a sample of shift workers. In: Johnson L C, Colquhoun W P, Tepas D I, Colligan M J (eds) Biological rhythms, sleep and shift work, p 357–369 Spectrum Publications, New York (Advances in sleep research, vol 7)
Guillerm R, Radziszewski E, Reinberg A 1975 Persisting and unaltered circadian rhythms of six healthy young men with a night-work shift every 48 hrs and a 2% CO_2 atmosphere during a 4-week span. Chronobiologia 2: 336–345
Haider M, Kundi M, Koller M 1981 Methodological issues and problems in shift work research. In: Johnson L C, Colquhoun W P, Tepas D I, Colligan M J (eds) Biological rhythms, sleep and shift work, p 145–163 Spectrum Publications, New York (Advances in sleep research, vol 7)
Hak A, Kampman R 1981 Working irregular hours: complaints and state of fitness of railway personnel. In: Reinberg A, Vieux N, Andlauer P (eds) Night and shift work. Biological and social aspects, p 229–236 Pergamon Press, Oxford (Advances in the biosciences, vol 30)
Hannunkari I, Järvinen E, Partanen T 1978 Work conditions and health of locomotive engineers. II. Questionnaire study, mortality and disability. Scandinavian Journal of Work, Environment and Health 4: Suppl 3: 15–28
Härmä M I, Ilmarinen J, Knauth P, Rutenfranz J, Hänninen O 1986 The effect of physical fitness intervention on adaptation to shiftwork. In: Haider M, Koller M, Cervinka R (eds) Night and

shiftwork: longterm effects and their prevention. Lang, Frankfurt am Main, Bern (Studies in industrial and organizational psychology, vol 3, p 221–228)

Harrington J M 1978 Shift work and health. A critical review of the literature. Her Majesty's Stationery Office, London

Herrmann H 1982 Bedeutung und Bewertung der Nachtarbeit — Gedanken zur Erstellung eines Berufsgenossenschaftlichen Grundsatzes –. In: Fliedner T M (ed) Kombinierte Belastungen am Arbeitsplatz. Der chronisch Erkrankte im Betrieb. Bericht über die 22. Jahrestagung der Deutschen Gesellschaft für Arbeitsmedizin eV, Ulm/Neu-Ulm, 27.–30.4.1982. Teil III: Arbeitsmedizinisches Kolloquium. S 61–70 Gentner, Stuttgart

Hildebrandt G, Breithaupt H, Döhre S, Stratmann I, Werner M 1977 Arbeitsphysiologische Bedeutung und Bestimmung der circadianen Phasentypen. Zeitschrift für Arbeitswissenschaft 31: (3 NF), 98–102

Irion H, Roßner R, Lazarus H 1983 Entwicklung des Hörverlustes in Abhängigkeit von Lärm, Alter und anderen Einflüssen. In: Stalder K (ed) Immunbiologische Aspekte in der Arbeitsmedizin. Bericht über die 23. Jahrestagung der Deutschen Gesellschaft für Arbeitsmedizin e.V, Göttingen, 4.–7.5.1983. S 355–356 Gentner, Stuttgart

Ishihara K, Saitoh T, Miyata Y 1983 Short-term adjustment of oral temperature to 8-hour advanced-shift. Japanese Psychological Research 25: 228–232

Jansen B, Van Hirtum A, Thierry H 1983 Unannehmlichkeiten der Schichtarbeit. Fallstudie auf der Grundlage des Modells für innovative Kompensationsfunktionen. Endbericht. Europäische Stiftung zur Verbesserung der Lebens- und Arbeitsbedingungen, Dublin

Kelly R J, Schneider M F 1982 The twelve-hour shift revisited: recent trends in the electric power industry. Journal of Human Ergology 11: Suppl. 369–384

Kiesswetter E, Knauth P, Schwarzenau P, Rutenfranz J 1985 Daytime sleep adjustment of shiftworkers. In: Koella W P, Rüther E, Schulz H (eds) Sleep '84. G Fischer, Stuttgart, New York

Kiesswetter E, Knauth P, Weier R, Theissen W, Rutenfranz J 1981 Reentrainment of rectal temperature and heart frequency during days with experimental night shifts and morning and afternoon sleep. In: Reinberg A, Vieux N, Andlauer P (eds) Night and shift work. Biological and social aspects, p 99–106 Pergamon Press, Oxford (Advances in the biosciences, vol 30)

Knauth P 1983 Ergonomische Beiträge zu Sicherheitsaspekten der Arbeitszeitorganisation. VDI-Verlag, Düsseldorf (Fortschr-Ber VDI-Z, Reihe 17, Nr 18)

Knauth P, Eichhorn B, Löwenthal I, Gärtner K H, Rutenfranz J 1983a Reduction of nightwork by re-designing of shift-rotas. International Archives of Occupational and Environmental Health 51: 371–379

Knauth P, Emde E, Rutenfranz J, Kiesswetter E, Smith P 1981 Re-entrainment of body temperature in field studies of shiftwork. International Archives of Occupational and Environmental Health 49: 137–149

Knauth P, Kiesswetter E, Ottmann W, Karvonen M J, Rutenfranz J 1983b Time-budget studies of policemen in weekly or swiftly rotating shift systems. Applied Ergonomics 14: 247–252

Knauth P, Landau K, Dröge C, Schwitteck M, Widynski M, Rutenfranz J 1980 Duration of sleep depending on the type of shift work. International Archives of Occupational and Environmental Health 46: 167–177

Knauth P, Rutenfranz J 1981 Duration of sleep related to the type of shift work. In: Reinberg A, Vieux N, Andlauer P (eds) Night and shift work. Biological and social aspects, p 161–168 Pergamon Press, Oxford (Advances in the biosciences, vol 30)

Knauth P, Rutenfranz J 1982 Development of criteria for the design of shiftwork systems. Journal of Human Ergology 11: Suppl 337–367

Knauth P, Rutenfranz J, Herrmann G, Poeppl S J 1978 Re-entrainment of body temperature in experimental shift-work studies. Ergonomics 21: 775–783

Knauth P, Schmidt K-H 1985 Beschleunigung der Schichtrotation und Ausweitung der regulären Betriebszeit auf das Wochenende. Zeitschrift für Arbeitswissenschaft 39: (11NF), 226–230

Kogi K 1981 Comparison of resting conditions between various shift rotation systems for industrial workers. In: Reinberg A, Vieux N, Andlauer P (eds) Night and shift work. Biological and social aspects, p 417–424 Pergamon Press, Oxford (Advances in the biosciences, vol 30)

Kogi K 1982 Sleep problems in night and shift work. Journal of Human Ergology 11: Suppl 217–231

Koller M 1983 Health risks related to shift work. An example of time-contingent effects of long-term stress. International Archives of Occupational and Environmental Health 53: 59–75

Koller M, Haider M, Kundi M, Cervinka R, Katschnig H, Küfferle B 1981 Possible relations of irregular working hours to psychiatric psychosomatic disorders. In: Reinberg A, Vieux N, Andlauer P (eds) Night and shift work. Biological and social aspects, p 465–472 Pergamon Press, Oxford (Advances in the biosciences, vol 30)

Koller M, Kundi M, Cervinka R 1978 Field studies of shift work at an Austrian oil refinery. I: Health and psychosocial wellbeing of workers who drop out of shiftwork. Ergonomics 21: 835–847

Küpper R, Rutenfranz J, Knauth P, Romahn R, Undeutsch K, Löwenthal I 1980 Wechselwirkungen zwischen lärmbedingten Störungen des Tagschlafs und der Häufigkeit verschiedener Beschwerden bei Schichtarbeitern. In: Brenner W, Rutenfranz J, Baumgartner E, Haider M (eds) Arbeitsbedingte Gesundheitsschäden — Fiktion oder Wirklichkeit. Bericht über die 20. Jahrestagung der Deutschen Gesellschaft für Arbeitsmedizin e.V, Innsbruck, 27.–30.4.1980. S 165–170 Gentner, Stuttgart

Lille F, Sens-Salis D, Ullsperger P, Cheliout F, Borodulin L, Burnod Y 1981 Heart rate variations in air traffic controllers during day and night work. In: Reinberg A, Vieux N, Andlauer P (eds) Night and shift work. Biological and social aspects, p 391–397 Pergamon Press, Oxford (Advances in the biosciences, vol 30)

Lortie M, Foret J, Teiger C, Laville A 1979 Circadian rhythms and behaviour of permanent nightworkers. International Archives of Occupational and Environmental Health 44: 1–11

Masuda T, Takagi K 1975 Influence of shift work on the meal taking. On the food intake of young work women. Journal of Science of Labour 51: No 6: 323–338

Matsumoto K 1978 Sleep patterns in hospital nurses due to shift work: an EEG study. Waking and Sleeping 2: 169–173

Matsumoto K, Matsui T, Kawamori M, Kogi K 1982 Effects of nighttime naps on sleep patterns of shiftworkers. Journal of Human Ergology 11: Suppl, 279–289

Matsumoto K, Sasagawa N, Kawamori M 1978 Studies on fatigue of hospital nurses due to shift work. Japanese Journal of Industrial Health 20: 81–93

Meers A 1975 Performance on different turns of duty within a three-shift system and its relation to body temperature — two field studies. In: Colquhoun P, Folkard S, Knauth P, Rutenfranz J (eds) Experimental studies of shiftwork, p 188–205 Westdeutscher Verlag, Opladen (Forschungsberichte des Landes Nordrhein-Westfalen, Nr 2513)

Meijman T F 1981 Analyse subjective de la recuperation apres les postes de nuit dans le cas de rotation lente (7 jours). Le Travail Humain 44: 315–323

Melton C E, McKenzie J M, Smith R C, Polis B D, Higgins E A, Hoffmann S M, Funkhouser G E, Saldivar J T 1973 Physiological, biochemical, and psychological responses in air traffic control personnel: comparison of the 5-day and 2-2-1 shift rotation patterns. Department of Transportation, Federal Aviation Administration, Office of Aviation Medicine, Washington, DC. National Technical Information Service, Springfield, Virginia (FAA-AM-73-22)

Mills J N, Minors D S, Waterhouse J M 1978 Adaptation to abrupt time shifts of the oscillator(s) controlling human circadian rhythms. Journal of Physiology (London) 285: 455–470

Minors D S, Waterhouse J M 1981 Anchor sleep as a synchronizer of rhythms on abnormal routines. In: Johnson L C, Colquhoun W P, Tepas D I, Colligan M J (eds) Biological rhythms, sleep and shift work, p 399–414 Spectrum Publications, New York (Advances in sleep research, vol 7)

Minors D S, Waterhouse J M 1983 Circadian rhythm amplitude — is it related to rhythm adjustment and/or worker motivation? Ergonomics 26: 229–241

Minors D S, Waterhouse J M 1985 Circadian rhythms in deep body temperature, urinary excretion and alertness in nurses on night work. Ergonomics 28: 1523–1530

Monk T H, Embrey D E 1981 A field study of circadian rhythms in actual and interpolated task performance. In: Reinberg A, Vieux N, Andlauer P (eds) Night and shift work. Biological and social aspects, p 473–480 Pergamon Press, Oxford (Advances in the biosciences, vol 30)

Monk T H, Knauth P, Folkard S, Rutenfranz J 1978 Memory based performance measures in studies of shiftwork. Ergonomics 21: 819–826

Moog R, Hildebrandt G 1982 Comparison between autorhythmometric methods and a baseline measurement of circadian rhythms in night-workers. Journal of Human Ergology 11: Suppl 385–391

Moog R, Hildebrandt G 1985 Morning-evening-types and tolerance to shift work. Journal of Interdisciplinary Cycle Research 16: 147–148

Moog R, Hildebrandt G 1986 Comparison of different causes of masking effects. In: Haider M, Koller M, Cervinka R (eds) Night and shiftwork: longterm effects and their prevention. Lang, Frankfurt am Main, Bern (Studies in industrial and organizational psychology, vol 3, p 131–140)

Mott P E, Mann F C, McLoughlin Q, Warwick D P 1965 Shift work. The social, psychological and physical consequences. The University of Michigan Press, Ann Arbor

Nachreiner F 1975 Role perceptions, job satisfaction, and attitudes towards shiftwork of workers in different shift systems as related to situational and personal factors. In: Colquhoun P, Folkard S, Knauth P, Rutenfranz J (eds) Experimental studies of shiftwork, p 232–243 Westdeutscher Verlag, Opladen (Forschungsberichte des Landes Nordrhein-Westfalen, Nr 2513)

Nachreiner F, Frielingsdorf R, Romahn R, Knauth P, Kuhlmann W, Klimmer F, Rutenfranz J, Werner E 1975 Schichtarbeit bei kontinuierlicher Produktion. Wirtschaftsverlag Nordwest, Wilhelmshaven (Bundesanstalt für Arbeitsschutz und Unfallforschung, Dortmund, Forschungsbericht Nr 141)

Nakano Y, Miura T, Hara I, Aono H, Miyano N, Miyajima K, Tabuchi T, Kosaka H 1982 The effect of shift work on cellular immune function. Journal of Human Ergology 11: Suppl 131–137

Nicholson N, Jackson P, Howes G 1978 Shiftwork and absence: an analysis of temporal trends. Journal of Occupational Psychology 51: 127–137

Northrup H R, Wilson J T, Rose K M 1979 The twelve-hour shift in the petroleum and chemical industries. Industrial and Labour Relations Review 32: 312–326

Orth-Gomer K, Knutsson A, Jonsson B, Freden K, Akerstedt T 1985 Direct and indirect evidence of ischemic heart disease in shift workers. In: Bertazzi P A (ed) 4th International Symposium Epidemiology in Occupational Health, Como, Italy, 10.–12.9.1985. Abstracts, p 43 Como, Italy (Ricerca scientifica ed educazione permanente, Suppl 45)

Ottmann W, Plett R, Knauth P, Gallwey T, Craig A, Rutenfranz J 1986 Combined effects of experimental shiftwork and heat stress on cognitive performance tasks. In: Haider M, Koller M, Cervinka R (eds) Night and shiftwork: longterm effects and their prevention. Lang, Frankfurt am Main, Bern (Studies in industrial and organizational psychology, vol 3, p 361–368)

Patkai P, Akerstedt T, Pettersson K 1977 Field studies of shiftwork: I. Temporal patterns in psychophysiological activation in permanent night workers. Ergonomics 20: 611–619

Pokorski J, Oginski A, Knauth P 1986 Work physiological field studies concerning effects of combined stress in morning, afternoon and night shifts. In: Haider M, Koller M, Cervinka R (eds) Night and shiftwork: longterm effects and their prevention. Lang, Frankfurt am Main, Bern (Studies in industrial and organizational psychology, vol 3, p 369–377)

Queinnec Y, De Terssac G, Dorel M 1984 Temporal organization of activities in process control. In: Wedderburn A, Smith P (eds) Psychological approaches to night and shift work. International research papers, seminar paper 20.1–20.7 Heriot-Watt University, Edinburgh, Scotland

Queinnec Y, Teiger C, de Terssac G 1985 Reperes pour negocier le travail poste. Service des Publications UTM, Toulouse (Travaux de l'Universite de Toulouse-Le Mirail, Serie B, Tome 2)

Reinberg A, Andlauer P, De Prins J, Malbecq W, Vieux N, Bourdeleau P 1984 Desynchronization of the oral temperature circadian rhythm and intolerance to shift work. Nature 308: No 5956, 272–274

Reinberg A, Andlauer P, Guillet P, Nicolai A, Vieux N, Laporte A 1980 Oral temperature, circadian rhythm amplitude, ageing and tolerance to shift-work. Ergonomics 23: 55–64

Reinberg A, Bourdeleau Ph, Andlauer P, Levi F, Bicakova-Rocher A 1986 Internal desynchronization of circadian rhythms and tolerance to shift-work. In: Haider M, Koller M, Cervinka R (eds) Night and shiftwork: longterm effects and their prevention. Lang, Frankfurt am Main, Bern (Studies in industrial and organizational psychology, vol 3, p 17–20)

Reinberg A, Chaumont A-J, Laporte A 1975 Circadian temporal structure of 20 shift workers (8-hour shift-weekly rotation): An autometric field study. In: Colquhoun P, Folkard S, Knauth P, Rutenfranz J. Experimental studies of shiftwork p 142–165 Westdeutscher Verlag, Opladen (Forschungsberichte des Landes Nordrhein-Westfalen, Nr 2513)

Reinberg A, Migraine C, Apfelbaum M, Brigant L, Ghata J, Vieux N, Laporte A, Nicolai A 1979 Circadian and ultradian rhythms in the feeding behaviour and nutrient intakes of oil refinery operators with shift-work every 3–4 days. Diabete & Metabolisme (Paris) 5: 33–41

Reinberg A, Vieux N, Ghata J, Chaumont A-J, Laporte A 1978 Is the rhythm amplitude related to the ability to phase-shift circadian rhythms of shift-workers? Journal de Physiologie (Paris) 74: 405–409

Rutenfranz J 1982 Occupational health measures for night- and shiftworkers. Journal of Human Ergology 11: Suppl 67–86

Rutenfranz J, Colquhoun W P 1979 Circadian rhythms in human performance. Scandinavian Journal of Work, Environment and Health 5: 167–177

Rutenfranz J, Knauth P, Angersbach D 1980a Arbeitsmedizinische Feststellungen zu Befindlichkeitsstörungen und Erkrankungen bei Schichtarbeit. Arbeitsmedizin Sozialmedizin Präventivmedizin 15: 32–40

Rutenfranz J, Knauth P, Küpper R, Romahn R, Ernst G 1980b Pilot project on the physiological and psychological consequences of shiftwork in some branches of the services sector. In: European Foundation for the Improvement of Living and Working Conditions, Dublin (ed) Effects of shiftwork on health, social life and family life of the workers. Dublin

Rutenfranz J, Neidhart B, Ottmann W, Schmitz B, Plett R, Knauth P, Klimmer F 1986 Circadian rhythms of physiological functions during experimental shift work with additional heat stress. In: Haider M, Koller M, Cervinka R (eds) Night and shiftwork: longterm effects and their prevention. Lang, Frankfurt am Main, Bern (Studies in industrial and organizational psychology, vol 3, p 347–359

Saito Y 1982 A permanent night work system in the electronics industry. Journal of Human Ergology 11: Suppl 399–407

Sakai K, Kogi K 1986 Conditions for three-shift workers to take night-time naps effectively. In: Haider M, Koller M, Cervinka R (eds) Night and shiftwork: longterm effects and their prevention. Lang, Frankfurt am Main, Bern (Studies in industrial and organizational psychology, vol 3, p 173–180)

Sakai K, Kogi K, Watanabe A, Onishi N, Shindo H 1982 Location-and-time budget in working consecutive night shifts. Journal of Human Ergology 11: Suppl 417–428

Schroeder D J, Goulden D R 1983 A bibliography of (on!) shift work research: 1950–1982. US Department of Transportation, Federal Aviation Administration, Office of Aviation Medicine, Washington, DC. National Technical Information Service, Springfield, Virginia (FAA-AM-83-17)

Seibt A, Friedrichsen G, Jakubowski A, Kaufmann O, Schurig U 1986 Investigations of the effect of work-dependent noise in combination with shift work. In: Haider M, Koller M, Cervinka R (eds) Night and shiftwork: longterm effects and their prevention. Lang, Frankfurt am Main, Bern (Studies in industrial and organizational psychology, vol 3, p 339–346)

Seibt A, Hilpmann Ch, Friedrichsen G 1983 Zur auralen Wirkung des hörschädigenden Lärms bei Schichtarbeit. Zeitschrift für die gesamte Hygiene und ihre Grenzgebiete. 29: 206–208

Sen R N, Kar M R 1978 Circadian rhythms in some groups of Indians working in shifts. Journal of Human Ergology 7: 65–79

Shift Work Committee, Japan Association of Industrial Health 1979 Opinion on night work and shift work (Yakin-kotai-kimmu ni kansuru ikensho). Journal of Science of Labour 55: No 8, Pt II, 1–36

Smith A, Miles C 1986 The combined effects of nightwork and noise on human function. In: Haider M, Koller M, Cervinka R (eds) Night and shiftwork: longterm effects and their prevention. Lang, Frankfurt am Main, Bern (Studies in industrial and organizational psychology, vol 3, p 331–338)

Smith M J, Colligan M J, Tasto D L 1982 Health and safety consequences of shift work in the food processing industry. Ergonomics 25: 133–144

Smith P 1979 A study of weekly and rapidly rotating shiftworkers. International Archives of Occupational and Environmental Health 43: 211–220

Tasto D L, Colligan M J, Skjei E W, Polly S J 1978 Health consequences of shift work. US Department of Health, Education, and Welfare, Cincinnati, Ohio. (DHEW (NIOSH) publication, no 78–154)

Taylor P J, Pocock S J 1972 Mortality of shift and day workers 1956–68. British Journal of Industrial Medicine 29: 201–207

Teiger C 1984 Overmortality among permanent nightworkers: some questions about 'adaptation'. In: Wedderburn A, Smith P (eds) Psychological approaches to night and shift work. International research papers, seminar paper 15.1.–15.34 Heriot-Watt University, Edinburgh, Scotland

Tepas D I 1982 Shiftworker sleep strategies. Journal of Human Ergology 11: Suppl 325–336

Tepas D I, Walsh J K, Moss P D, Armstrong D 1981 Polysomnographic correlates of shift worker performance in the laboratory. In: Reinberg A, Vieux N, Andlauer P (eds) Night and shift work. Biological and social aspects, p 179–186 Pergamon Press, Oxford (Advances in the biosciences, vol 30)

Thierry H, Hoolwerf G, Drenth P J D 1975 Attitudes of permanent day and shift workers towards shiftwork — a field study. In: Colquhoun P, Folkard S, Knauth P, Rutenfranz J (eds) Experimental studies of shiftwork, p 213–231 Westdeutscher Verlag, Opladen (Forschungsberichte des Landes Nordrhein-Westfalen, Nr 2513)

Tilley A J, Wilkinson R T, Drud M 1981 Night and day shifts compared in terms of the quality and quantity of sleep recorded in the home and performance measured at work: a pilot study. In: Reinberg A, Vieux N, Andlauer P (eds) Night and shift work. Biological and social aspects, p 187–196 Pergamon Press, Oxford (Advances in the biosciences, vol 30)

Torsvall L, Akerstedt T, Gillander K, Knutsson A 1986 24 h recordings of sleep/wakefulness in shift work. In: Haider M, Koller M, Cervinka R (eds) Night and shiftwork: longterm effects and their prevention. Lang, Frankfurt am Main, Bern (Studies in industrial and organizational psychology, vol 3, p 37–41)

Torsvall L, Akerstedt T, Gillberg M 1981 Age, sleep and irregular workhours. A field study with electroencephalographic recordings, catecholamine excretion and self-ratings. Scandinavian Journal of Work, Environment and Health 7: 196–203

Ulich E, Baitsch C 1979 Schicht- und Nachtarbeit im Betrieb. Probleme und Lösungsansätze. Korrigierter Nachdr. der 2. Aufl. gdi-Verlag, Rüschlikon/Zürich

Vieux N, Ghata J, Laporte A, Migraine C, Nicolai A, Reinberg A 1979 Adjustment of shift workers adhering to a three- to four-day rotation (study 2). Chronobiologia 6: Suppl 1: 37–42

Vokac A, Gundersen N, Magnus P, Jebens E, Bakka T 1981 Circadian rhythm of urinary excretion of mercury. In: Reinberg A, Vieux N, Andlauer P (eds) Night and shift work. Biological and social aspects, p 425–431 Pergamon Press, Oxford (Advances in the biosciences, vol 30)

Vokac Z, Rodahl K 1975 A field study of rotating and continuous night shifts in a steel mill. In: Colquhoun W P, Folkard S, Knauth P, Rutenfranz J (eds) Experimental studies of shiftwork, p 168–173 Westdeutscher Verlag, Opladen (Forschungsberichte des Landes Nordrhein-Westfalen, Nr 2513)

Vokac Z, Vokac M Sleep throughs as indicators of endo- and exogenous components of circadian rhythms. In: Proceedings of the 17th International Conference of the International Society for Chronobiology, 3.–6.11.1985, Little Rock, Arkansas (preprint)

Walsh J K, Tepas D I, Moss P D 1981 The EEG sleep of night and rotating shift workers. In: Johnson L C, Colquhoun W P, Tepas D I, Colligan M J (eds) Biological rhythms, sleep and shift work, p 371–381 Spectrum Publications, New York (Advances in sleep research, vol. 7)

Webb W B 1983 Are there permanent effects of night shift work on sleep? Biological Psychology 16: 273–283

Wedderburn A A I 1981 Is there a pattern in the value of time off work? In: Reinberg A, Vieux N, Andlauer P (eds) Night and shift work. Biological and social aspects, p 495–504 Pergamon Press, Oxford (Advances in the biosciences, vol 30)

Werner E, Borchardt N, Frielingsdorf R, Romahn H 1980 Schichtarbeit als Langzeiteinfluß auf betriebliche, private und soziale Bezüge. Westdeutscher Verlag, Opladen (Forschungsbericht des Landes Nordrhein-Westfalen, Nr 2974)

Williamson A M, Sanderson J-W 1986 Changing the speed of shift rotation: a field study. Ergonomics 29: 1085–1095

Wolf S, Almy T P, Bachrach W H, Spiro H M, Sturdevant R A L, Weiner H 1979 The role of stress in peptic ulcer disease. Journal of Human Stress 5: No 2, 27–37

Wynne R F, Ryan G M, Cullen J H 1986 Adjustment to shiftwork and its prediction: results from a longitudinal study. In: Haider M, Koller M, Cervinka R (eds) Night and shiftwork: longterm effects and their prevention. Lang, Frankfurt am Main, Bern (Studies in industrial and organizational psychology, vol 3, p 101–108)

SECTION 5

Controversy

17. Occupational bronchitis

C. A. Soutar

INTRODUCTION

Occupational bronchitis is a vague but convenient term used to describe the chronic inflammatory or degenerative effects on the lung of dusts, fumes or gases. The term excludes pneumoconiosis, asthma and neoplasms. Fortunately individual components of this group of syndromes can be defined precisely for the purposes of measurement. The measurable components of occupational bronchitis described in this chapter include:

1. impairment of lung function or accelerated rate of loss of function;
2. symptoms of chronic productive cough (chronic bronchitis);
3. necropsy evidence of emphysema or bronchial or bronchiolar disease.

Other inflammatory or degenerative responses could be included within the definition (for instance, a susceptibility to respiratory infections) but are not discussed in this chapter, mostly for lack of new or any evidence on the topics. Abnormalities of the chest radiographs related to occupational bronchitis are, however, discussed briefly. This chapter will review the evidence relating components of occupational bronchitis to exposure to dusts, fumes or gases at work. Most of the evidence relates to dusts, and consists of two main types of comparisons.

(a) Comparisons of the frequency of disease between populations in dusty and non-dusty occupations. (A dusty occupation, for the purposes of this discussion, is one where exposure to dust is obvious to the casual observer; a non-dusty occupation is one where exposure to dust appears to be no greater than average domestic conditions).
(b) Comparisons of the frequency of disease with quantitative estimates of exposure (exposure/response relationship).

The main advances in recent years have been in the study of exposure/response relationships, and these are given more attention in this chapter than the descriptions of frequency of disease in exposed populations.

FREQUENCY OF INFLAMMATORY OR DEGENERATIVE LUNG DISEASE IN DUSTY OCCUPATIONS

It is well accepted that workers in many dusty occupations suffer more impairment of lung function, more chronic productive cough (chronic bronchitis), more spells of sickness absence from chest illnesses and higher mortality from respiratory disease than average. There has however, been much discussion on whether this is a consequence of exposure to dust or of other factors related to socio-economic differences. No attempt will be made in this chapter to reproduce the debate of the last thirty years, or review all the evidence on which it was based. The reader is referred to reviews by Higgins (1970), Enterline (1967), Lowe (1968) and Morgan (1978).

Comparisons of the prevalence of disease in large populations of workers in dusty and non-dusty occupations are invaluable in identifying and quantifying possible health problems, and, if positive, may raise strong suspicions, but cannot conclusively demonstrate, that the dust is responsible for the excess of disease. Other factors such as smoking habit or other differences related to social behaviour or circumstances may contribute to the apparent differences between occupational groups in frequency of disease, and confirmation of the effect of dust rests on the demonstration of exposure/response relationships.

Discussions of exposure/response relationships form the most important part of this chapter. Firstly, however, note is made of those dusty industries in which excesses of respiratory disease have been demonstrated.

Impairment of lung function

Impairment of lung function, irrespective of the underlying structural abnormality, is the most important component of occupational bronchitis, since it potentially leads to breathlessness, disability and premature death. Impairments of lung function, relative to other occupational groups, have been demonstrated in workers in many dusty industries; underground coalminers in Britain (Higgins et al, 1956, 1959; Lloyd et al, 1986), West Germany (Reichel & Ulmer, 1978) and the United States (Enterline, 1967; Kibelstis et al, 1973; Higgins & Whittaker, 1981); goldminers in South Africa (Sluis Cremer et al, 1967a,b), foundry workers (Higgins et al, 1959); cement production workers in Yugoslavia (Kalacic, 1973; Saric et al, 1976); grain workers in the USA (Dosman et al, 1980) and in Canada (Becklake et al, 1979) and coke oven workers (Walker et al, 1971). Additionally, though not strictly a dusty industry, pulp mill workers exposed to sulphur dioxide in Sweden (Skalpe, 1964) and Utah (Smith et al, 1977) have also demonstrated impairments of lung function greater than those found in other workers.

Chronic productive cough

Chronic productive cough (chronic bronchitis), though often not considered a serious symptom by its sufferers, is useful as an indication that the lungs are responding to some insult, usually external, and for its associations with poor lung function and respiratory infections (Fletcher et al, 1976). It is not invariably associated with impaired lung function however, and the association between these two is much less apparent when smoking habit, which causes both these features indpendently, is taken into account (Fletcher et al, 1976).

Increased prevalences of chronic productive cough have been demonstrated in all the groups of workers in dusty industries listed in the above section on functional impairment, though occasionally the symptom complex examined differed slightly in detail (for instance, a history of recent chest illness was included in some studies). These results confirm independently the evidence provided by the lung function studies that workers in dusty industries suffer more respiratory disease than other men.

Necropsy evidence of emphysema

Pathology studies are naturally limited by the availability of satisfactory necropsy material, but two studies of emphysema in coalminers have shown an excess of this condition in the lungs of miners compared with non-miners. The first (Ryder et al, 1970) was criticised for its case selection procedure which could have induced bias in the results. The second study (Cockroft et al, 1982) neatly satisfied the requirements for unbiased selection by studying the lungs of miners and non-miners who died in hospital from a disease unrelated to exposure to dust (ischaemic heart disease), and confirmed that miners in South Wales have more emphysema at death than non-miners, thus supporting the conclusions of the original study.

Interpretation of findings

The results of these various studies strongly suggest that exposure to dust at work can impair lung function and cause respiratory symptoms, but other, non-occupational factors could in theory be responsible for the observed differences. The most important of those other factors are social class and smoking habit. Lung disease, smoking habit and social class are strongly linked (see the reviews quoted earlier in this chapter), and it is difficult to be sure on the basis of the above studies that the excess of disease in dust exposed workers is the consequence of exposure to dust and not some other, non-occupational difference in social behaviour (it should not be forgotten of course that social class is defined by occupation, so it would not be surprising if occupationally related illnesses occurred more often in some social classes than others).

To overcome these difficulties, comparisons of the frequency of lung disease with measured or estimated quantitative exposures to dust fumes or gases are necessary to demonstrate exposure/response relationships.

EXPOSURE/RESPONSE RELATIONSHIPS

Exposure/response relationships are analogous to dose/response relationships in other contexts, and their power in scientific proof of an association rests on the demonstration of an increase in risk of disease over a series of increments of exposure.

What does exposure mean?

To compare exposure to dust with biological response it is necessary to be able to express exposure in quantitative terms. Most workers in dusty conditions experience fluctuating concentrations of airborne respirable dust, and their exposure may be discontinuous. It is difficult to express all the features of such an experience in simple quantitative terms.

However, it has been suggested on theoretical grounds (and this supports the traditional empirical view) that for agents which are cleared slowly and tend to cumulate in the body (such as inorganic dusts) it is the total cumulative dose which most influences the biological effect (Rappaport et al, 1985). This view is supported by the results of many studies over the years, which have consistently demonstrated relationships between total cumulative exposure to dust and biological response.

It is recognised that exposure is not the same thing as dose, the latter being affected by such variables as breathing patterns, wearing of respirators, partitioning of dust between nose, pharynx, lung and stomach, and clearance from the lung, but exposure is relatively easy to measure, and is immediately relevant to the prevention of disease, since dust control measures are monitored primarily by measuring airborne concentrations of dust.

A man's lifetime cumulative dust exposure would therefore consist of the sum of the respirable dust concentrations he has experienced, each multipliedby the duration of exposure to that concentration. It can be regarded as a concentration-weighted time, and its units might be, for instance, grammes of respirable dust per cubic metre of air, multiplied by hours ($g \cdot h/m^3$). To aid the reader in interpreting dust exposure quantities, in a recent study (Soutar & Hurley, 1986) of miners working in the coal industry in the 1950s, the lifetime dust exposures of these men by the time of a medical survey in the late 1970s averaged $182\,g \cdot h/m^3$, and 4% of them had experienced $400\,g \cdot h/m^3$ or greater.

Respirable dust also requires some definition. There is currently a great deal of interest in which fractions (by aerodynamic size) of airborne inhalable dust are relevant to different kinds of disease (Vincent & Mark, 1981). However, many of the established relationships between dust and disease are based on dust concentration measurements resulting from an inspired proposal by the British Medical Research Council Working Group at the Johannesburg Conference on Inhaled Particles in 1959 (Walton, 1960). The fraction of airborne dust thought by virtue of its aerodynamic properties to be able to penetrate to the alveolar region of the lung was to be selected by the sampling instrument according to those aerodynamic properties. The proposed fraction was defined by:

> 'a sampling efficiency curve which depended on the falling velocity of the particles and which passed through the following points: effectively 100% efficiency at 1 micron and below, 50% efficiency at 5 microns and zero efficiency for particles of 7 microns and upwards; all the sizes refer to equivalent diameters' (Orenstein, 1960).

This definition of respirable dust has enabled the demonstration of many important relationships between disease and exposure to dust.

Impairment of lung function in relation to exposure

Studies of miners in Britain, West Germany, the USA and South Africa have all shown inverse relationships between estimates of exposure to dust and lung function. The methods of estimating exposure differed in sophistication according to the availability of suitable data, but all showed broadly similar results. These are listed in Table 17.1.

The British studies of coalminers were based on an extraordinarily detailed prospective strategy of measurements of respirable dust concentrations undergound, and compatible records of jobs and place of work undergound, over more than 20 years. A series of papers has been published, based on different British coalmining populations, which confirms that lung function is inversely related to exposure to dust, after allowing for the effects of age and smoking.

The original estimate of Rogan and colleagues (1973), based on a selected group of 4000 faceworkers, was a deficit of -60 ml of forced expired volume in one second (FEV_1) for every 100 units of dust exposure ($g \cdot h/m^3$). Subsequent study of a less

Table 17.1 Quantitative estimates of cumulative dust exposure on lung function, grouped as far as possible by the units in which the dust/function relationship was expressed

Study group	Effect of dust on lung function FEV_1(ml)/100 g . h/m^3	Dust exposure of group (g . h/m^3; mean unless stated otherwise)	Source
3581 Coalface workers in Britain	−60	175 g . h/m^3	Rogan et al (1973)
1867 Coalminers in Britain (currently employed)	−68	174 g . h/m^3	Soutar & Hurley (1986)
1023 Coalminers in Britain (ex-miners)	−89	167 g . h/m^3	Soutar & Hurley (1986)
1305 Coalminers in Britain (who died before later medical survey)	−78	185 g . h/m^3	Gauld et al (1985)
1486 Coalminers in Britain (who did not attend a later medical survey)	−78	147 g . h/m^3	Gauld et al (1985)
818 Workers in a polyvinylchloride factory	−53	13	Soutar et al (1980)
152 Coalminers in W Germany (moderate smokers)	FEV_1 − 140 ml lower for men with exposures greater than 100 g . h/m^3 than for men with lower exposures	Not explicitly stated	Reichel & Ulmer (1978)
180 Coalminers in W Germany (heavy smokers)	FEV_1 − 89 ml similarly	Not explicitly stated	Reichel & Ulmer (1978)
2209 Gold miners in South Africa	Significant negative effect of exposure on FEV_1 (also see Fig. 17.2)		Wiles & Faure (1977)
	PEFR (ml/s)/ years of exposure		
4318 Coalminers in USA (smokers)	−57	Ranged from 4 to over 30 years	Hankinson et al (1977a)
1696 Coalminers in USA (non-smokers)	−35		

selected population of miners, originally working in the coal industry in the 1950s, and followed-up 22 years later (Soutar & Hurley, 1986) confirmed the approximate magnitude of the relationship between dust exposure and lung function in 1867 men still working in the industry (estimated at -68 ml $FEV_1/100\,g.h/m^3$ exposure) and in 1023 men of a similar range of ages who had left the industry by the time of follow-up examination (estimate -89 ml $FEV_1/100\,g.h/m^3$ exposure).

A similar effect of dust on the FEV_1 was found in these men when their FEV_1 measured 11 years earlier was compared with their dust exposure up to that time (estimate $-78\,ml/100\,g.h/m^3$) (Gauld et al, 1985), and furthermore the estimated effect of dust was of very similar magnitude in an additional 2791 men studied at that time, both in 1305 of them who subsequently died before the follow-up examination and in 1486 others who remained alive but did not attend the follow-up examination (Gauld et al, 1985).

Thus, an inverse relationship between dust exposure and FEV_1, after allowing for smoking habit, has been demonstrated independently in five different, large groups of British miners, and in each case the average estimated effect of dust on FEV_1 has been of similar magnitude. The universality of the effect of dust in exposed men is illustrated again by the data shown in Fig. 17.1, which shows the mean FEV_1 of 4059 miners and ex-miners (Soutar & Hurley, 1986), grouped by age and smoking habit, in relation to dust exposure. The effects of dust are obvious in almost all groups of men.

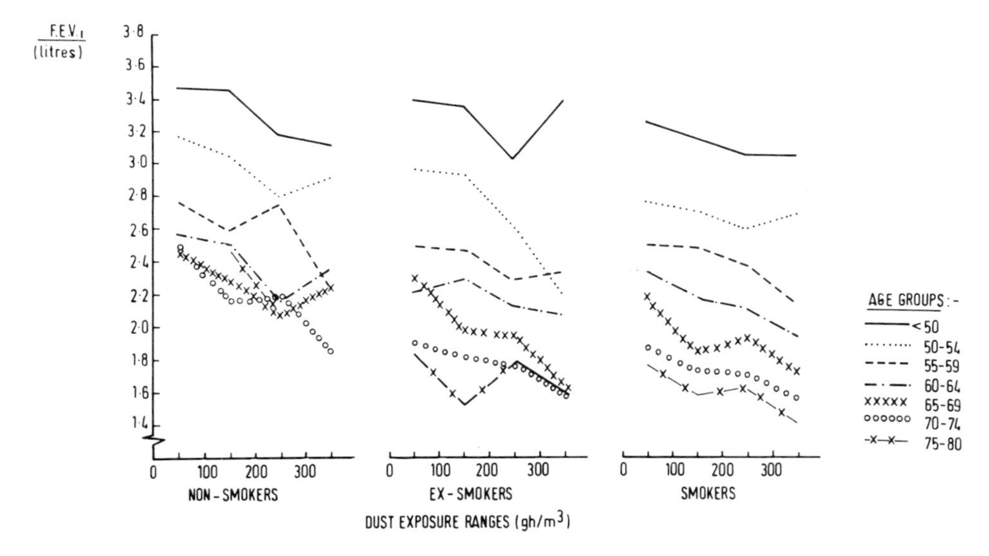

Fig. 17.1 Mean observed FEV_1 by dust exposure ranges for 4059 coalminers and ex-miners. Groups consisting of one man, without men in adjacent dust exposure groups, have been omitted (from Soutar & Hurley 1986), (reproduced by permission of the Editor, British Journal of Industrial Medicine).

In West German studies the dust exposure estimates have been based on less detailed records of occupations and dust concentrations kept for dust control and health surveillance purposes, over periods of up to 16 years (Reisner et al, 1978). Men with dust exposures greater than $100\,g.h/m^3$ (an arbitrarily chosen figure) had, on average,

FEV_1 values 140 ml lower than men with exposures less than this, after allowing for age and smoking (Reichel & Ulmer, 1978). While this result (without more detailed analysis) cannot be directly compared in quantitative terms with the results in British miners, the effect of dust on the FEV_1 appears to be of the same order in West Germany and in Britain.

In the USA, dust concentration data have not been adequate for calculating individual dust exposures and years worked in dusty conditions has therefore been used as an exposure index. Even so, lung function has been shown to be inversely related to exposure after allowing for age (Hankinson et al, 1977a). A more recent study of two USA mining communities confirmed that FEV_1 was inversely related to years of work at the coalface, though small numbers of men and differences in smoking habit contributed to some inconsistencies in the data (Higgins & Whittaker, 1981).

In gold miners in South Africa, estimates of dust exposure were estimated retrospectively on the basis of records of occupation and a cross-sectional programme of dust concentration measurements (Beadle, 1971). Figure 17.2 shows the mean FEV_1 of 2020 gold miners according to their estimated dust exposure (Wiles & Faure, 1977). No account was taken of age in this analysis, though mean ages were similar in all groups. Similar trends were seen in non-smokers as well as in smokers. There is an obvious and statistically significant inverse relationship between dust exposure and FEV_1.

An inverse relationship between dust exposure and lung function has also been demonstrated in another (non-mining) dusty industry. Workers exposed to polyvinylchloride dust demonstrated an inverse relationship between FEV_1 and an estimate of dust exposure based on retrospective occupational histories and cross-sectional measurements of respirable dust concentrations (Soutar et al, 1980).

Studies of workers in granite sheds in Vermont have also suggested an inverse relationship between FEV_1 and estimates of exposure to granite dust (Theriault et al, 1974a,b), although subsequent discovery of inaccuracies in the lung function measurements causes some doubt about these results (Graham et al, 1981).

Accelerated loss of lung function

The finding that impairment of lung function is related to exposure to dust implies that there has been an accelerated rate of loss of function over a period which has resulted in the impairment. Study of rates of loss of function could therefore provide a partially independent confirmation, if confirmation were needed, that exposure to dust impairs lung function. A study of loss of function in 1677 British coalminers over an 11-year period confirmed that loss of FEV_1 was increased by exposure to respirable dust (Love & Miller, 1982). The estimated effect of dust exposure on the loss of FEV_1 over an 11-year period was 36 ml of $FEV_1/100\,g\,.\,h/m^3$ of lifetime exposure, in excess of that attributable to ageing and smoking. Very similar relationships were demonstrated in a longitudinal study of USA miners (Attfield, 1985).

Relationships between dust exposure and chronic productive cough

The frequency of chronic productive cough in young British coalminers has been shown to be influenced by their exposure to respirable dust, though this relationship was not apparent in older men (Rae et al, 1971), and in South African gold miners these symptoms were also strongly related to dust exposure (Wiles & Faure, 1977).

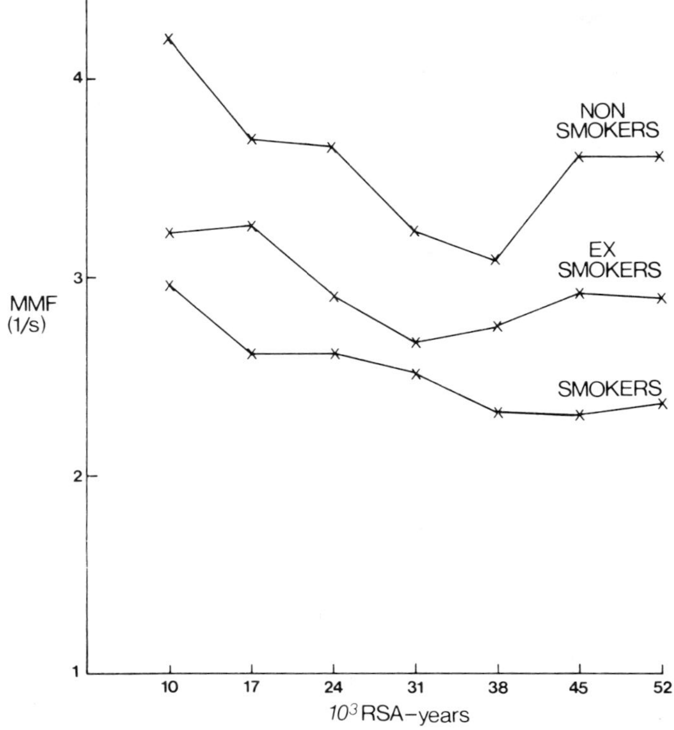

Fig. 17.2 Mean maximal mid-expiratory flow rates (a simple test of lung function) of goldminders in South Africa, related to dust exposure ranges. The units of dust exposure are respirable surface area times years $\times 10^3$ (10^3 RSA-years), based on particle count measurements of airborne dust (data from Wiles & Faure, 1977).

Interpretation of findings: relationships between dust exposure and lung function and chronic productive cough

In the face of so many independent and careful studies clearly demonstrating an inverse relationship between exposure to dust and lung function, supported by demonstrated relationships between dust exposure and respiratory symptoms, it would take the most determined sceptic to deny the relationship. While the demonstration of a relationship does not of course necessarily imply causation, the thesis that dust causes the loss of lung function would appear to be the simplest and most obvious explanation for the relationship.

Further questions which arise include whether the effect of dust on lung function is ever severe enough to cause disability, and to what extent the smoking habit may have either confused the results or be acting as a synergist with the effects of dust.

Influences of smoking habit

Smoking has such a widespread influence on lung function that failure to allow for its affects adequately may obscure the effects of other factors. All the studies mentioned above took account of smoking habit, in some cases in considerable detail. Estimates of pack-years of smoking were usually not available, but this is unlikely to have affected the estimates of the effects of dust much, in view of the recognised inaccuracies of retrospective smoking histories (Todd, 1966), and the insensitivity of estimates of the effects of dust to the way smoking habit is taken account of in the analyses (Wiles & Faure, 1977; Soutar & Hurley, 1986). There is no *a priori* reason to suppose that smoking would cause a false overestimate of the effect of dust; it is more likely to obscure such dust effects, if smoking were not taken adequately into account, by increasing the residual variability of the functional data. Additionally, in most of the above studies, similar effects of dust were observed in non-smokers and in smokers.

There is no evidence in general of synergism between smoking habit and dust exposure on lung function and respiratory symptoms. In the studies quoted above where dust exposure estimates have been available, the effects of dust exposure and smoking have been additive not multiplicative, and the estimated effects of dust have been as great in non-smokers as in smokers. The only exception to this was the study of workers exposed to PVC dust, in whom the data suggested that the effects of dust on FEV_1 might be greater in smokers than in non-smokers (Soutar et al, 1980).

Is occupational bronchitis disabling?

The average effect of the average dust exposures in the studies quoted above would appear to be quite modest in quantitative terms. For instance, in the recent study of British miners and ex-miners (Soutar & Hurley, 1986), the average deficit of FEV_1 related to the average exposure for the group ($182\,g\,.\,h/m^3$) was 138 ml, a relatively modest amount when compared with the average FEV_1 for the group (2.4 litres). However, there are two reasons why some men may suffer a greater loss than the average. One is simply that some men may experience much greater exposures to dust than the average. For instance, nearly 4% of men in the latter study had experienced $400\,g\,.\,h/m^3$ (the mean exposure plus two standard deviations) or more, and the average estimated deficit of FEV_1 occurring as a result of this exposure would be 300 ml or greater. This amount of damage, when added to the effects of age and smoking, could contribute to disability.

The other reason is that, as in all biological systems, the response to dust exposure will vary from person to person, and while there may on the one hand be many men unaffected by their dust exposure, there may also be men with much greater responses to their dust exposure than average. Until recently there had been little evidence on whether the extreme end of the range of responses is sufficient to cause a clinically important loss of lung function.

Since in larger groups of men, any extreme responses to dust are likely to be obscured by the blander responses occurring in the majority, one way of investigating extreme responses is to study smaller groups of men, who for *a priori* reasons might include preferentially those with unusually severe responses.

One such group of miners has recently been identified (Hurley & Soutar, 1986). It consisted of men who had been miners but had left the industry before normal retiring age, and had then taken other jobs. Only men with chronic productive cough were included in the selection, since this symptom helps to identify men with more severe responses to dust, and may have helped to identify men who left the industry because of ill-health. In this group the estimated effect of dust was $-200\,ml/100\,g.\,h/m^3$ exposure, more than three times the average effect. This would represent a loss of −800 ml FEV_1 or greater in response to an exposure of 400 g . h/m^3 or greater (2% of this group experienced this exposure). The observed FEV_1 of this group of men, compared with dust exposure and age, are shown in Fig. 17.3. A

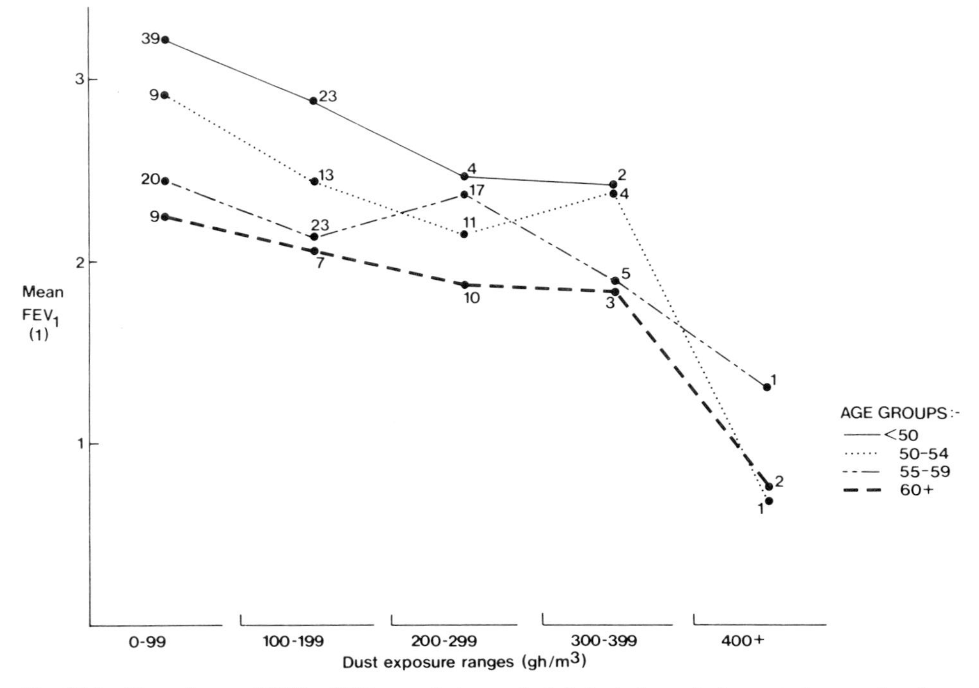

Fig. 17.3 Mean observed FEV_1 of 199 ex-coalminers who left the industry before retiral age, and took other jobs, and had chronic productive cough. Numbers of men in groups are shown (from Hurley & Soutar, 1985).

loss of 800 ml of FEV_1 is substantial and likely to contribute importantly to disability. The clinical effect will be more obvious in smokers than non-smokers since the functional loss is additional to the effects of smoking as well as ageing. The result is consistent with the hypothesis that the effects of dust exposure can in a few men be severe. It is noteworthy that the dust effect in the men in the group who had given up smoking were even more pronounced, estimated at −300 ml FEV_1/100 g . h/m^3. Presumably these men had given up smoking in an attempt to limit any further damage to their health.

This is the first epidemiological morbidity study to identify miners without progressive massive fibrosis who have nevertheless been severely affected by their dust exposure. This result suggests that the range of functional responses to dust exposure may be similar in one respect to the effects of smoking, that is, the effects are in most men mild, but in a minority they are severe. There is nothing to suggest that increased susceptibility to the effects of dust is an all or nothing characteristic however, and this also is comparable to the ill effects of smoking (Fletcher et al, 1976).

Mortality

Such severe effects of dust on lung function should be reflected in mortality, and a recently reported mortality study of 25 000 British miners over a 22-year period (Miller & Jacobsen, 1985) has indicated that mortality from all non-violent causes was related to lifetime cumulative exposure to respirable dust, and that this was accounted for not only by deaths from pneumoconiosis but also from the certified causes chronic bronchitis and emphysema. Figure 17.4 shows that the estimated percentage mortality from chronic bronchitis and emphysema increases independently with both age and dust exposure. While recognising the diagnostic uncertainties of death certification, it is highly likely that these men were suffering from the chronic inflammatory or degenerative conditions of the lung that are the subject of this chapter.

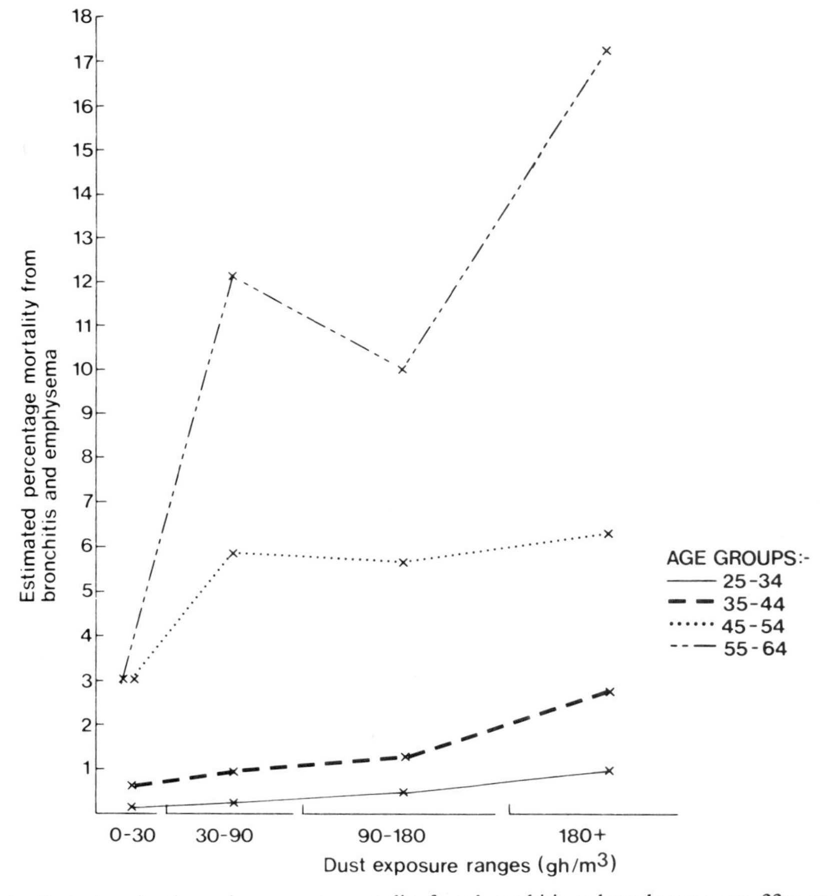

Fig. 17.4 Coalminers' estimated percentage mortality from bronchitis and emphysema over 22 years, according to dust exposure and age, both at the beginning of the period of observation. The data are adapted from survival data published by Miller and Jacobsen (1985), and are based on observed deaths from these causes expressed as a percentage of men still alive at the time of each of these deaths.

Thus, both mortality and morbidity studies indicate that a small proportion of miners are disabled by and die either from occupational bronchitis caused by exposure to respirable dust, or from the addition of the effects of dust and the effects of smoking on the lung.

Comparisons of pathology with dust exposure

At least one component of the structural abnormalities causing the loss of lung function resulting from exposure to dust is centriacinar emphysema. A post-mortem study of the lungs of 450 British coalminers whose lifetime exposure to respirable dust was known, has shown that the higher a man's exposure to dust the more likely he is to have centriacinar emphysema (Ruckley et al, 1984). This relationship only holds if pneumoconiotic simple nodular fibrosis or progressive massive fibrosis is present as well (visible to the pathologist), suggesting that the fibrosis in some way mediated the emphysema. Interestingly, relatively high proportions of quartz in the dust exposure were associated with a reduction in the risk of having emphysema.

Dust exposure has also been shown to correlate with bronchial mucous gland hypertrophy (Douglas et al, 1982), a finding in keeping with the relationship between chronic productive cough and dust exposure (Rae et al, 1971).

Conclusions from morbidity, mortality and pathology studies

Thus a morbidity study has shown that exposure to dust is related to severe impairment of lung function in a minority of men, a mortality study has shown that mortality from chronic bronchitis and emphysema is increased by exposure to dust, and a necropsy study has shown that the likelihood of having emphysema is related to dust exposure.

Taken together this is powerful evidence that exposure of miners to coalmine dust can sometimes cause an occupational bronchitis which is disabling.

Character of the functional impairment related to dust

There is evidence that the type of functional impairment associated with exposure to inorganic dusts is different from the airflow obstruction caused by smoking. One study of USA coalminers (Hankinson et al, 1977b) showed that in non-smokers the presence of chronic productive cough (assumed to be the consequence of exposure to dust) was associated with changes in the forced expiratory flow/volume curves suggesting, in the authors' view, airflow obstruction in large airways. Certainly the type of defect was different from that seen in smokers.

No doubt airflow obstruction at some site in the bronchial tree does occur in response to dust. Further evidence of this derives from an increase in the residual volume/total lung capacity ratio in this study group and in another group of USA miners (Morgan et al, 1971). However, evidence that the functional defect is not pure airflow obstruction is provided by studies of British miners, in which dust exposure was found to be related to a parallel reduction of FEV_1 and FVC, and had only a slight effect on the ratio between them (Soutar & Hurley, 1986). This suggests an element of restrictive defect as well as obstructive, and this pattern is also evident in the study of USA miners referred to above (Hankinson et al, 1977b). Table 17.2 shows the effect on FEV_1, FVC and FEV_1/FVC ratio of the presence of chronic productive cough (occupational bronchitis), and of smoking in their study group. It is obvious that, while smoking and the presence of cough in smokers are associated with a greater reduction of FEV_1 than FVC, and substantial reduction of FEV_1/FVC ratio, the presence of chronic productive cough in the non-smokers is associated with parallel reductions of FEV_1 and FVC, and hardly any reduction of the FEV_1/FVC ratio. These differences, though not discussed by the authors, are also consistent with the presence of a dust-induced restrictive lung functional component as well as a degree of obstruction.

Table 17.2 Effects in USA coalminers of the presence of chronic productive cough (occupational bronchitis) and of smoking on lung function. Levels of FEV_1, FVC and FEV_1/FVC ratio for non-smokers without cough are shown, and the average levels for other groups, expressed as mean differences from the first group. The data were obtained from Hankinson et al (1977b)

	Non-smokers		Smokers	
	Mean level	Mean difference	Mean difference	Mean difference
Chronic productive cough	Absent	Present	Absent	Present
FEV_1(1)	3.77	−0.17	−0.28	−0.43
FVC(1)	4.89	−0.18	−0.12	−0.23
FEV_1/FVC(%)	76.8	−0.3	−3.7	−5.5

This is a similar pattern to that observed in workers exposed to PVC dust (Soutar et al, 1980). It seems possible that exposure to mineral and other dusts in general causes a mixed restrictive and obstructive type of functional defect.

However, there is evidence that when the miner disabled by dust is a smoker, the final result is classical airflow obstruction (Soutar & Hurley, 1986). Since most miners smoke, the clinical presentation of miners disabled by exposure to dust is likely to be as an airflow obstruction indistinguishable from the effects of smoking. While the obstructive component of the functional defect may well be the result of emphysema, as discussed in an earlier section, the structural defect responsible for the restrictive component is not known, though the finding by Churg et al (1985) of fibrosis of the respiratory bronchioles in subjects who have been exposed to various mineral dusts may possibly indicate that this is the site and nature of the lesion. The results of Churg's work, though based on relatively small numbers of subjects, also tend to suggest that the lesions in the respiratory bronchioles are associated with an element of restrictive functional defect as well as some airflow obstruction.

Studies which did not demonstrate a relationship between estimates of occupational exposure and lung disease

Not many studies of occupational bronchitis in which good estimates of exposure to dusts or gases have been available have failed to show an effect of dust. However, a study of steelworkers in South Wales, Lowe et al (1970) did not demonstrate a relationship between exposure to particulates and symptoms or lung function, possibly because airborne concentrations of the dust were quite low compared with those in mines. Nor was any effect of exposure to sulphur dioxide demonstrated in this group.

Studies of possible effects of estimated exposure of coalminers to oxides of nitrogen on symptoms and lung function have so far drawn negative results (Robertson et al, 1984; Reger et al, 1985).

Radiographic abnormalities in occupational bronchitis

The ILO scheme for recording the chest radiographic appearances of pneumoconiosis enables the distinction between small opacities which are rounded in shape and those which are irregular (International Labour Office, 1980). Until recently studies of coalworkers' pneumoconiosis have considered only rounded opacities, though there is now evidence that irregular opacities also are related to dust exposure (Amandus et al, 1976; Cockroft et al, 1983; Dick et al, 1984). These opacities are different from rounded opacities in that they are related to an impairment of lung function (Cockroft et al, 1982), and have been shown to be related to the presence of emphysema and fibrosis in necropsy material (Cockroft et al, 1981). Confirmation of the relationship of these opacities with functional impairment in a different context is provided by a study of polyvinylchloride workers, where irregular opacities (related to age but not dust exposure in this population) were associated with airflow obstruction (Soutar et al, 1980).

It would appear that the presence of irregular opacities may be another non-specific indicator of occupational bronchitis.

Prevention of occupational bronchitis

Dust induced occupational bronchitis should be prevented by controlling exposure to respirable dust. At present it is not clear what the control limits should be to avoid important loss of lung function (and these would not necessarily be the same for dusts of different materials). The main aim should be the avoidance of important loss of lung function.

The current respirable dust control limit in British mines corresponds to a maximum possible exposure of 245 g . h/m^3 in the unlikely event of a man working continuously in the worst possible conditions for 35 years (Jacobsen, 1984). In practice, exposures are much less than this, though precise information on the actual cumulative exposures currently being experienced is not yet available. Using this maximum figure as the worst-possible case, together with estimates of the effects of dust exposure on FEV_1 taken from the most recent British study, enables some very approximate estimates of the numbers of men affected by dust to various defined extents (Table 17.3). Some judgement would need to be made on what constitutes an acceptable loss of FEV_1 in relation to occupation. A maximum acceptable loss of 300 ml FEV_1 might be a reasonable figure, bearing in mind, for example, that an average smoking miner in South Wales aged 65, would have a predicted FEV_1 of 2.2 litres if he had not been exposed to dust. The data in Table 17.3 indicate that the average effect of the maximum possible exposure is well within this limit, but that unusually susceptible men would not be sufficiently protected.

Table 17.3 Estimated effects of current maximum possible dust exposure (245 g . h/m^3), over a working lifetime of 35 years, in all men and in susceptible subgroups of miners. In practice not many men will experience exposures near this amount

Estimate of effect of dust on FEV_1 (ml FEV/g . h/m^3 exposure)	Proportion of susceptible men	Loss of FEV_1 in susceptible men related to maximum possible exposure (ml)	Source
0.76	All (4059)	−186	Soutar et al, 1984
2.00	4.9% (199/4059)	−490	Hurley & Soutar, 1986
3.10	0.9% (35/4059)	−760	Hurley & Soutar, 1986

More information is required on the cumulative exposures currently being experienced, in order to estimate whether current regulations protect sufficient men. Clearly it would be difficult to protect by general measures the individual who is extremely susceptible, but it should be possible to identify such individuals before they have developed disabling lung damage. At present it is not possible to identify susceptible individuals before they have developed some lung damage, but they can be identified by serial simple spirometric measurements, which enable the recognition of an unusually rapid decline of function, even before substantial damage has occurred (Pern et al, 1984).

Those responsible for the health of workers in other dusty industries will wish to take note of these considerations. So far there is no information on the relative toxicity of various dusts in the context of occupational bronchitis, the only known determinant being the physical quantity (in most cases, mass) of the respirable dust to which men are exposed. There is no reason at present to exempt any dust from suspicion of risk of occupational bronchitis, if the airborne concentrations are high enough.

CONCLUSIONS

The conclusion that exposure to dust can cause loss of lung function in the absence of pneumoconiosis is irresistable. Furthermore there is sufficient evidence, in the author's view, that the loss of function is sometimes disabling, to require recognition of this as a potential risk by those responsible for the health of workers in dusty jobs. There is at present no reason to exonerate any type of dust from suspicion of risk of lung damage.

It is probable that the structural abnormality causing the functional loss is (in the case of coalminers at least) partly emphysema, and partly some other abnormality, possibly fibrosis of the respiratory bronchioles. Prevention of occupational bronchitis should be by dust control measures, supported by regular medical surveillance of exposed workers in order to identify unusually susceptible individuals before severe lung damage has occurred. More information is needed on quantitative estimates of risk of specified degrees of functional loss to enable appropriate decisions to be made on control limits for airborne dust concentrations.

REFERENCES

Amandus H E, Lapp N L, Jacobson G, Reger R B 1976 Significance of irregular small opacities in radiographs of coalminers in the USA. British Journal of Industrial Medicine 33: 13–17

Attfield M D 1985 Longitudinal decline in FEV_1 in United States coalminers. Thorax 40: 132–137

Beadle D G 1971 The relationship between the amount of dust breathed and the development of radiological signs of silicosis: an epidemiological study in South African gold miners. In: Walton W H (ed) Inhaled Particles III. Unwin Bros, Old Woking 2: 953–966

Becklake M R, Jodoin G, Lefort L, Rose B, Mandl M, Fraser R G 1980 A respiratory health study of grain handlers in St Lawrence River ports. In: Dosman J A, Cotton D J (eds) Occupational pulmonary disease: focus on grain dust and health. Academic Press, New York

Churg A, Wright J L, Wiggs B, Pare P D, Lazar N 1985 Small airways disease and mineral dust exposure. Prevalence, structure and function. American Review of Respiratory Diseases 131: 139–143

Cockroft A, Berry G, Cotes J E, Lyons J P 1982 Shape of small opacities and lung function in coalworkers. Thorax 37: 765–769

Cockroft A, Lyons J P, Andersson N, Saunders M J 1983 Prevalence and relation to underground exposure of radiological irregular opacities in South Wales and workers with pneumoconiosis. British Journal of Industrial Medicine 40: 169–172

Cockroft A, Seal R M E, Wagner J C et al 1982 Post mortem study of emphysema in coal workers and non-coal workers. Lancet ii: 600–603

Cockroft A E, Wagner J C, Seal E M E, Lyons J P, Campbell M J 1981 Irregular opacities in coalworkers pneumoconiosis — correlation with pulmonary function and pathology. Annals of Occupational Hygiene 26: 767–787

Dick J A, Jacobsen M, Gauld S, Pern P O 1984 The significance of irregular opacities in the chest radiographs of British coalminers. In: Proceedings of the VIth International Pneumoconiosis Conference, Bochum 1983. International Labour Office, Geneva, p 283–299
Dosman J A, Cotton D J, Graham B L, Li K Y R, Froh F, Barnett D 1980 Chronic bronchitis and decreased forced expiratory flow rates in lifetime non-smoking grain workers. American Review of Respiratory Diseases 121: 11–16
Enterline P E 1967 The effects of occupation on chronic respiratory disease. Archives Environmental Health 14: 189–200
Fletcher C, Peto R, Tinker C, Speizer F E 1976 The natural history of chronic bronchitis and emphysema. An eight-year study of early chronic obstructive lung disease in working men in London. Oxford University Press, Oxford
Gauld S J, Hurley J F, Miller B G 1985 Differences between long term participants and non-responders in a study of coalminers' respiratory health and exposure to dust. Proceedings of the Sixth International Symposium on Inhaled Particles, September 1985. British Occupational Hygiene Society. Pergamon Press, Oxford
Graham W G B, O'Grady R V, Dubuc B 1981 Pulmonary function loss in Vermont granite workers: a long-term follow-up and critical reappraisal. American Review of Respiratory Diseases 123: 25–28
Hankinson J L, Reger R B, Fairman R P, Lapp N L, Morgan W K C 1977a Factors influencing expiratory flow rates in coalminers. In: Walton W H (ed) Inhaled particles IV. Pergamon Press, Oxford, p 737–755
Hankinson L, Reger R B, Morgan W K C 1977b Maximal expiratory flows in coalminers. American Review of Respiratory Diseases 116: 175–180
Higgins I T T, Oldham P D, Cochrane A C, Gilson J C 1956 Respiratory symptoms and pulmonary disability in an industrial town. British Medical Journal 2: 904–910
Higgins I T T, Cochrane A L, Gilson J C, Wood C H 1959 Population studies of chronic respiratory disease: a comparison of miners, foundry workers and others in Staveley, Derbyshire. British Journal of Industrial Medicine 16: 255–268
Higgins I T T 1970 Occupational factors in chronic bronchitis and emphysema. In: Bronchitis III. Orie N G M, Van der Lende R (eds) Proceedings of the Third International Symposium on bronchitis, Groningen, the Netherlands 1969. Charles C Thomas, Royal Vangorcum, the Netherlands.
Higgins I T T, Whittaker D E 1981 Chronic respiratory disease in coalminers: follow-up study of two mining communities in West Virginia. US Department of Health and Human Services. Public Health Service Centres for Disease Control. National Institute for Occupational Safety and Health. Morgantown, West Virginia, 26505
Hurley J F, Soutar C A 1986 Can exposure to coalmine dust cause a severe impairment of lung function? British Journal of Industrial Medicine 43: 150–157
International Labour Office 1980 Guidelines for the use of ILO international classification of radiographs of pneumoconiosis. Revised edition 1980. Occupational Safety and Health Series. International Labour Office, Geneva
Jacobsen M 1984 Coal workers pneumoconiosis: results from epidemiological studies in Britain. In: Proceedings of the VIth International Pneumoconiosis Conference, Bochum 1983. International Labour Office, Geneva
Kalacic I 1973 Ventilatory lung function in cement workers. Archives of Environmental Health 26: 84–85
Kibelstis J A, Morgan E J, Reger R, Lapp N L, Seaton A, Morgan W K C 1973 Prevalence of bronchitis and airway obstruction in American bituminous coalminers. American Review of Respiratory Diseases 108: 886–893
Lloyd M H, Gauld S J, Soutar C A 1986 Respiratory ill-health among coalminers and telecommunication workers in South Wales. British Journal of Industrial Medicine 43: 177–181
Love R G, Miller B G 1982 Longitudinal study of lung function in coalminers. Thorax 37: 193–197
Lowe C R 1968 Chronic bronchitis and occupation. Proceedings of the Royal Society of Medicine 61: 89–102
Lowe C R, Campbell H, Khosla T 1970 Bronchitis in two integrated steel works III. Respiratory symptoms and ventilatory capacity related to atmospheric pollution. British Journal of Industrial Medicine 27: 121–129
Maestrelli P, Simonatu L, Bartolucci G B, Gemignani C, Maffessanti M M 1979 Distribuzione della pneumoconiosi e della bronchite cronica negli addetti alla produzione del cemento. Medicina del Lavoro 3: 195–202
Miller B G, Jacobsen M 1985 Dust exposure, pneumoconiosis and mortality of coalminers. British Journal of Industrial Medicine 42: 723–733
Morgan W K C 1978 Industrial bronchitis. British Journal of Industrial Medicine 35: 285–291
Morgan W K C, Burgess D B, Lapp N L, Seaton A 1971 Hyperinflation of the lungs in coal miners. Thorax 26: 585–590

Orenstein W H (ed) 1960 Recommendations adopted by the conference. Proceedings of the Pneumoconiosis Conference, Johannesburg. J A Churchill, London, p 617–621
Pern P O, Love R G, Wightman A J A, Soutar C A 1984 Characteristics of coalminers who have suffered excessive loss of lung function over 10 years. Bulletin European de Physiopathologie Respiratoire 20: 487–493
Rae S, Walker D D, Attfield M D 1971 Chronic bronchitis and dust exposure in British coalminers. In: Walton W H (ed) Inhaled Particles III. Unwin Bros, Old Woking, Surrey, p 873–881
Rappaport S M, Spear R C, Selvin S 1985 The influence of exposure variability on dose-response relationships. Proceedings of the Sixth International Symposium on Inhaled Particles, Cambridge. British Occupational Hygiene Society. Pergamon Press, Oxford
Reger R, Hankinson J, Piacitelli G, Gamble J, Ames R. 1985 Effects of exposure to diesel emissions amongst coalminers: a prospective evaluation. Proceedings of the Sixth International Symposium on Inhaled Particles. British Occupational Hygiene Society (in press)
Reichel G, Ulmer W T 1978 Results obtained by the various investigation centres: coal mine; active staff. In: Research Report: chronic bronchitis and occupational dust exposure: cross-sectional study of occupational medicine on the significance of chronic inhalative burdens for the bronchopulmonary system. Deutsche Forschungmeinschaft, p 237–247. Harald Boldt Verlag KG, Boppard
Reisner M T R 1978 Dust in coal mines. In: Research Report: chronic bronchitis and occupational dust exposure: cross-sectional study of occupational medicine on the significance of chronic inhalative burdens for the bronchopulmonary system. Deutsche Forschungmeinschaft, p 105–109. Harald Boldt Verlag KG, Boppard
Robertson A, Dodgson J, Collings P, Seaton A 1984 Exposure to oxides of nitrogen: respiratory symptoms and lung function in British coalminers. British Journal of Industrial Medicine 41: 214–219
Rogan J M, Attfield M D, Jacobsen M, Rae S, Walker D D, Walton W H 1973 Role of dust in the working environment in development of chronic bronchitis in British coalminers. British Journal of Industrial Medicine 30: 217–226
Ruckley V A, Gauld S J, Chapman J S, Davis J M G, Douglas A N, Fernie J M, Jacobsen M, Lamb D 1974 Emphysema and dust exposure in a group of coal workers. American Review of Respiratory Diseases 129: 528–532
Ryder R, Lyons J P, Campbell H, Gough J, 1970 Emphysema in coalworkers' pneumoconiosis. British Medical Journal 3: 481–487
Saric M, Kalacic I, Holetic A 1976 Follow-up of ventilatory lung function in a group of cement workers. British Journal of Industrial Medicine 33: 18–24
Skalpe I O 1964 Long-term effects of sulphur dioxide exposure in pulp mills. British Journal of Industrial Medicine 21: 69–72
Sluis-Cremer G K, Walters L G, Sichel H S 1967a Ventilatory function in relation to mining experience and smoking in a random sample of miners and non-miners in a Witwatersrand Town. British Journal of Industrial Medicine 24: 13–25
Sluis-Cremer G K, Walters L G, Sichel H S 1967b Chronic bronchitis in miners and non-miners: an epidemiological survey of a community in the gold-mining area in the Transvaal. British Journal of Industrial Medicine 24: 1–12
Smith T J, Peters J M, Reading J C, Castle C H 1977 Pulmonary impairment from chronic exposure to sulphur dioxide in a smelter. American Review of Respiratory Diseases 116: 31–39
Soutar C A, Copland L H, Thornely P E, Hurley J F, Ottery J, Adams W G F, Bennett B 1980 Epidemiological study of respiratory disease in workers exposed to polyvinylchloride dust. Thorax 35: 644–652
Soutar C A, Hurley J F 1986 Relationships between dust-exposure and lung function in minders and ex-miners. British Journal of Industrial Medicine 43: 307–320
Soutar C A, Hurley J F, Gurr D C 1984. The relationship between dust exposure and lung function in miners and non-miners. In: Sixth International Pneumoconiosis Conference. BOCHUM 1983. International Labour Office, Geneva
Theriault G P, Peters J M, Fine L J 1974a Pulmonary function in granite shed workers of Vermont. Archives of Environmental Health 28: 18–22
Theriault G P, Peters J M, Johnson W M 1974b Pulmonary function and roentgenographic changes in granite dust exposure. Archives of Environmental Health 28: 23–27
Todd G F 1966 Reliability of statements about smoking habits. Supplementary report. Tobacco Research Council, London
Vincent J H, Mark D 1981 The basis of dust sampling in occupational hygiene: a critical review. Annals of Occupational Hygiene 24: 375–390
Walker D D, Archibald R M, Attfield M D 1971 Bronchitis in men employed in the coke industry. British Journal Industrial Medicine 28: 358–363

Walton W H 1960 Questions put by the dust/engineering group and the discussion of these. In: Orenstein A J (ed) Proceedings of the Pneumoconiosis Conference, Johannesburg. J A Churchill, London, p 612–616

Wiles F J, Faure M H 1977 Chronic obstructive lung disease in goldminers. In: Walton W H (ed) Inhaled particles IV. Pergamon Press, Oxford, p 727–735

18. Medical fitness to drive

J. Taylor

INTRODUCTION

Accident studies involving on the spot assessment of the cause of road accidents both in the UK and in the USA, reveal human factors as the overwhelming cause of road accidents. Transport and Road Research Laboratory's survey in Great Britain was published in 1977.

ALCOHOL CONSUMPTION AND ROAD ACCIDENTS

Paramount amongst human factors is the role of alcohol and Borkenstein et al published the results of their City of Grand Rapids survey (Borkenstein et al, 1964). This clearly illustrated the linear relationship between accident risk and the amount of drink consumed. Borkenstein's study has led to most nations prescribing legal limits to the amount of alcohol present in the breath or blood of motor vehicle drivers. In Great Britain on 6 May 1983 the evidential breathalyser (Fig. 18.1) was introduced in place, in most cases, of a specimen of blood or urine. The last 5 years has also seen the progressive replacement of simple chemical breathalysers by catalytic hand-held devices (Fig. 18.2) with a high degree of accuracy. Whilst most often used by the police, these do have a clinical use in detecting recidivism in problem drinkers. May 1983 also saw the launching, in Great Britain, of the high risk offender procedure. A person convicted of a first drink/drive offence or refusal to give a specimen in Great Britain, normally on conviction is banned from driving for 12 months, whilst a second offence or refusal to give a specimen involves a period of 3 years disqualification. However in Great Britain persons having two convictions within 10 years exceeding 2½ times the legal limit (80 mg of alcohol per 100 ml of blood or 35 μg of alcohol per 100 ml of blood), will not have their driving licence restored at the end of the period of disqualification unless they can satisfy the Department of Transport medical examiner that they have not been suffering from an alcohol problem likely to cause them to be a source of danger to the public driving. The pilot project will involve the assessment of some 2000 to 3000 high risk offenders per year coming on to main stream from May 1987 (due to the 4-year disqualification lead-in time). On being categorised as a high risk offender, the driver receives a letter from the Department warning him that restoration of his licence is not automatic at the end of the period of disqualification, that he has been categorised as a high risk case, and he is urged to consider seeking help either through his doctor or a local treatment agency. He is also offered telephone advice on the location of these. Then, some 3 months prior to the end of the period of disqualification, he is invited to attend a medical examination which comprises a structured interview involving specific problem drinking-related questions, and also to a haematological and biochemical assay. It is estimated that

Fig. 18.1 The evidential breathalyser being used.

these combined procedures should result in a 99% detection of problem drinkers. Whilst at the present time there is no 100% specific detection chemical test available (gamma glutamyl transferase is only positive in about 33% of problem drinkers), current research at the Wolfson Research Laboratory at Birmingham suggests that problem drinkers have a specific triangular defect in their red cells (triangulocyte; Homadan et al, 1984). This can be detected on electron microscopy but the screening process takes about 3 hours in each case. The Birmingham team are hoping to detect a membrane biochemical marker common to the triangulocyte. Once an accurate chemical test has been established, it should be possible to extend screening at the time of arrest, not only to detecting body levels of alcohol, but also the existence of a drink problem.

In tandem with the efforts to establish a simple screening detection of problem drinkers which are of obvious importance not only in the context of driving but in the whole sphere of occupational medicine, are efforts to structured education programmes and effective literature in combating problem drinking. The British high

Fig. 18.2 A catalytic hand-held breathalyser in use.

risk offender project is to be closely scientifically monitored and it is hoped that the value of educational techniques can be assessed. A survey of British road users killed in road accidents reveals that one-third killed during the 24-hour period exceeded the legal limit, and two-thirds are over the limit between the hours of 10 pm and 4 am (Sabey & Staughton, 1980). The precise clinical role of chemical addiction to alcohol and road accidents has yet to be assessed and it seems likely that social and heavy drinking represents the bulk of alcohol-related accidents. To what extent these are indicative of an underlying personality problem, is entirely speculative.

SUDDEN COLLAPSE AT THE WHEEL

Road accidents by their very nature are multi-factoral in aetiology and it is only in the acute, sudden precipitate collapse at the wheel that health factors form the sole causal agent. Acute collapse is a rare event and a study of 10 000 road accidents by Grattan revealed only one of these with a direct medical cause (Grattan & Jeffcoate, 1968). The part played however by chronic illnesses in road accidents is much more difficult to determine. Waller (1965) found that patients with chronic illnesses averaged twice as many accidents per million miles travelled as drivers in an age adjusted comparison group. However Ysander (1970) suggested that persons with chronic ill health had a lower accident rate per distance travelled than healthy controls. These differences in the two most important epidemiological studies perhaps reflect the element of chance and also the fact that drivers innocent of all blame may be involved in accidents.

Epilepsy

A survey of 2000 police-reported personal injury accidents to the Licensing Centre, in which drivers had survived the accident but in which eye witness statements confirmed sudden collapse, revealed that 39% of these were due to generalised tonic clonic epileptic seizures. 70% of these were suffering from epilepsy and had failed to notify their condition in respect of their driving licence. In this series, 21% were witnessed to lose consciousness but subsequent investigations failed to reveal a cause. Diabetics taking insulin treatment accounted for 17% in the series, again with a 70% failure to inform the authorities, and only 10% were due to all forms of heart condition, and 7% to cardiovascular accidents. It is estimated that sudden collapse is responsible for about 0.1% of all road accidents. Many countries restrict driving of people with epilepsy, some banning it completely. In Britain, people with epilepsy were totally barred from driving from 1930 onwards until 1969 when a 3-year period of freedom from attacks when awake was required, but a special concession was introduced for people who had epilepsy confined to sleep for at least 3 years. In 1982 the current legislation came into effect and applies only to people who are suffering from epilepsy. This was defined many years ago by Russell Brain as 'a continuing liability to seizures'. A person suffering from epilepsy in order to be allowed to drive, must be free from any attack of epilepsy for a period of 2 years or where the attacks were confined to sleep, shall have had attacks asleep but not awake for at least 3 years. In addition a person being granted a licence with epilepsy must not be likely to be a source of danger to the public.

Much tougher standards are applied within the EEC to commercial and bus drivers who are barred from this occupation if there is a past history of epilepsy. Specific British law precludes the grant of a heavy goods or public service vehicle driving licence to a person who has suffered 'any attack of epilepsy since attaining the age of 5 years'.

The steady relaxation in medical standards for driving licence holders has been justified on grounds that nowadays it is possible to monitor compliance with anti-convulsant medication and the modern anti-convulsants are less liable to toxic impairment than the earlier ones. Also, there has been an endeavour internationally to unify standards but wide differences still persist from the total ban in Italy to 6 months freedom from attacks in certain parts of Scandinavia. At the time of going to press, negotiations are proceeding in Brussels for a Second Directive towards the objective of a Common Community Driving Licence. This goal of the Commission will certainly not be achieved before 1990. The major problem however in regard to epilepsy and driving is the failure of people with epilepsy to notify their condition to the Licensing Authorities. In Holland, van der Lugt (1975) showed that only 14% of young men who sought exemption from National Service on account of epilepsy, had declared their condition in relation to their driving licence. In Great Britain the legal obligation to notify rests with the driver and is said to take legal effect as soon as the driver has been made aware that his condition can adversely affect driving either now or in the future, by a medical practitioner. In some parts of the USA, legislation binds on the medical practitioner to notify. Most European ethical professional medical codes enable a practitioner exceptionally and where the safety of the public is being placed in jeopardy, to breach his patient's secrecy and inform the authorities. A significant effect of failure of an individual to notify a health condition to the Licensing Authorities, can be the effect on motor insurance. British motor insurers are likely to refuse to meet liability where a driver has made a false declaration to obtain a driving licence. A medical practitioner who connives with a patient in making a false driving licence statement in Great Britain, runs the risk of having to meet the costs of the accident if he is found by a court to be guilty of an act or omission likely to harm his neighbour. Many practitioners feel there is a dilemma and many give no specific advice to patients, but in so doing may be running the risk of falling foul of the tort of negligence each time their patient drives illegally.

In some countries, a distinction is made between epileptic attacks which impair attention, and those which do not. This distinction is not an easy one to establish. The majority of patients with absence seizures were convinced that they remained fully conscious throughout their brief seizures and temporal lobe discharges (complex partial seizures) are frequently associated with states of altered consciousness. A 24-hour EEG and video monitoring is proving a useful new diagnostic tool and it is evident that it is rarely the case that minor seizures do not represent a danger driving. It is important to realise that a routine 20-minute electroencephalogram may well show a normal trace between seizures, and therefore a negative trace does not necessarily preclude a diagnosis of epilepsy. This is not to say that modern electrophysiological techniques cannot establish or confirm a diagnosis. Jennett (1985) has drawn attention to the need to advise against driving where there is a high prospective risk of epilepsy following head injury or craniotomy. In general, patients who have undergone a major supratentorial craniotomy or suffered severe brain trauma, are advised not to drive privately for 12 months and to give up professional driving on a permanent basis.

Diabetes mellitus

The commonest reason for withdrawing a private driving licence from a person suffering from diabetes in Great Britain, is defective vision. The current trend towards tighter physiological control and a broader use of insulin in maturity onset (type 2) diabetes has not, in Great Britain, led to a significant increase in the number of reported acute collapses at the wheel, and this may well be due to wider blood glucose monitoring by patients themselves. Sadly, however, a significant proportion of patients fail to inform the authorities about their diabetes whilst applying for a driving licence, and may subsequently discontinue regular hospital clinical attendance. Such patients may be deprived of the advantages of photocoagulation treatment to combat diabetic retinopathy. The EEC Directive of January 1983, now binding on all Community members, requires that diabetics must be granted short term private driving licences so that their health can be periodically reviewed. In Great Britain this has resulted in the reintroduction of many patients to their diabetologists. Unfortunately through the non-declaration of their diabetes, many patients deprive themselves of this review procedure and in consequence blindness may ensue. The Community Directive imposes a ban on member nations granting driving licences to commercial and bus drivers. As mentioned above, 17% of the 2000 police reported accidents associated with witnessed collapse at the wheel, were on account of diabetes. All of these were on account of insulin, bar one which was attributable to glibenclamide.

Patients were questioned as to the reason for their collapse and Table 18.1 details the results of those enquiries. The commonest attributable cause of collapse was due to missing a meal and the second commonest due to unusual effort. In Britain insulin is generally self-administered by subcutaneous injection, usually into the thigh. Physical effort and muscular action may accelerate the absorption and lead to an overdose

Table 18.1

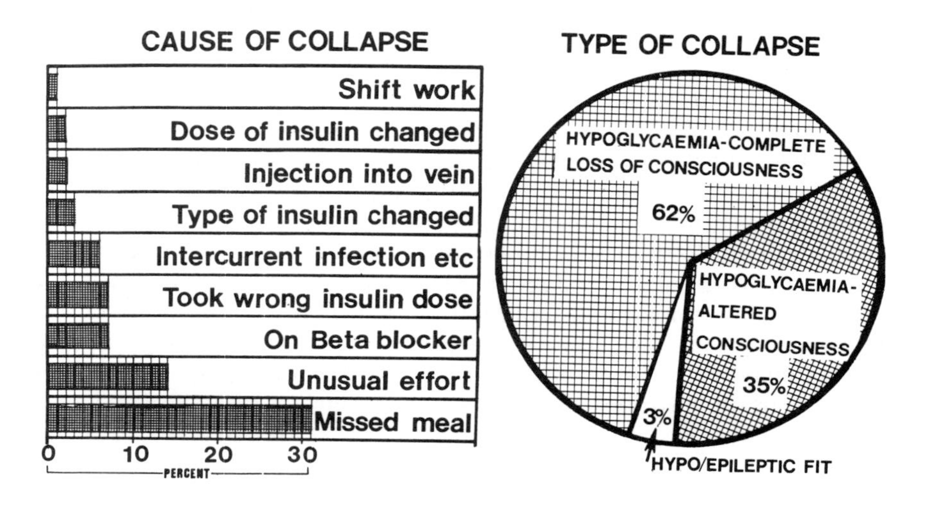

effect. A classical case was that of a coal lorry driver who had to deliver a large quantity of coal to a flat at the top of a converted old house. He had recently been commenced on insulin treatment. After making the delivery he returned to his diesel lorry, started the engine, put it into gear and then acutely and precipitately collapsed over the wheel. The vehicle moved slowly forward at tick-over rate and killed 2 people sitting in a parked car in front. In the series mentioned earlier, 62% of the patients on insulin collapsing suddenly, did so with immediate complete loss of consciousness, whilst 35% suffered a state of altered consciousness in which they were physically able to drive the motor vehicle albeit mindlessly. One was stopped by the police after driving 20 minutes the wrong way up a motorway. Another had driven 22 miles and was stopped driving the wrong way up a dual carriageway at 20 mph. Whilst most diabetics are well versed and trained in what to do if they suffer early warning hypoglycaemic symptoms, altered consciousness sometimes prevents them from taking carbohydrate. Notwithstanding this however Clark & Ward (1980) showed that a significant number of patients attending the clinic did not carry carbohydrate with them when they were driving. Overdose situations apart, the individual dose of insulin in insulin accident cases showed no relevance and indeed the highest risk related to relatively small doses in new patients, particularly maturity onset patients.

AGE OF THE DRIVER

A number of British motor insurers still require drivers over the age of 65 to submit annual medical certificates of fitness to drive. This is despite protestations by the British Medical Association who feel that medical practitioners may be placing themselves at risk of being held liable in the event of an accident. It is therefore important that any certificate issued should be based on a full medical examination and should relate only to fitness at the time that the examination was conducted. However a survey of drivers involved in accidents in Great Britain reveals that the greatest motor accident involvement is between the ages of 17 and 19 years, when males have 3.3 times the average traffic accident driver risk and females 1.9 times. Drivers are at above average risk until about the age of 34 years, and thereafter until the age of 72 years, they are below average risk. The British driving population beyond 70 years is still relatively sparse, and accident rates have not been explored yet. Although the British population over the age of 65 years has been increasing progressively, by the turn of the century it is expected to fall although one would anticipate a continuing increase in the driving population in that age group if only due to the increasing number of women drivers.

HEART DISEASE

Although all forms of heart disease are a relatively small factor in road accidents, coronary artery disease is a high cause of morbidity. A person who has recovered from a first cardiac infarction, has a significantly higher risk of a second attack than matched controls with a negative cardiac history. For this reason the consensus view of British cardiologists until quite recently, has been that post-cardiac infarction patients should not drive very heavy vehicles; buses or heavy lorries. In Great Britain heavy goods vehicles exceed 7.5 metric tonnes laden weight and are responsible for 84% more deaths per mile travelled than cars. Public service vehicles (nine or more passengers carried for hire or reward), probably have three times the killing capacity of heavy goods vehicles due to the fact that they operate primarily in urban and city areas. Although in Britain we have seen a spate of public service vehicle accident catastrophes, the risk of being killed per 100 000 km travelled by bus remains the same as that by train. More recently, the British cardiological consensus view has been that advances have been made in prognosis to the point that specific parameters can be laid down to allow persons who have had coronary artery bypass surgery or cardiac infarction, to resume public service vehicle or heavy goods vehicle driving (Oliver & Somerville, 1985). The first test applied is a resting cardiograph and the presence of persisting Q waves debars from vocational driving. In the absence of those, the patient is subjected to an exercise test and providing ST depression is 2 mm or less, the patient is then subjected to a coronary angiography and may be granted a licence subject to the results of this, on an annual review basis.

THE TWO-WHEELED VEHICLE

If the trend in oil prices continues to increase, this may well result in more people resorting to two rather than four wheels for economy reasons. Motorcyclists in Great Britain represent 2% of registrations and 20% of driver deaths. A mitigating effect of seat belts on car occupant serious injuries and deaths, does not apply to two-wheeled vehicles. Whilst driver motorcycle training courses and the new two part licence seems to have had some beneficial accident value, it is perhaps important to advise medical high risk accident groups such as people with epilepsy and diabetics on insulin, to avoid using two-wheeled machines.

REFERENCES

Borkenstein R F, Crowther R P, Shumate W B, Ziel W B, Zylman R 1964 The role of the drinking driver in traffic accidents. Bloomington: Department of Police Administration, Indiana University 1964

Clarke B, Ward J D 1980 Hypoglycaemia in insulin-dependent diabetes. British Medical Journal 586

Grattan E, Jeffcoate G O 1968 Medical factors and road accidents. British Medical Journal 1: 75

Homadan F R, Kricka L J, Whitehead T P 1984 Morphology of red blood cells in alcoholics. Lancet i: 913–914

Jennett B 1985 Epilepsy (2) — After head injury and craniotomy. In: Oliver M F, Somerville, W (eds) Medical aspects of fitness to drive, 4th edn. Medical Commission on Accident Prevention, ch 3, p 35–37

Oliver M F, Somerville W 1985 Cardiac conditions, Medical aspects of fitness to drive, 4th edn. Medical Commission on Accident Prevention

Sabey B, Staughton G 1980 The driving road user in Great Britain. SR 616 Leaflet 912. Transport and Road Research Laboratory, Crowthorne

TRRL 1977 On the spot accident investigation. Leaflet LF392. Transport and Road Research Laboratory, Crowthorne

van der Lugt P J M 1975 Is an application form useful to select patients with epilepsy who may drive? Epilepsia 16: 743–746

Waller J A 1965 Chronic medical conditions and traffic safety. New England Journal of Medicine 273: 1413–1420

Ysander L 1970 Diabetic motor-vehicle drivers without driving licence restrictions. Acta Chirurgica Scandinavia Supplementum 409: 45–53

19. Hazards of working with VDUs

C. Mackay

I think that I shall never see
A calculator made like me.
A me that likes Martinis dry
And on the rocks, a little rye.
A me that looks at girls and such,
But mostly girls, and very much.
A me that wears an overcoat
And likes a risky anecdote.
A me that taps a foot and grins
Whenever Dixieland begins.
They make computers for a fee,
But only moms can make a me.
Hilbert Schenck Jr (1960)

Non e vero, ma e bene trovato!
Italian proverb

INTRODUCTION

It is now a quarter of a century since the verse by Schenck was published. At the same time a paper by Licklider (1961) coined the term 'man (sic)—computer symbiosis' to articulate the then currently prevalent view that anticipated very close co-operation between the human being and the computer as a prominent feature of the development and subsequent spread of such systems. During those 25 years technical development, enabling miniaturisation and economy of production of crucial components, has progressed at a staggering rate, far surpassing the predictions of the foremost proponents of information technology, as it has latterly become known.

For the most part the physical manifestation of such changes has been the appearance of the visual display unit (VDU) referred to in North America as the video display terminal (VDT). During the 1960s the use of VDUs was largely confined to specialised user groups. Beginning in the 1970s and accelerating into the 1980s all that has changed. Such devices are no longer the exclusive tool of the specialist. They have become both commonplace and indispensable. They are used by an astonishingly heterogeneous group of users for an equally broad range of tasks. Forecasts suggest that by the early 1990s around 40 million working people in the USA will be using such devices.

The change to new technology has undoubtedly brought with it enormous benefits both to organisations who implement it and to individual users. In other instances problems have arisen largely because of inadequate planning and implementation policies including job redesign, poor design of equipment and poor working environments. More particularly the VDU has been rightly or wrongly blamed for health concerns ranging from mild headaches to perinatal mortality. Much of the debate about alleged VDU-related illness has taken place in the popular press: some of the accounts imply that almost every bodily system or organ is a potential target. The medical press has not been immune from such discussions, albeit on a more objective level (Lee, 1985; Frank, 1985) and most countries have or are in the process of issuing guidance, developing minimum standards or enacting legislation on VDU related health issues. Meanwhile controversy exists over whether VDUs can be considered a hazard to human health (the term hazard being broadly conceived). Consensus on some issues exists whilst others, including the reliability and validity of the data on which their existence is based, are hotly disputed. Are VDUs the cause of the complaints heard apparently from large numbers of workers or, as the physical embodiment of the new systems, are they a convenient target for hostility and reaction against fundamental changes introduced into the work process, particularly where such changes are perceived as being threatening to livelihood and status? In many industries the VDU health debate has been the subject of much discussion, negotiation and anguish with the result that there has been a proliferation of company–union agreements, some of which have been extremely wide-ranging in their coverage and technically very detailed; others less so. The debate has impinged upon the occupational physician in several ways including dealing with issues only marginally related to mainstream occupational medicine, being the final arbiter in bargaining contexts, being asked to pronounce on the desirability of screening for VDU users (in the case of eyesight examinations) and grappling with allegations of miscarriage and birth defect, whilst simultaneously trying to allay users' (legitimate) fears and worries about such effects.

The VDU health debate has generated an enormous number of papers, monographs, volumes and conference proceedings. A number of objective reviews of the subject are now available (Berquist, 1985; Marriott & Stuchly, 1986) and it is not the present author's intention or remit to produce a further review: rather to highlight some of the areas where controversy still exists, discuss the most recent research findings and suggest how some of the implications following from the most up-to-date and reliable studies can be of help to the occupational health practitioner.

A SCHEME FOR CONSIDERING HEALTH EFFECTS

Since we are considering the user of the system, the nature of the device he is using as a tool (in this case the VDU) can be viewed from two aspects. The first concerns the physical processes which enable the system to operate to fulfil its principal role (to transmit information). Some have suggested that these processes, certainly in the case of cathode ray tube (CRT) technology may as a by-product produce harmful radiation emissions. The second concerns the interaction of the user with the system via the device. As far as the user is concerned access to the system is achieved through the medium of an interface. The interface is any hardware or software feature which the user may have to interact. As such, the performance of the interface, i.e. its usability, both colours the perception of the user to the system as a whole, and if deficient may impede performance and generate discomfort. In the past, the health debate has tended to focus more on the technology (the physical manifestations and possible hazardous processes within the device) rather than the information handling requirements of the task and the related information processing limitations of the user. This would seem to be a reflection of perception of risk on the part of users rather than objective estimations of the level of hazard. These two general aspects of physical processes and usability are best conceived schematically as in Figure 19.1.

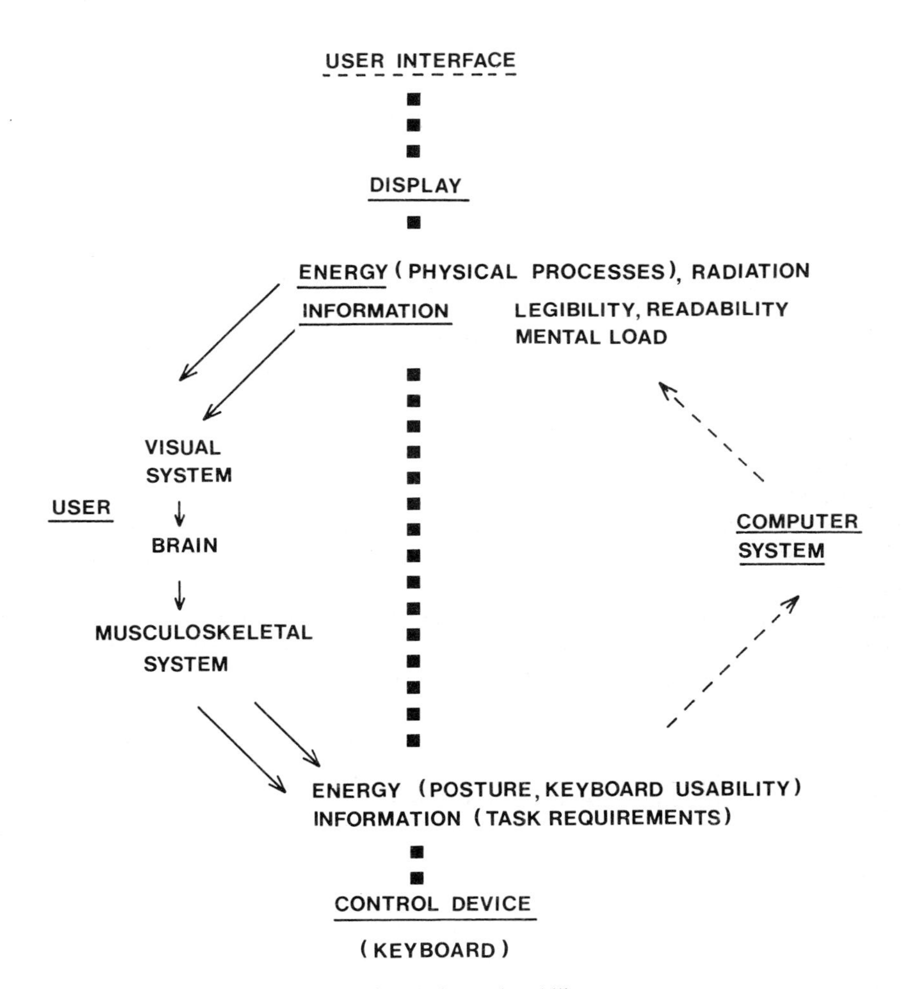

Fig. 19.1 Health aspects of VDU use: interface design and usability.

PHYSICAL PROCESSES AND ELECTROMAGNETIC RADIATION EMISSIONS

Information displayed on a CRT-based VDU is generated in almost identical manner to that used in a television receiver. An electron beam is accelerated by a high voltage field and projected onto a fluorescent screen. Fluorescence is produced by a coating of phosphorescent material (known as the phosphor) deposited in a thin film on the inner surface of the screen. The beam is made to scan the screen at a rapid but pre-determined rate. Excitation of the phosphor by the electrons results in visible light being emitted. Variations in beam current affect the intensity of brightness of the images on the screen. Theoretically, all the energy of incident electrons should be converted into light. In practice, however, the process is not perfect and radiation of other wavelengths could be emitted. Electromagnetic radiation comprises ionising and non-ionising types. The former is that in which the photon energy or quantum exceeds that which is required to break molecular bonds. Additionally, mechanical energy in the form of acoustic radiation may also be emitted. In this context acoustic radiation refers to the propogation of sound and ultrasound.

Sources and nature of radiation from visual display units

In assessing the risks of exposure to radiation it is important to take a critical look at the types of radiation emitted by VDUs and determine their sources, their intensity and the biological effects of each individual type. Moreover, in instances where more than one type of radiation is occurring simultaneously it is important to consider the combined effects of the spectrum as a whole from the point of view of biological effects. Also the use of cosmetics or medication which may lower the threshold for harmful radiation effects and similarly environmental potentiating factors must also be considered.

Ionising radiation

In CRT-based VDUs the only significant potential source of X-rays is the tube itself. VDUs usually operate at relatively low CRT voltages (about 12 kV for monochrome units and about 25 kV for colour). Nevertheless, the relatively energetic election beam inside the CRT is capable of X-ray generation when rapidly decelerated by striking internal metal or glass components. However, the energy levels are reduced very markedly by appropriate circuit design and the high lead content of the glass envelope. (Insofar as the material from which the VDU is constructed may contain radioactive materials these must also be considered). Further, modern VDU design is such that a significant increase of the CRT voltage with a resulting increase in the production of X-rays, is very unlikely. In monochrome CRTs the display becomes unstable over beam currents of about 15 kV. In colour CRTs, where higher voltages are used, circuitry is used which ensures the VDU will fail safe when the voltage exceeds a pre-set level.

A large scale study of X radiation from VDUs and TV receivers carried out in the USA indicated that the likely emissions under normal and severe test conditions are 0.1 mR/h or less (Bureau, 1981). Numerous tests of VDUs in use invariably showed that X-ray emissions were below the natural background, i.e. about 10 mR/h (National Institute for Occupational Safety and Health, 1981; Environmental Health Directorate, 1983; Terrana et al, 1980; Wolbarsht et al, 1979). In a few cases excessive levels of X-rays found were attributable to faulty measurements (National Institute for Occupational Safety and Health, 1977; Environmental Health Directorate, 1983). Some models, which in pre-market testing were found to produce X-rays about 0.5 mR/h, were not allowed onto the market (Bureau, 1981). Finally, tests performed in a low-background facility with the natural background radiation reduced showed that X-ray emissions for nearly 70 different models were less than 0.003 mR/h. This is an extremely low level and it corresponds to the background level of the test facility employed. No X-rays could be attributed to operation of the VDU since the measured radiation levels were the same whether or not power to the test VDUs was switched on or off (Pomroy & Noel, 1984).

In view of the aforementioned studies, the conclusions and recommendations of various government agencies seem well justified. The main conclusion is that X-ray emissions from VDUs are either non-existent or so low that they do not pose a health hazard to the operator. Furthermore, there is no need for periodic testing for X-rays of VDUs whose models meet appropriate government regulations, as there is nothing inherent in their design and operation that can cause an increase in production of X-rays (Marriott & Stuchly, 1986).

Non-ionising radiation (NIR)

This area covers both optical radiation (non-ionising UV, visible light and infrared (IR) and those parts of the spectrum below 300 GH (referred to as Hertzian radiation; primarily the microwave region, radiofrequency and low frequency). Clearly substantial levels of the former are found from both natural and man-made sources. Concerning the later, this frequency range is used extensively as an energy source for a wide range of household and industrial products including radio, television, microwave ovens, radar, microwave transmitters as well as the ubiquitous electric fields associated with the power-line frequency (50/60 Hz) in such as incandescent and fluorescent lighting, electric tools, hi-fi, electric blankets and other electrical appliances.

The generation of non-ionising radiation by video display terminals is most easily understood if one looks at the way in which these devices function. In the cathode ray tube, electrons are accelerated by the application of a current of 10 to 25 kV, depending on the type of the CRT. The accelerated electrons strike the screen of the CRT which is covered with luminescent materials that govern the emission of ultraviolet, visible light and infrared radiation. Infrared is generated by the cathode heater as well. Various electrodes and coils deflect and focus the electron beam which is carried by the rapid oscillations of a current, ranging from 15 000 to 22 000 oscillations per second, depending on CRT type. This produces radiofrequency radiation. The horizontal deflection system is considered to be the primary source of radiofrequency radiation, as well as of sound and ultrasound emissions. Because of the complex deflection signals imparted to the electronbeam, miltiples (harmonics) of the fundamental frequency of 15 to 22 kHz have to be expected. The transformers and coils contained in a VDU are sources of low-frequency magnetic fields, but the emission of microwave radiation by VDUs is made virtually impossible by their design.

All non-ionising radiation is capable of gross thermal effects (bulk tissue heating) at sufficiently high energy densities. Acute biological effects are possible when there is a significant tissue temperature rise in the case of microwave and radiofrequency, such power levels are only possible in very close proximity to high-power sources such as transmitters, radar or heat sealers. At medium power levels bulk tissue heating by di-electric absorption is proportionately less severe, but can still cause significant biological effects. The classical example is microwave radiation of the eye leading to cataract formation. Neither of these power levels is appropriate in the case of

VDUs and increasing attention is now being paid to lower frequency ranges including VLF and ELF (0 to ~500 Hz including power line frequencies).

Ultraviolet, optical and infrared radiation

A number of studies have examined radiation emissions in this portion of the spectrum. Visible light is emitted by a VDU in order for it to function. With the screen full of characters and at maximum brightness the risk to the eye is negligible. Overall for UV, IR and optical radiation typical measurements show that emission levels are at least two orders of magnitude below recommended limits (Murray et al, 1981; Terrana et al, 1980; Roy et al, 1984).

Radiofrequencies

The main source of emissions within the 10 kHz to 10 GHz region is from the fly-back transformer usually located towards the rear of the VDU. The nature and extent of these emissions have been well documented in a number of recent studies (Stuchly et al, 1983b; Weiss & Petersen, 1979; Weiss, 1983; Wolbarsht et al, 1979). Because of the complex waveform of the scanning signal up to ten harmonics of the fundamental frequency are present (Harvey, 1984; Ontario Hydro, 1984). Outside the frequency range of about 15–22 kHz the emissions of radio and microwave frequencies are extremely low. When considering radio and lower frequency portions of the electromagnetic spectrum it is important to realise that emissions are manifested in the form of electrical and magnetic fields. The temporal and spatial distributions of such fields are exceedingly complex and often present formidable measurement difficulties. When placed in such fields the body acts as an electrical conductor and current flows through it to earth. In determining possible biological effects it is important to assess the coupling of such fields to the operator and the resultant anatomical distribution of body currents. Marriott & Stuchly (1986) have summarised the position regarding the likely field strengths impinging upon an operator 30 cm from the screen. At this distance maximum electric fields strengths are in the order of 7 V/m and the maximum values were derived from measurements carried out on 165 different models of VDU and thus can be considered as representative. Because of the positioning of the fly-back transformer localised maximum fields strengths at the surface of the VDU casing may reach 40 V/m and 1.1 A/m for the electric and magnetic fields respectively with 15 V/m and 0.5 A/m being the more typical values (Ontario Hydro, 1984; Radiation, 1984). Comparison of these data (at normal viewing distances) with recommended limits for RF exposures suggest that even the highest exposures are within the most stringent standards. The majority of exposures (65%) are below the current limits by a factor of at least 10 (i.e. the electric field strength is below 1.5 V/m) (Marriott & Stuchly, 1986).

Extremely Low Frequencies (ELF)

Because of the nature of the physical processes within VDUs they, like any other electrical and electronic device, produce stray electric and magnetic fields at the power-line frequency of its harmonics (50 or 60 Hz). So far measurements have revealed a maximum electric field strength of 10 V/m (RMS) (Harvey, 1984; Ontario Hydro, 1984) and magnetic field strengths of 0.22–0.56 A/m (Stuchly et al, 1983a; Ontario Hydro, 1984). These fields are of the same order as those found from ambient levels in laboratories and homes and are typically lower than those found around other commonly used domestic electrical devices (electric shavers, electric blankets).

NIR from VDTs: possible biological effects
Until recently most of the emphasis in examining NIR effects associated with VDUs has been concerned with characterising the nature and magnitude of the field strengths. Little attention has been paid to the dosimetry aspects; the quantification of induced currents and absorbed energy in those using VDUs and like devices (Guy, 1986). Within the frequency range of interest in connection with VDUs the only well established NIR interactions with biological material are electrical stimulation or shock, spark discharge, elevation of tissue temperature, burns and possible stimulated bone growth. Only exposures orders of magnitude above those measured in the vicinity of VDUs can elicit the first four of the above. In some circumstances current densities of the same order of magnitude as those associated with VDUs may be capable of bone growth induction. However in other respects the fields known to have such effects differ markedly from VDU field characteristics, notably waveshape. In the last few years it has been shown that weak electromagnetic fields are capable of interactions with biological systems at specific frequencies and intensities (Adey, 1981; Bioelectromagnetics, 1984). Both the electric and the magnetic field appear to be capable of such interactions, but their mechanisms are not known. The effects reported occur, however, at frequencies between 1 and 1000 Hz and they are pertinent rather in evaluating emissions at extremely low frequencies (ELF) than radiofrequencies. Suggestions have been made that the waveform of the fields which can cause biological effects may be of importance. However, this can only be considered as an interesting hypothesis yet to be scientifically tested.

Some recent work on the effects of pulsed magnet field on chick and mouse embryos has attempted to demonstrate a possible biological mechanism whereby VDT like fields may be teratogenic. These will be dealt with in a following section dealing with reproductive hazards. Overall the conclusions drawn from the literature on VDU field characteristics, dosimetry and known biological effects is that such fields are well below the exposure standards applicable in the West and are below levels shown to reliably cause harmful bioligical effects.

Electrostatic fields
A quasistatic (varying very slowly in time) electric field is produced by the electric charge on the screen of the CRT (Harvey, 1984). The strength of the static electric field decreases rapidly with distance away from the screen. The charge on the CRT screen depends on the display brightness, the number of characters on the screen, the rate at which the writing beam is turned on and off and the operating history of the unit (Harvey, 1984). The design of the unit and the ambient conditions (humidity) are also likely to play a role, as are measurement technique and individual differences in electrostatic charge on the user.

Measurements performed on 54 units which comprised 27 different models showed that typical static electric field strengths were between – 150 V/m and 150 V/m, and a maximum of 1500 V/m was found 30 cm from the screen (Ontario Hydro, 1984). Electrostatic field strengths up to 30 kV/m at a distance of 30 cm from the screen were reported in a study of 44 models of VDUs. Some models, however, did not produce a static field (Paulsson et al, 1984).

DISPLAY CHARACTERISTICS

To enable the VDU user to extract information from alphanumeric text and graphical symbols, the form, structure and content of such material needs to be considered. Much is now known about what makes a good display (thereby enhancing human performance) and there are various techniques for measuring display parameters and evaluating image quality. The latter largely determines the ease by which the user can see (legibility) and read (readability) material presented on the screen, although other factors such as task requirements, ambient lighting and eyesight are important and tend to interact in complex ways in influencing performance and comfort. Although much of what is summarised below is derived from work on the existing and most widespread technologies (i.e. CRT-based) requirement for the newer devices (plasma panels, LCD) are likely not to be dissimilar apart from some specific exceptions.

In terms of display characteristics the legibility of an individual picture element is determined by its luminance with respect to the background against which it is displayed, its size, and a variety of time dependent variables of screen luminance. Legibility of single letters is affected by character generation methods, resolution and the sharpness of individual strokes or dots and the shape of individual characters (aspect ratio, stroke width). Legibility of single words and passages of text are further influenced by inter-character and inter-line spacing and text format. These are discussed briefly below in the context of the requirements for comfort and visual performance.

Luminance, contrast and character structure

Increases in display luminance have several direct effects on visual physiological and optical responses, and visual performance. Generally, increases in display luminance will cause decreases in pupil size, which in turn leads to increases in the optical depth of field and improvement in optical quality. The result is that the visual acuity of the normal healthy eye is enhanced. Thus displays having higher luminance permit a user to see finer details on the display. In so far as image brightness depends upon individual preference, control of either display luminance and/or contrast is desirable. There remains some dispute as to whether positive or negative contrast displays are better for user performance. With positive contrast displays one may expect an increase in visual acuity of about 15% because of higher average luminance (NRC, 1984). The data of Bauer & Cavonius (1980) and Radl (1980) suggest that positive contrast displays yield greater legibility and are preferred by users over negative polarity. However, these experimental data are not sufficiently clear cut to allow a definite decision to be made. Rupp (1981) has suggested that whichever of the two items has the higher luminance, will essentially control the adapting luminance thus removing any effect on pupil size due to positive or negative contrast displays. Both positive and negative polarities have purported advantages. Negative contrast may help to increase flicker perception thresholds whilst negative ones may help to reduce the effects of veiling reflections on the screen. Negative contrast may also ameliorate losses in visibility that can occur as a result of transient adaptation (the fluctuation in eyes sensitivity due to rapid background luminance changes, as occurs for example, when positive contrast displays are used with negative contrast source documents). This issue is by no means settled.

Various groups have set out very specific recommendations for luminance values of the background and characters. On negative contrast these vary from 10–20 cd.m^{-2} for the former and 25–160 cd.m^{-2} for the latter. These suggestions must be seen in the context of visual performance generally: the human eye can adapt itself automatically to operate sufficiently well over a very large range (Campbell & Durden, 1983). For a general discussion of contrast and visual performance see Campbell (1980) and Campbell & Maffei (1974). So long as VDUs have contrast/brightness controls available to individual users, requirements for individual preferences and variations in ambient workroom lighting conditions should be met.

The modulation transfer function (MTF) is a useful parameter for evaluating image quality. The MTF essentially describes the amount of signal output for a given input. It is used in the context of VDUs to describe the contrast attainable on a CRT and relates directly to the sharpness and clarity of the alphanumeric image so displayed (Task, 1979). Maximum legibility and ease of reading are obtained for contextual characters when contrast modulation ($L_c - L_b/L_b + L_c$) is approximately 70%. For non-contextual or single letters about 95% is required for adequate performance. Apart from luminance and contrast, character structure is known to have a crucial impact on visual performance and comfort. Character generation methods most widely used at the time of writing are dot matrix and stroke generation. Different information processing requirements and application dependency account for the equivocal guidance given in the literature (Snyder & Maddox, 1980; Bouma, 1982). Acceptibility, identifiability and distinctness are among the important criteria. Optimum font design is crucial for non-contextual presentations (data entry versus word processing). With dot matrix system both decreasing the space between dots and increasing the number of dots improves reading speed and reduces error rates.

The need for upper and lower case characters (capitals and small letters) again depends upon the operator's task. Both are required for tasks involving continuous text, but in other instances upper case characters may be adequate. If dot matrix generation techniques are used a 5 × 7 matrix is adequate for a character set of upper case letters and digits. A larger matrix (7 × 9) is desirable when lower case letters are to be used as well. Generally, upright (i.e. not slanted) characters are to be preferred, but where it is the practice to allow italic script to be used for coding purposes, this should not be discouraged. However it may not be possible to accommodate italics within an orthogonal 5 × 7 or 7 × 9 matrix and produce the desired legibility.

Visible raster structure decreases legibility and therefore imperceptible scan lines improve legibility. If the raster is too pronounced the line structure may become apparent and therefore reduce effective resolution and contrast (and incidentally cause annoyance). Characters with sharp edges lead to greater legibility than those with blurred ones. Because of limitations of current technology most characters have graded luminance distributions. The minimum desirable physical dimensions of a character required for legibility depend upon a number of factors. An optimum size may only be determined for one particular set of circumstances. However the subtended angle should not be less than 16 minutes of arc because below this value legibility decreases rapidly. Provided there is no dissociation in the elements (strokes, segments or dots) of which the character is constructed, and assuming good contrast, a value somewhat above this figure is, in fact, comfortable for most people. Larger symbol size, although allowing the display to be viewed at greater distances and providing for some improvements in legibility, may limit the amount of text that can be displayed, and may also affect the ease with which text can be read at shorter viewing distances.

Time-dependent variables of image quality

Many current display units based upon the most widely used CRT technologies are susceptible to a range of time-dependent display instabilities. These phenomena are important because they influence comfort and probably performance. Under operational conditions there are a number of ways in which image stability can be adversely affected. The appearance of flicker on visual display units is dependent upon relevant design characteristics of the display and upon personal factors of the individual operator. The former include: phosphor type and regeneration rate; character luminance, size and colour; area of screen illuminated; viewing angle. The latter include large individual differences in both the threshold at which flicker is perceived and sensitivity to it, in terms of annoyance and distractibility. These individual factors must be borne in mind if complaints arise from some users but not others. Nevertheless, because of its importance in minimising fatigue and promoting legibility, stable character image should be an important consideration when selecting a VDU.

Other, less common forms of display instability may also occur. These include the apparent tendency for characters to periodically jump or jitter, and a slow, rhythmic wavering effect known as swim.

Where they happen, fluctuations in the stability of the image are largely a reflection of current VDU design, often beyond the scope of VDU maintenance, and largely outside the control of the operator. It is to be hoped that progress in display technology will overcome these problems. Should display instabilities occur on individual displays, assistance should be sought from the supplier.

Colour

There are few definite data to support the choice of one phosphor/screen colour compared with another. Most phosphors have been of a yellow/green appearance. These have been suggested to be preferred perhaps because they are the commonest. Colour *per se* is not a crucial determinant of legibility but highly saturated primary colours immediately adjacent to one another may cause perceptual difficulties of apparent depth because they are focused at slightly different distances. Used appropriately colour is also useful for coding purposes but consideration must be given to deficiencies in colour vision in the user population.

CONTROL DEVICES: INFORMATION INPUT

It has been suggested (Card, 1983) that to cope with the complex workstation now available (which may incorporate several control devices) it would be useful if the human being could now be re-designed with four upper limits instead of two; but until reliable systems based upon voice input/speech recognition become available less direct forms of information input must suffice. Less direct means the physical manipulation of some form of control device. For the most part either keying or pointing are involved. Smooth reflexive keying depends upon a user's ability to generate a pattern of motor signals and to receive information back via the fingers (and other types of feedback) so that the perceived and intended activity are matched. A range of factors are important in order to ensure operator efficiency and comfort.

These include training (skill levels), the nature of the task, posture and keyboard design. These four factors tend to interact in complex ways and a knowledge of the physical and cognitive limits of the human user is important.

The design of specific control devices has a number of implications for the users' health and performance effectiveness. One approach to keyboard design suggests that as more and more individuals are becoming familiar with keyboards it is important to have one standard approach. The opposite view suggests that as the use of keyboards is proliferating rapidly amongst previously naive users, together with the opportunity to spend more time and resources on the design of the interface (because of reducing microprocessor costs) a technology/training window is occurring which would allow new forms of keyboard design to be rapidly assimilated. Ergonomic requirements for existing (i.e. QWERTY type key arrangements) are well established and the author does not propose to dwell on them here (Snyder, 1983; Alden et al, 1972). As to controversy concerning key layouts, attempts have been made to improve the efficiency and comfort aspects of keyboards by balancing the loading over the two hands, increasing lateral hand movements whilst decreasing longitudinal ones, and minimising overall finger travel. There are clear advantages in some of the new layouts concerning productivity and comfort. Until recently little progress has been made in developing these novel keyboard designs because of the investment in training and re-training time. Nobody has been trained on the newer keyboards since they have not been widely available (and/or compatible) and there have been no simplified keyboards available; nobody has been trained to operate them. It is likely in the near future that a number of well-developed new keyboards will become available on the market with features such as alternative key layouts, dished and split plinths and with wholly programmable keys. The development of such a keyboard is described by Beusen (1984). So long as such devices can meet standards and requirements for usability (which are currently being developed by ISO) there is no reason to think that they will not become much more common and thus acceptable training requirements will follow.

HEALTH EFFECTS

The debate about health effects from VDUs can be regarded as an example of the polarisation of opinion which seems to inevitably greet technological change. One extreme (the null hypothesis) implies that harm cannot be assumed until reliable evidence has accumulated. The opposing view is that harm attributable to the technology must be assumed until there is evidence that is not present. Those in the first camp are accused of complacency, exploitation, and an interest in technology at any cost. Those in the second are labelled either as Luddites or blind to, or dismissive of the scientific approach to risk assessment. In connection with the latter, demonstrating with absolute certainty that a particular product or process never produces harm is logically and practically impossible.

Clearly some form of middle ground position is desirable in which society desires certain limits or restrictions on the design or use of the technology consistent with existing good practice. But these guides as to usage must be grounded upon scientifically validated data rather than upon mere speculation or assertions. This approach of course presupposes an existing database on which claims can be assessed and guidance generated. Unfortunately in this controversial area the database is patchy and competing hypotheses as to likely risk and to risk mechanisms cannot be easily tested. Furthermore anxiety on the part of users is generated by inappropriate perception of causality which is often reinforced by media reports of physical hazards. Indeed the media role in this particular issue has been profound. Perhaps because the industry itself had not been immune to the controversy and turmoil surrounding the introduction of new technology. Those readers who are familiar with some of the background to the alleged health hazards will recall that the original impetus for many of these was associated with VDU users in newspaper offices (Cataracts at the New York Times; birth defects at the Toronto Star; repetitive strain injury at the Melbourne Age). On page two of The Times for Friday 17 May 1985 was a passage labelled 'Computers no danger to pregnant women'. On the same day a reader who also turned to the foreign news on page 4 would have found a piece headed 'VDU linked to birth problems'.

Several health risks arising from the operation of VDUs have been postulated and have been discussed widely. Much of the concern has been expressed about the possible direct effects of VDU operation upon the individual operator, particularly hazards such as radiation, epileptogenic properties and effects on reproduction. Two factors contribute to the anxiety felt by operators: typically they spend the majority of each working day in close proximity to the screen, and thus feel they would be subject to long-term exposure to these hazards, and furthermore as these cannot be perceived as such they represent a hidden threat. Some reports continue to link radiation with impairments of eyesight and eyestrain, and understandably, the two have become inextricably linked in the minds of many operators.

The remainder of this chapter deals with each of the alleged health outcomes associated with the use of VDUs.

Models for considering health effects

One of the major difficulties in assessing types of hazards and levels of risk to health possibly associated with VDUs is the lack of any clear conceptual framework for hypothesis building. Some have proposed that argument by analogy is perhaps most appropriate and it is often suggested that VDU operation is in many ways similar to viewing television and, as such, should be associated with the same low level of health risk. This analogy is not a particularly useful one. The viewing distance, nature of the information displayed (text rather than pictures), duration of exposure, postural constraints and motivational and environmental factors are clearly all different.

These factors indicate that a variety of different models may be appropriate; a biological one for considering radiation hazards; a psychophysical one for considering impact of screen characteristics, a biomechanical one for posture, a cognitive model for considering information processing requirements of VDU tasks and socio-technical/psychosocial approaches for delineating stress effects and the impact of new technology. Until recently the debate has concentrated on the micro level as opposed to macro level approaches.

VISUAL SYSTEM

In so far as much of the information transmission involved in VDU work is in the visual mode, much of the concern about possible health effects has concentrated upon vision and the visual system. Essentially these can be divided into short term phenomena often referred to as visual fatigue, eyestrain or asthenopia as well as possible chronic effects involving ocular pathology.

Short-term effects

Many of the unpleasant symptoms experienced by the users of VDUs have been referred to as eyestrain or visual fatigue. This is a generic, loosely defined term which may include a cluster of symptoms such as burning, pain in the eyes, blurring of vision, headache and general fatigue. Visual fatigue is also used to refer to decrement of performance of visual tasks as indicated by impaired reaction time and increased error rates. Changes in visual function (e.g. accommodation, convergence, eye movements, blinking, pupillary response) have been observed during prolonged near visual work and these are often evidenced as indicants of visual fatigue.

Symptoms

Numerous field studies have indicated that VDU users (as a group) report high prevalence of visual disturbance of mainly the ocular (irritation, redness and soreness) and visual (difficulties in accommodation, blurring, photophobia) types (Dainoff et al, 1981; Gunnarsson & Soderberg, 1983; Knave et al, 1985a; Meller & Moberg, 1983). Most studies report a higher prevalence of symptoms in VDU users than in corresponding non-VDU users (Belluci & Mauli, 1984; Laubli et al, 1981; Ong et al, 1981; Rey et al, 1982) but more recent and perhaps better designed studies have failed to reveal clear differences (Gould & Grischkowscky, 1984; Lewis et al, 1982; Turner, 1982; Sauter, 1984; Starr, 1984; Howarth & Instance, 1985). The visual complaints described in these studies appear to be qualitatively similar to those found in previous studies of visual fatigue and are thus not unique to VDU work. It is possible that some specific characteristics of VDU design are responsible (such as display instability) but the individual effects of such parameters cannot be derived from currently available studies. Adequate research that would establish whether there is anything inherent in VDT tasks that can unavoidably cause visual difficulties not encountered in comparable non-VDT tasks has not been conducted (NRC, 1983).

In some of the studies referred to above, the prevalence of reported discomfort was quite high. On this basis one may expect objectively verifiable changes in visual functioning to occur as a result of VDU screen characteristics or task demands. Thus transient changes in accommodation including transient myopia (Haider et al, 1980; Ostberg, 1980; Ostberg et al, 1980; Mourant et al, 1981; Jaschinski-Kruza, 1984) convergence (Gunnarsson & Soderberg, 1980) and apparent visual acuity (Haider et al, 1980) have been shown to occur during or following VDU work. Such changes have been evidenced as manifestations of visual fatigue and have been attributed to the particular visual demands of VDU work. On the face of it these changes in measures of visual function are supportive of the notion of a VDU-related impairment. However two important caveats must be borne in mind. First methodological deficiencies, specifically the nature of the visual work undertaken in the VDU as opposed to non-VDU groups, makes interpretation difficult. For example the operator's familiarity with the information being processed affects the pattern of eye movements and may be related to motivational factors as well. This illustrates the need for studies which incorporate detailed breakdowns of job and tasks requirements rather than the all embracing term VDU work. Second, and perhaps more crucially, the connection between fatigue in the oculomotor system and temporary changes in these visual functions has not been established despite much effort. So far all of the reported changes are transient in nature and may contribute to the symptoms reported by some VDU users. Alternatively they may be only side effects of the factors that produce the discomfort or may even be a consequence of it (Brown et al, 1982). No evidence exists to suggest that such changes are the precursors of more serious pathological disturbances in the visual system.

Ocular pathology

Because of the possible radiation emissions from VDUs, the close proximity of users to the screen and the need for prolonged viewing concern about long term, irreversible damage to the visual system has been expressed. Anecdotal reports of cataract in two young copy editors from the New York Times fuelled this debate (Zaret, 1984). Zaret diagnosed their cataracts as radiant energy cataracts caused by VDT emitted microwaves. Exposure to high levels of ionising or microwave radiation is known to be risk factor for cataract and ultraviolet radiation may also be a causal factor. The threshold dose of ionising radiation for cataractogenesis in humans is generally regarded to be from 200–500 rads for a single exposure and 1000 rads for exposure over several months. This equates with a lifetimes exposure (40 years) for a full-time user of less than 1 rad. Similar arguments can be advanced for UV, IR and microwave portions of the spectrum as emitted from VDUs as being typically orders of magnitude below those levels shown to be cataractogenic in experimental studies (Michaelson, 1986). Radiation-induced cataracts are posterior subcapsular and corporal cataracts, types seen idiopathically without any recognised environmental exposure. Follow-up investigation by three independent physicians determined that the VDTs were not the cause of the lens changes in the two men referred to above. A review of claims that Dr Zaret has diagnosed 10 other cases of VDT-induced cataracts was reviewed by the National Research Council with the following results: Six of the cases actually had inconsequential opacities that did not appreciably reduce visual acuity. The other four patients either had exposures to cataractogenic agents prior to working with VDTs or were genetically predisposed to cataract development. Moreover statistical estimates (Weale, 1981) suggest that for VDUs to be implicated as a causal factor the number of suspected cases would need to be ten times greater than those already seen.

A number of epidemiological studies have been indertaken in order to examine the prevalence of lens opacities in VDU users. Two of these are based on questionnaire reports of visual health (Frank, 1983; Canadian Labour Congress, 1982) and suffer from other methodological shortcomings. Smith et al (1982) and more recently Boos et al (1985) carried out ophthalmological examinations. Neither showed an elevated level of opacities in VDU users. In the latter study the largest differences were found in the results from the three ophthalmologists taking part. Similarly, negative findings have been found in the routine prospective studies undertaken in the British (Weale, 1982) and Dutch (De Groot & Kamphuis, 1983) telecommunications sectors.

THE MUSCULOSKELETAL SYSTEM AND POSTURAL PROBLEMS

From the outset it was recognised that the introduction of VDUs with keyboards into offices could be a cause of biomechanical problems. Two of these were inter-related and concerned the nature of the tasks to be performed and the posture needed to carry them out satisfactorily. Constrained postures caused by continuous work with keyboards, together with badly designed workplaces were thought to lead to a range of problems. Even before the introduction of VDUs, typists and others with a sedentary posture were know to suffer a range of musculoskeletal problems (Zipp et al, 1983). It is clearly evident that the finger muscles are the only actual effectors in keyboard work, whilst the body as a whole functions as a support, holding the fingers in their working position and it has long been recognised that rapid repetitive movements of the fingers, hands and arms may lead to a variety of disorders including tendinitis, tensynovitis, carpal tunnel syndrome, myositis, bursitis and ganglionic cysts (Arndt, 1983; see also Chapter 5 of this volume for a review). Previous work has indicated that the very rapid repetitive keying patterns involved in some forms of keyboard work are associated with such disorders (Maeda et al, 1983; Duncan & Ferguson, 1974; Birkbeck & Beer, 1975; Gainer & Nugent, 1977). Flexion, extension and deviations at the wrist joint caused by poor keying style, poor keyboard design and poor workstation design are also relevant in the context of such disorders. Furthermore, the static tension produced while maintaining the sedentary posture in typewriting over long periods is thought to give rise to strain and fatigue (Rhomert, 1960), especially in the shoulders, neck and back (Grandjean, 1984). In keyboard work such fatigue is expressed in several ways (Mackay, 1980). Biomechanical fatigue in the cervical and upper thoracic spine results in aching and dull pain in the head, neck, lower back and between the shoulders. This may result from the tendency of operators to lean forward in order to view the screen more easily, often as a result of visual fatigue. Constraint of the lower parts of the body usually results in pain in the joints and muscles and is particularly concentrated in the knees. Compression of the main weight-bearing tissues results in impaired blood flow. One consequence of this may be a transient decrease in circulation to the extremities, the common result being a loss of sensation in various parts of the body. Minor adjustments of posture, fidgeting and flexing of the limbs helps to prevent this in fit individuals. For those with even minor physical disabilities that prevent easy movement, poor workplace design is likely to represent more of a problem.

Prolonged extension, rotation and bending of the neck can lead to minor neurophysiological difficulties. These appear to be much less common symptoms than those described above. A temporary loss of strength in the arm and hand may result together with a transient decrease in the ability to perform fine manipulative skills. When this occurs it has obvious implications for keyboard use, especially when high data entry rates are involved.

Essentially there are two points of issue concerning musculoskeletal disorders in VDU work. The first concerns the likelihood of permanent damage or injury, or long-lasting disablement as a result of prolonged keyboard work. Such a suggestion has been highlighted by recent experiences in Australia where there has been an apparent epidemic of so-called Repetition Strain Injuries (RSI) in keyboard workers (see Chapter 5). The second issue concerns the validity of existing requirements for posture as incorporated into guidance on current good practice in the light of recent empirical studies on preferred posture and comfort.

There is a substantial body of literature which has been devoted to examining postural problems and discomfort in VDU work. For reviews see Arndt (1983) and

Berquist (1984). Problems range from complaints of discomfort to pain and occasionally medical disability. The back, neck and shoulders are the most frequent sources of complaints: the arms, wrists and hands less frequently. Most studies have relied solely upon the use of subjective reporting. In some instances these have been supplemented by physical examination (anamnesis and palpitation of painful pressure points) and electrophysiological methods. Overall studies completed thus far tend to indicate an over-reporting of symptoms in VDU work compared with non-VDU jobs, although some studies have been unable to show such differences. Matching of control groups has not always been appropriate. The nature and duration of the work itself seem to be important determinants as does the male:female ratio in the populations under study.

Risks in keyboard work

It is the author's view that terms such as repetitive strain injury do not help in trying to identify and prevent problems in keyboard work. Fatigue does seem to be as prevalent in this sort of work as in others where similar sorts of postures need to be adopted. The current data such as they are do not indicate that the classical injury types (tenosynovitis, peritendinitis, carpal tunnel syndrome) are particularly prevalent even in full-time keyboard users (South Australian Health Commission, 1984). Back and neck problems of the type referred to as occupational cervicobrachial disorders appear to be more common in full-time keyboard users. Poor job design and ergonomic factors are largely to blame.

Some of the suggested problem areas include:

(a) abduction of the hands;
(b) pronation of the forearm, often close to the anatomical limit;
(c) raising the shoulders;
(d) lack of appropriate support for the back;
(e) prolonged work spells;
(f) lack of support for hands and wrists (Sauter et al, 1986; Nakeseko et al, 1985);
(g) sharp edge of desk cutting into wrists or forearm.

Prevention

Five main areas are involved:

(i) improvements in keyboard design;
(ii) altered work practices including job redesign;
(iii) attention to postural requirements through workplace design;
(iv) selection and training to improve or modify technique;
(v) occupational health care.

REPRODUCTIVE EFFECTS

Over the past few years there has been specific worldwide concern over the possible reproductive effects from working with VDUs. Concern originally arose because of anecdotal reports of apparently high rates of spontaneous abortion and birth defect in groups of pregnant users. Until recently most of the evidence for a possible link between VDU usage and reproductive problems stemmed from reports of such clusters and a handful of small-scale and largely uncontrolled field studies. The high profile debate about possible links between VDU use and pregnancy outcome has naturally led to apprehension and worry among female users. In assessing the possible risk to reproduction three pieces of evidence need to be considered. First, the nature of the cluster phenomena; second, possible biological mechanisms and purported risk factors and third, the conclusions to be drawn from reliable epidemiological studies.

Clusters

Most of the evidence of a possible link between VDU usage and reproductive problems has arisen because of anecdotal reports of birth defects and miscarriages in small groups of operators. Clusters are by definition unusual occurrences and anecdotal reports in the press preceded scientific evaluation in most cases by months if not years. Four clusters initially provoked the bulk of concerned comment; in the classified advertising department of the Toronto Star four cases of birth defects were reported in seven pregnancies; at Dorval Airport check-in counter, employees of Air Canada reported seven miscarriages in 13 pregnancies; in a Dallas office of the Sears Roebuck Company seven miscarriages and a premature death in 12 pregnancies; finally, in a USA Defence Logistics Agency near Atlanta, three birth defects and seven miscarriages were reported in 15 pregnancies. Additionally, the author understands that NIOSH are currently investigating further clusters. There have been other reports from European countries and doubtless there will be more. Because the cases have occurred in the way they have (anecdotal clusters of *unusual* events), rather than from routine pregnancy outcome surveillance studies, the very high level of concern about VDUs and pregnancy is based upon data which constitute extremely weak evidence of the likelihood of a causal mechanism operating rather than merely the existence of an association.

Problems stem both in trying to specify what constitutes a cluster and, taking the population of VDU users as a whole, what would we expect to observe by chance. Generally, the more widely used the agent (or suspected hazard) and the more common the outcome event the more often they will be associated merely by chance and the more difficult it will be to establish a causal relationship between them.

Human malformations, defined as 'macroscopic abnormalities of structures attributable to faulty development and present at birth' (McKeown & Record, 1960) occur at a frequency of around 25 per 1000 births in most large studies (e.g. Hook, 1971). These defects are clearly successful anomalies in terms of survival but represent a small percentage of abnormal conceptions (Berry, 1980). At 3% of live births therefore the likelihood of a defect as defined above is by no means uncommon. Based on this 3% estimate Roberts (1983) has calculated the effect of variation in observed birth defect rates due to the chance selection of a small number of births (sampling variation). Statistical probability theory allows us to predict the natural sampling variability we would expect to see in birth rate from community to community based upon, say, the last 20 births in a sample of 190 hospitals. (The calculations were based upon the number of maternity hospitals in Ontario, Canada). Assuming the underlying rate of about 3%, over half the hospitals would report no defects in their last 20 births, about one third would report one defect, about 10% would report two defects, and finally four hospitals (2%) would report three or more defects. It would not be unexpected for one of the hospitals to report four cases of birth defects in their last 20 live births which represents an incidence rate of 20%, almost seven times greater than the population average.

Over and above these statistical estimates must be superimposed variations due to genetic susceptibility (familial history of problems), medical problems during labour together with other known risks for birth defect. These will have the effect of increasing the natural variability in observed birth defect rate and will make it more likely that extreme rates (clusters) will be observed. Similar arguments are equally applicable in the case of other adverse outcomes reported in clusters, notably miscarriages and complications of delivery. Reported rates of miscarriage from well conducted cohort studies show wide variability from 10% to perhaps 70% (Roberts, 1983) with the majority typically in the range 14–20/30% (World Health Organisation, 1983). Data from the Hospital In-Patient Enquiry Maternity Tables (OPCS, 1980) for deliveries in England and Wales indicate that over half of all women delivered in hospital suffer from one or more complications (56% of cases had complications 1.4 complications per case).

Investigations of a number of the known clusters have, or are, taking place. As far as the author is aware detailed examination of the equipment concerned for radiation emissions at levels currently thought to be unacceptable have proved negative. Only one report of an investigation of one of the alleged cluster has appeared in the literature (Landrigan et al, 1983). No association (in terms of a dose-response relationship) was found between duration of exposure or proximity to VDUs and adverse pregnancy outcome. Prospective surveillance disclosed no further problems. One year after initial investigation four new pregnancies had been reported. All resulted in term births. As individual clusters are subjected to formal epidemiological study further information should be accumulated on the problem. However, although it is important to respond to reports of clusters they do not in themselves provide strong evidence of risk. If clusters are of sufficient size and there is sufficient variation in the sample as to degree of exposure the most valuable contribution their investigation can make is to establish the presence of dose-response relationships. To provide more reliable empirical evidence of a causal link one must look to formal epidemiological studies of VDU use and pregnancy outcome combined with assessments based upon possible risk factcors inherent in VDU work and their known association with pregnancy outcome.

Possible risk factors

Elsewhere the author has dealt at some length with the putative risk factors associated with VDU use and possible biological mechanisms affecting reproduction (Mackay, 1984). What follows is a short summary of current thinking concerning likely candidates namely: radiation; physical/ergonomic constraints; psychological stress.

Because of the general concern in the past about irradiation from VDUs much of the debate about reproductive hazards has centred on harmful radiation emissions to both the fetus and the mother. There is universal agreement that ionising radiation emissions are not relevant in the context of VDUs. As far as non-ionising radiation is concerned, although some residual concern has been expressed, a consensus does exist. Considerable discussion of possible health effects of exposure to radiofrequency fields from VDTs has recently ensued following research reports on the biological effect of weak electromagnetic fields (BEMS, 1984). Such fields are capable of interactions with biological systems at specific frequencies and intensities (Adey, 1981). Both the electric and magnetic fields appear to be capable of such interactions, but their mechanisms are not known. Suggestions have been made that the waveform of the fields and in particular the rate of change of the magnetic field may be of importance in inducing biological effects. These sugestions followed an uncorroborated study of chick embryo teratology (Delgado et al, 1982; Ubeda et al, 1983). An international effort is now under way to try to replicate these findings.

In these and subsequent experiments, square pulses of primarily 100 Hz and various field strengths have been investigated for effects, mostly on chicken embryos (Maffeo et al, 1984; Juntilainen & Saali, 1986; Sandstrom et al, 1986) and also on pregnant mice (Tribukait et al, 1986). In the main, attempts to replicate the original findings have proved difficult. In the Juntilainen et al study exposure to 100 Hz, 1A/m magnetic fields increased the number of abnormal embryos significantly. However in so far as the supposed effects are thought to be specific in terms of waveshape, the fact that sinusoidal (as opposed to triangular waveforms from VDUs) fields showed the effect the relevance to VDUs remains in doubt. The recently reported preliminary data of Tribukait et al (1986) on the effects of triangular pulse shaped fields (VDU-like) on mice embryos has therefore received much attention since there appeared to be a significant excess of malformed fetuses following exposure to such fields. However, since the significance of the data seemed to depend crucially upon the inclusion of dead fetuses the reliability of the findings are not clear. Further replications are necessary.

Work with VDUs is usually of a sedentary nature particularly when full-time or nearly full-time operation is involved. Physical activity particularly whilst standing undoubtedly causes the diversion of blood away from the abdominal viscera, and the uterus is not spared in this respect (Hytten, 1984). The fetus may show cardiographic signs of distress when the mother exercises, and in conditions where the mother habitually works hard in a standing position there is convincing statistical evidence of reduced fetal growth (Briend, 1980).

Typically the mechanical requirements of VDU work (including prolonged data entry) do not equate with the levels of gross physical activity mentioned above, nor does it seem to involve the level of metabolic cost associated with standing. Whether seated in a constrained posture in badly designed seating for long periods without a break leads to impaired circulation, and hence reduced blood supply to the fetus, is unclear. In the recent WHO report on women and occupational health risks the problem of constained posture was highlighted as a priority area and requiring further elucidation. Whilst being plausible physiologically, the extent to which it is relevant to reproductive problems in seated persons (including VDU operators) is, at the present time, unclear. Similarly unclear is the extent to which gross repetitive movements imposed by work routines are relevant.

Presumably such demands are less prevalent in typical VDU tasks than in production line tasks where such requirements are more common. In the context of miscarriage in the early stages of pregnancy, the requirement for manual handling and the transport of loads should be minimised. Increased intra-abdominal pressure can significantly increase the risk of miscarriage. The extent to which a particular posture adopted whilst sitting significantly elevates intra-abdominal pressure, and thus acts as a possible risk factor for miscarriage, is not known. Kashiwazaki et al (1979) have shown that miscarriage rates in clerical, nursery-nurses and school teachers differs significantly (7.6%, 18.2% and 16.0% respectively) and has attributed this to a combination of standing and lifting. In that these data are based upon a self-completion questionnaire and the individual response rates, nor possible biases, can be assessed, they must be treated with caution. The scope for more detailed investigations in the area of ergonomic aspects is clearly evident (MacKay & Bishop, 1984).

The notion that psychosocial stress affects reproduction receives support from many fields, including experimental work on animals and studies in humans (Bjorseth et al, 1985). Low birthweight, preterm labour, the need for obstetric intervention in delivery and disturbed early mother–child relationships are among those pregnancy outcomes in which stress has been implicated (Shaw et al, 1970; Cohler et al, 1975; Erikson, 1976; Yang et al, 1976; Lederman et al, 1978; Newton et al, 1979; Wolkind, 1981). Stress reduction also appears to be helpful. There is mounting evidence that social support mechanisms may positively influence health status (Broadhead et al, 1983) and a number of studies have documented the buffering effect of social support on the experience of stress. Both the provision of emotional support (Sosa et al, 1980) and instrumental support (Sokal et al, 1980) have been shown to reduce the likelihood of adverse reproductive outcomes. It is unlikely that any given psychosocial process or stressor is aetiologically specific for any particular reproductive problem. Rather the neuroendocrine response to stress (in which the growth promoting anabolic hormones and processes of immunity are supressed) acts to increase the likelihood of adverse effects. Also stress, may be associated more directly with fetotoxic substances as a result of behavioural attempts to cope, in so far as it depresses or modifies nutritional intake, causes sleep difficulties and leads to increased consumption of alcohol, caffeine and nicotine in users of these substances. There is some evidence that both maternal smoking and drinking are responses to, or attempts to cope with, stress (Graham, 1977; Farrant, 1980). In the case of VDUs however, stress is not necessarily a factor since it is a function of the job rather than the equipment. Indeed one of the criteria for introducing new technology is to reduce undesirable and unnecessary job demands but clearly some forms of work organisation in which VDUs are used violate well established job design principles (Eason & Sell, 1984; MacKay & Cox, 1984). Stress may therefore be a problem for *some* users.

Studies of reproductive performance in female users

Until recently only a few small-scale studies had been performed on the link between VDU use and the likelihood of adverse pregnancy outcome and only some of these are available in the open literature. From 389 volunteer workers at 13 Australian firms using VDTs (Lewis et al, 1982) identified 30 women who had had a spontaneous abortion. Each of these case women was matched by maternal age and date of delivery to two other women in the survey who had no reported history of spontaneous abortion (controls) and the two groups were compared on exposure to VDTs during pregnancy. The proportion of cases who reported such exposure (13%) was not statistically significantly different from the proportion of controls who did so (8%).

Beyond the problem of small sample size, this study has several methodological limitations that make interpretation of the findings difficult. Among these are possible selection biases (only volunteers were studied and several cases with spontaneous abortion were eliminated for unspecified reasons), chronological bias (it is not clear when the pregnancy outcomes actually occurred) and recall biases (the study women knew the reason for the study; in fact, concern among VDT workers had apparently motivated the study). In addition, exact exposures and their relation to the timing of pregnancy are not provided nor is it clear that possible confounding factors (smoking, previous reproductive history, etc.) were appropriately controlled.

A retrospective, cross-sectional self-report study of health complaints by Frank (1983, unpublished) also examined reproductive outcomes. Reproductive questions were asked at some length. Unfortunately, when the study population was subdivided by sex, age, and marital status, relatively few individuals remained for a thorough evaluation of many of the questions of concern. The results obtained from this questionnaire survey are unable to support either the view of a harmful or of no harmful effect on pregnant women.

The question of birth defects could not be thoroughly evaluated. Of the 11 female VDT users noting that they had had a child born with a birth defect, three had had the abnormal birth within the past 5 years, whilst on their current VDT using job. One case was reported from 40 years previously. Of note was the fact that although males were a minority of the total study population, they reported more birth defects among their offspring than did the working women. Five of the 18 reported birth defects among the offspring of males had occurred within the past 5 years while the users were on their current VDT job.

There was insufficient data to support or reject a relationship between VDT exposure and miscarriages. From the total study population, 62 miscarriages were reported by 57 different individuals. Among the women, 39 reported miscarriages, as did 18 males in their wives. As a percentage of the total female population of this study, some 7% of women reported a miscarriage. This is not readily comparable with other data since this includes unmarried women, and women older than childbearing years. There were no marked differences between any of the six study locations. Miscarriages in this population were reported by females as having occurred between the years 1948 and 1982. Of these, 15 were reported since 1977 and from these, seven had been among VDT user group while three had not, with the remainder being unknown for some of the specific details.

Among this group, there was no evidence of premature birth or infant mortality related to VDT exposure. There also did not appear to be any effect upon the menstrual cycle with VDT use. Clearly this study, as with the Lewis et al (1982) study, suffers from many methodological problems, which, even if significant effects had been found, would make interpretation difficult.

A Japanese study (Kajiwara, 1984) on 1591 VDU workers included data on 50 pregnancies. Thirteen were described as abnormal pregnancies or abnormal deliveries. Of these four (8%) ended in abortion whilst six others (12%) were described as threatened abortion. These data appear to be similar to spontaneous abortion statistics quoted in other Japanese sources. Conclusions cannot be drawn because of the small size of the sample.

A retrospective, questionnaire based study has been described by Lee & McNamee (1984). Questionnaires were sent to women currently and previously employed in a data preparation unit. The questionnaire asked about past reproductive history.

An exposed pregnancy was defined as one in which the woman had worked with VDUs for at least 10 hours per week during the 3-month period prior to conception plus 3 months afterwards. Although the miscarriage rate for the exposed women was within the usual range (14.5%) the reported frequency in the non-exposed was unusually low (5.3%). Although the differences could be due to natural sampling variation this is difficult to evaluate. More likely the results reflect a number of selection and response biases common with this type of approach, as pointed out by the authors, together with some other methodological difficulties. These include the tendency for members of the exposed groups, knowing themselves to be at risk, to be more rigorous in their reporting of reproductive problems and sometimes exaggerate their exposure to the hazard of concern, coupled with a probable selective under-reporting of miscarriage in the non-exposed; a common finding in such studies.

A study of birth defect and VDU use has been described by Kurppa et al (1984, 1985). The investigation was an extension of the standard operating procedure of the Finnish national register of congenital malformations. Detailed information has been collected about possible occupational and leisure-time exposures of the mothers. Included were 365 consecutive defects of the central nervous system, 581 orofacial clefts, 360 structural defects of the skeleton, and 169 selected cardiovascular malformations.

The scrutiny revealed 386 discordant pairs with respect to potential VDT exposure by occupational title such as clerical worker, secretary, typist, or automated data processing employee. 183 of the potentially VDT exposed were case mothers and 203 were referent mothers. The interview forms of the mothers with potential exposure were then perused for VDT information by someone unaware of the case/referent status. VDT work during early pregnancy was explicitly described in questionnaire forms of 108 pairs out of which 50 were case mothers and 58 referent mothers. (The age structure, parity, previous miscarriages or stillbirths, drug consumption, and alcohol intake during pregnancy were comparable for the VDT exposed and non-exposed mothers). The comparison of the mothers exposed to VDT for at least 4 hours a day during early pregnancy with those not exposed at all showed an overall odds ratio of 1.0 (95% confidence limits; 0.6, 1.6) indicating no increased risk.

McDonald et al (1984) reported the preliminary result of the Montreal study of 51 200 deliveries and 4300 spontaneous abortions.

Among the current pregnancies some preliminary results did suggest an association between a rising level of spontaneous abortion and increasing numbers of hours worked per week on VDUs. Of the 3799 current pregnancies there was a 5.7% rate of spontaneous abortion among those doing no VDU work, 8.2% rate of spontaneous abortion among those doing up to 15 hours per week VDU work; and 9.3% among those doing more than 15 hours per week. The same trend was apparent in all the five occupational groups.

McDonald et al suggested that some bias in the responses to the enquiries made and exaggeration of the extent of VDU use among those who had experienced spontaneous abortion might explain this apparent trend. Bearing in mind that the low overall rate of spontaneous abortions in the five occupational groups studied, McDonald et al considered that the balance of evidence was against there being any true association as the preliminary results had suggested.

Further analyses based upon a greater number of pregnancies (McDonald, 1985) showed that in past pregnancies (3881) the abortion rate was greatly increased when the interval before the current pregnancy was short. Excluding conceptions before 1980, the spontaneous abortion rate was 15.4–15.3% for 443 women who had used VDTs and 15.4% for the 2287 who had not. Allowing for known risk factors relating to previous abortion, age and smoking, there was no difference in abortion or defect rates in users or non-users.

In current pregnancies, ascertainment of spontaneous abortion was incomplete: early cases did not reach hospital and, of those which did, only 75% were interviewed. The abortion rate in the study group was 6.3% compared with 6.7% for all other pregnancies: in VDT users the rate was 8.4%, in non-users 5.1%. There was no systematic exposure-response relationship as was apparent in the earlier analyses. There was evidence from a subsidiary questionnaire inquiry that the use of VDTs early in full-term pregnancies was under-reported by at least 10%. Despite the equivocal findings in current pregnancies, which could have been due to biases referred to earlier it was concluded that the absence of any association in previous pregnancies between VDT use and abortion or congenital defect probably reflects the truth.

A recent Swedish study carried out for the National Board of Occupational Safety and Health (Arbetaskyddsstyrelsen) and the National Social Welfare Board (Socialstyrelsen) used the Swedish miscarriage register to examine pregnancy outcome in occupational groups thought to have differing exposures to VDUs. Low exposure included post office assistants, librarians and bank cashiers; medium exposure, insurance company secretaries; and high exposure, computer personnel, travel agency clerks and social insurance clerks. Validation of exposure was carried out with mailed questionnaires. A comparison was made between reproductive experience of these groups during 1976–1977 compared with more recent times when overall exposure in the workplace was likely to be higher. None of the outcome measures of interest (perinatal deaths, significant malformations, low birth weight and spontaneous abortions) were related to gross level of exposure (low, medium, high usage) nor did more detailed examination of dose-response effects indicate a relationship between number of hours exposure per week and adverse outcomes. In this particular population there were however significant effects of self-reported stress and smoking which were in turn inextricably linked with prolonged usage of VDUs (Ericson & Kallen, 1986a, 1986b).

The conclusion from these most recent studies is that whilst a link cannot be absolutely ruled out, the data accumulated thus far remove all cause for concern as far as the VDU itself is concerned. Nevertheless a vexed question remains, as before, concerning the most appropriate guidance to be given to those who are pregnant, or who are thinking of becoming so.

A number of organisations, in both Europe and North America, accept worry about dangers of VDU work as a justifiable reason for the woman concerned to be assigned to work away from the VDU during (or until the end of) a current pregnancy. These agreements do not imply an implicit acceptance of VDUs as dangerous during pregnancy; the explicit motivation is the woman's worry. Our own view is that at the present time the assembled data, such as they are, do not suggest that for a normal healthy female (pregnant and non-pregnant) there is an added risk to health from operating a VDU over and above that for comparable non-VDU tasks. On the basis of existing information therefore a strong case for transferring pregnant operators to non-VDU work cannot be made. As in any other job, however, it is recognised that additional medical reasons, including possible complications of pregnancy, require consideration on an individual basis (Kuntz, 1980; Carney, 1980). Although many pregnant VDU operators have, and continue to be, reassured by the evidence against there being a link, thus far accumulated, there are some who still remain very anxious about possible harm. It is this perceived risk which also must be taken into account, in so far as high levels of anxiety in themselves may be problematic (Kaffman et al, 1982). Apart from the provision of antenatal care generally, thoughtful employers could therefore remain flexible in responding to the concerns of individual operators. The extent to which such individual cases can be accommodated will depend very much upon local factors, including the opportunities for alternative work and the nature of such work. Apart from other medical factors which are contraindicated, any possible reduction in anxiety (which may not invariably occur) must be weighed against the problem of the change to a new line of work and any cost borne in adapting to that change, together with any objective risks to pregnancy inherent in the new job to which the woman is transferring.

SKIN DISORDERS

The possibility that VDU work may in some way be related to the appearance of facial dermatitis was raised following reports from Norway of dermatological examinations of 35 VDU operators. The case material has been variously reported (Tjonn, 1984; Nilsen, 1982; Linden & Rolfsen, 1981). The case reports described a fairly uniform clinical picture. The symptoms were transitory itch, redness, desquamation and sometimes papules on the cheeks. Their appearance seemed to follow the onset of VDU work and to subside on non-working days. Dermatological examination ruled out allergic contact dermatitis, photosensitivity and rosacea. Neither familial or childhood atopic dermatitis was noted in the case material.

A number of similar cases have been reported in the UK (Rycroft & Calnan, 1984) of a similar clinical picture to those described above but with some resemblance to rosacea. Cases continue to be reported to the Employment Medical Advisory Service in the UK. In many of these the appearance of skin problems is unrelated to VDU work even though there may be clustering of sufferers within the same office. Occasionally the VDU may serve as a focus for resolving long standing difficulties expressed in the form of psychologically generated skin disorders (dermatitis artefacta) (Rycroft, personal communication). In a field survey of VDU users in Sweden Knave et al (1985) found skin problems to be more frequently reported by users (36%) than by controls. The 96 individuals currently reporting skin disorders were subsequently examined (Liden & Wahlberg, 1985a). On an *a posteriori* basis the available material was grouped into four different diagnostic categories. In only one of these (seborrhoeic dermatitis, acne, rosacea and perioral dermatitis) was VDU use more prevalent. None showed the clinical picture described in the earlier studies but in so far as examinations were restricted to those originally complaining of skin problems a measure of response bias must be assumed. In a further study the same authors (Liden & Wahlberg, 1985b) examined the prevalence of VDU use in a series of patients with rosacea or perioral dermatitis (controls were not used). There was a suggestion of an aggrevation of the condition in some of the VDU users.

The role of the VDU in generating skin problems remains unclear. The anecdotal reports referred to earlier suggest one or more factors in the working environment in so far as studies report that rashes subside on leaving the working environment and reappear some hours after restarting work with the VDU. In none of the studies has ionising or non-ionising radiation from the VDU been detected at levels known to provoke skin disorders. A fairly consistent observation is the presence of electrostatic phenomena in the vicinity of the VDUs involved coupled with low ambient humidity. In some studies (Knave et al, 1985) no relationship between electrostatic fields and reported symptoms were found although in the Liden & Wahlberg (1985b) study significant differences in body potentials were found in the VDU related skin problems.

PHOTOSENSITIVE EPILEPSY AND VISUAL DISCOMFORT

Certain patients with epilepsy are photosensitive and suffer seizures induced by flickering lights and patterns of striped lines. When they are exposed to visual stimuli of this kind, epileptiform EEG activity (e.g. a photoconvulsive response) may be induced. The spatial and temporal characteristics of stimuli that induce epileptiform activity are surprisingly specific (Wilkins et al, 1981).

Estimates of the prevalence of photosensitive epilepsy in the population range from 1 in 5000 to 1 in 10 000. The incidence is, however, age-dependent (Harding, 1980). The age of onset is usually between the ages of 9 and 15 years and 90% will have had their first seizure before they reach the age of 22 years (Harding, 1986). Approximately 50% of individuals with this condition experience their first fit whilst watching television (Wilkins et al, 1979). Although the likelihood of the onset of photosensitive epilepsy being precipitated by a VDU is low, the use of microcomputers in schools is likely to lead to an increase in the number of seizures reported from VDU use as opposed to television viewing.

From a theoretical viewpoint and from experimental studies there appear to be a number of factors inherent in VDU work that increase the risk of paroxysmal activity in those individuals who are known to be, or may be, photosensitive relative to the epileptogenic effects of a domestic television. These include: a large screen, 25 Hz interlacing, large amounts of bright text and prolonged viewing at close range (Wilkins, 1978). Even with this information and given an individual's previous history of seizures, it is doubtful that an accurate assessment of the risk could be made given the lack of clinical statistics relating to the epileptogenic effects of VDUs.

More recent work however presents a rather different picture concerning relative risk, and it has been suggested that some of these parameters may indeed lessen the likelihood of a seizure (Jeavons et al, 1985). In a test series performed by Binnie et al (1985) a VDU with a short persistence phosphor consistently failed to cause epileptiform activity in photosensitive subjects, whilst conventional monochrome TV receivers provoked epileptiform discharges in many of the subjects.

STRESS, WORK DESIGN AND ORGANISATIONAL FACTORS

One of the supposed advantages of a VDU-based system is the activity for the user to interact intelligently with a computer system in ways compatible with human limitations in information processing capacity. Essentially this approach envisages the VDU as an intelligent tool working at the behest of the user. However, a cursory task analysis of many existing VDU based jobs indicates that such jobs contain undesirable features characteristic of some forms of repetitive work, which fail to provide work which fit existing patterns of skill (or where training is inadequate) and in so doing fail to meet the individual's need for challenging and interesting work.

The handful of studies undertaken so far on occupational stress factors inherent in some forms of VDU-based work supports these general assertions. Some studies have shown high levels of reported psychological distress amongst VDU operators (Elias et al, 1982; Ghinghirelli, 1982). Others report no significant differences between users and controls (Binaschi et al, 1982). It is most likely that in most of these studies, VDU work is confounded with undesirable job characteristics. The problem seems not to be the technology but the nature of work being undertaken (Sauter et al, 1983). Thus Gunnarsson & Ostberg (1977) showed that monotony experienced during VDU operation was clearly related to perceived feelings of lack of controllability and low levels of variety. Smith et al (1980) report that rigid work procedures, high production standards and constant pressure for performance were reflected in measures of self-reported stress and work demands which were substantially in excess of established norms (Caplan et al, 1975). Their respondents also complained of negative effects on their emotional health as well as musculoskeletal and visual problems. However, their data from other VDU sites suggest that perceived flexibility, autonomy and control over how work is to be carried out act as attenuating factors in the experience of stress. In these operators the greatest problems were those concerned with ambiguity over career development and future job activities.

One of the crucial variables in determining stress-related symptoms in VDU operators is linked with the perception of control by the system (or conversely, lack of control by the operator). Such control is evident in a number of ways. First, in some systems the processing power of the machine is exploited to such an operator. In many instances this information is used to determine levels of remuneration via piece-rate payment systems. Not unnaturally this level of control is often resented, is regarded with suspicion by many operators and, understandably, is associated with feelings of fatigue and stress. Second, very long response times from the computer, or those which are variable in length, create uncertainty and frustration. Third, technical disturbances and breakdowns, if they occur frequently, serve only to exacerbate these problems (Wallin et al, 1983; Dainoff et al, 1981). All these factors substantially increase the mental load upon the operator and inevitably lead to fatigue. Thus Johannson (1979) and Johansson & Aronsson (1981) have shown that these aspects of machine control lead to marked psychoneuroendocrine mobilisation in VDU operators as evidenced by increased urinary catecholamine levels. These effects do not go away when the person has stopped working but have a very insidious influence. They can affect one's ability to cope in other situations where one has little control.

VDU task-design must therefore seek to minimise repetitive elements in the operator's task by introducing variability in workload throughout the day; instead of long periods of concentrated work, whilst ensuring that the load is predicatable. This should be coupled with job-design features which allow the individual to have some discretion in how work is allocated over work periods and, by so doing introduce feelings of personal control and cater for individual differences in the need for brief pauses in work. Apart from these quantitative aspects of the VDU task, the qualitative features should also be examined. All too often, data entry tasks require only the use of simple psychomotor skills, where only minimal exercise of intellectual abilities is possible. Thus work should be designed to be mentally challenging, but within the scope of individual operator abilities. Ideally, the shift should be towards the use of the VDU as a tool for carrying out a much larger enriched job, rather than regarding the VDU user as solely a machine operator. Where more or less continuous work is unavoidable some form of work-rest break schedule must be considered.

Rest pauses

In most tasks, natural breaks or pauses occur as a consequence of the inherent organisation of the work. These informal breaks help to maintain performance by preventing the onset of fatigue. In some VDU work, for example those data entry tasks requiring continuous and sustained attention and concentration, together with high data entry rates, such naturally occurring breaks are less frequent. In situations where this type of task cannot be organised in any other way, and where natural breaks in work do not occur, the introduction of rest pauses should help attention and concentration to be maintained. It is difficult to be specific about guidance on rest pauses, since it is likely that if strictly laid down rest pauses are adhered to, they will often be found to be unnecessarily prolonged and frustrating for some, and, under other circumstances, too short to prevent the onset of fatigue. The most satisfactory length of pause can only be determined by consideration of the individual operator's job but some general statements can be made.

(a) Some of the symptoms reported by operators are often the result of the effort expended in order to maintain performance in the face of accumulating fatigue. Rest pauses should therefore be arranged so that they are taken prior to the onset of fatigue, not as a recuperative period from it. Rest should be therefore introduced when performance is at a maximum, just before a reduction in productivity. The timing of rest is more important than the length of the rest period, although optimal test period lengths can also be determined for individual jobs (Ghiselli & Brown, 1948).

(b) Short, frequently occurring pauses appear to be more satisfactory than longer ones taken occasionally.

(c) Ideally the break should be taken away from the VDU.

(d) Rest periods may be more useful for relatively ineffective workers; better workers seem to develop more efficient procedures and therefore have less need for rest.

(e) Rest periods are most effective with work requiring concentration than on jobs that are more or less automatic and leave the employee free to daydream, converse with others, or follow similar monotony-reducing strategies.

(f) A report issued by the Department of Health of New Zealand finds (Coe et al, 1980) that although fatigue-like complaints about the eyes are not alleviated by mandated formal breaks, they are alleviated by informal breaks, that is, time spent not viewing the screen, which may include time spent performing other work tasks. These findings are consistent with recent work in occupational stress which emphasises the need to allow individual operators discretion in the way tasks are carried out. Individual control over the nature and pace of work enable effort to be distributed optimally throughout the working day.

The job should be designed therefore to permit natural breaks, or changes in patterns of activity, as an integral part of the tasks to be performed. This may involve, for example, a mix of VDU-based and non-VDU-based work.

When the operator's job consists of a variety of tasks, some of which may involve occasional use of a VDU, minor faults in the ergonomics of the equipment may not be crucial. Conversely, when intense and continuous operation is required, the need for optimum workplace and screen characteristics become critical. However, solutions based entirely upon attention to ergonomic factors are not a panacea for low motivation and poor morale; work design and organisational factors are likely to be more important in determining the operator's overall acceptability of the computer. Thus in determining the acceptability of the system and minimising possible indirect health effects we would link work design and organisational factors with primary prevention.

The switch from orthodox office handling routines to VDU-based systems provides an opportunity for job-enrichment and enlargement. Job-flexibility and some control over work-allocation should enable the promotion and utilisation of individual skills. Whilst these are ideal outcomes stemming from the introduction of a computer system, they are only achieved by careful and thoughtful planning and implementation. The approach which appears to offer the most advantages and the most likely to achieve these objectives is that in which designers and users of the system, particularly VDU operators themselves, are jointly involved throughout the various phases of design and implementation.

We are beginning to understand something of the psychological demands inherent in VDU operation. We are not yet at a stage where very detailed recommendations can be given in order to optimise VDU task design, partly because of incomplete knowledge of possible health effects, either in the short or long term, and partly because what existing knowledge is available has been slow to be implemented.

REFERENCES

Adey W R 1981 Tissue interactions with non-ionizing electromagnetic fields. Physiological Review 61: 435–513

Alden D J, Daniels R W, Kanarick A F 1972 Keyboard design and operation — a review of the major issues. Human Factors 14: 275–293

Arndt R 1983 Working posture and musculoskeletal problems of video display terminal operators — review and reappraisal. American Industrial Hygiene Association Journal 44: 437–446

Bauer D, Cavonius C R 1982 Improving the legibility of visual display units through contrast reversal. In: Grandjean E, Vigliani E (eds) Ergonomic aspects of visual display terminals. Taylor & Francis, London, p 137–142

BEMS 1984 Sixth Annual Meeting of Bioelectromagnetics Society. Program and abstracts. Atlanta, Georgia, 15–19 July. BEMS, 1 Bank Street, #307, Gaitnersburg, MD 20878

Belluci R, Mauli F 1984 The effects of visual ergonomics and visual performance upon ocular symptoms during VDT work. In: Grandjean E (ed) Ergonomics and health in modern offices. Taylor & Francis, London, p 346–351

Berquist U 1984 Video display terminals and health. Scandinavian Journal of Work Environment and Health 10: Supp. 2, 1–87

Berry C 1980 The examination of embryonic and fetal material in diagnostic histopathology laboratories. Journal of Clinical Pathology 33: 317–326

Binashi S, Albonico G, Gelli E, Morelli Di Popolo M R 1982 Study on subjective symptomatology of fatigue in VDU operators. In: Grandjean E, Vigliani E (eds) Ergonomic aspects of visual display terminals. Taylor & Francis, London, pp 219–225

Binnie C D, Kasteleign-Nolst Terenite D G A, de Korte R et al 1985 Visual display units and risk of seizures. Lancet, i, p 991

Birkbeck M Q, Beer T C 1975 Occupations in relation to the Carpal Tunnel Syndrome. Rheumatology and Rehabilitation 14: 218–221

Bjorseth A, Warncke M, Ursin H 1985 Stress hos gravide: Konsekvenser for mor og barn (Stress in pregnancy: Consequences for mother and child). Institute of Physiological Psychology, University of Bergen

Bouma H 1980 Visual reading processes and the quality of text displays. In: Grandjean E, Vigliani E (eds) Ergonomic aspects of visual display units. London, Taylor and Francis

Boos S R, Calissendorff B M, Knave B G, Nyman K G, Voss M 1985 Work at video display terminals. An epidemiological health investigation of office employees. III. Opthalmological examination. Scandinavian Journal of Work Environment and Health 6: 475–482

Briend A 1980 Maternal physical activity, birth weight and perinatal mortality. Medical Hypotheses, 6: 1157–1170

Broadhead W E, Kaplan B H, Shermon A J, Wagner E H, Schoenback V J, Grimson R, Heyden S, Tibbling G, Gehlbach S H 1983 The epidemiological evidence for a relationship between social support and health. Americal Journal of Epidemiology 117: 521–537

Brown B S, Dismutes K, Rinalducci E J 1982 Visual display terminals and vision of workers. Summary and overview of a symposium. Behaviour and Information Technology, 2: 121–140

Buesen J 1984 Product development of an ergonomic keyboard. Behaviour and Information Technology 3: 387–390

Bureau of Radiological Health 1981 An evaluation of radiation emission from video display terminals. Bureau of Radiological Health, US Department of Health and Human Services, 1981 Publication FDA 81-8153

Campbell F W 1980 Recent attempts to link psychophysics with neurophysiology in vision research. Transactions of the Ophthalmic Society of the UK 99: 326–332

Campbell F W, Durden K 1983 The visual display terminal issue: A consideration of its physiological, psychological and clinical background. Ophthalmic and Physiological Optics 3: 175–192

Campbell F W, Maffei L 1974 Contrast and spatial frequency. Scientific American 213: 106–111

Canadian Labour Congress 1982 Labour Education and Studies Centre. Towards a more humanized technology; exploring the impact of Video Display Terminals on the health and working conditions of Canadian office workers Ottawa, Quebec

Caplan R D, Cobb S, French J R P J, Van Harrison R, Pinneau S R J 1975 Job demands and worker health, NIOSH Research Report, HEW Publication 75–160 US Dept of Health Education and Welfare, Washington

Card S K 1984 Human limits and the VDT computer interface. In: Bennett J, Case D, Sandelin J, Smith M (eds) Visual Display Terminals. Prentice-Hall. Eaglewood Cliffs N J

Carney P 1980 Working in pregnancy. How long? How hard? What's your role? Contemporary Obstetrics and Gynaecology

Coe J B, Cuttle K, McClellan W C, Warden N J 1980 Visual display units. A review of the potential health problems associated with their use. Wellington NZ Regional Occupational Health Unit, New Zealand Department of Health.

Cohler B J, Gallant D H, Grunebaum H U, Weiss J L, Gamer E 1975 Pregnancy and birth complications among mentally ill and well mothers and their children. Social Biology 22: 269–278

Cox E A 1984 Radiation emission from visual display units. In: Pearce B G (ed) Health hazards of VDTs? John Wiley & Sons, Chichester, p 25–37

Dainoff M J 1982 Occupational stress factors in visual display terminals (VDT) operation: A review of empirical research. Behaviour Information Technology 1: 141–176

Dainoff M J, Happ A, Crane P 1981 Visual fatigue and occupational stress in VDT operators. Human Factors 23: 421–438

De Groot J P, Kamphuis A 1983 Eyestrain in VDU users: Physical correlates and long-term effects. Human Factors 25: 409–413

Delgado J M, Leal J, Montegudo J L, Gracia M G 1982 Embryological changes induced by weak, extremely low frequency electromagnetic fields. Journal of Anatomy 134 (3)

Duncan J, Ferguson D 1974 Keyboard operating posture and symptoms in operating. Ergonomics 17: 651–662

Eason K D, Sell R G 1981 Case studies in job design for information processing task. In: Corlett E N, Richardson J (eds) Stress, work design and productivity. Wiley Chichester

Elias R, Cail F, Tisserand M, Christmann H 1982 Investigations in operators working with CRT display terminals: relationships between task content and psychophysiological alterations. In: Grandjean E, Vigliani E (eds) Ergonomic Aspects of Visual Display Terminals. Taylor and Francis, London, pp 211–217

Environmental Health Directorate 1983 Investigation of radiation emissions from video display terminals. Environmental Health Directorate, Health and Welfare Canada, Publication 83-EHD-91

Erickson M T 1976 The influence of health factors on psychological variables predicting complications of pregnancy, labour and delivery. Journal of Psychosomatic Research 20: 21–24
Ericson A, Kallen B 1986a An epidemiological study of work with video screens and pregnancy outcome: I A registry study. American Journal of Industrial Medicine 9: 447–457
Ericson A, Kallen B 1986b An epidemiological study of work with video screens and pregnancy outcome: II A case-control study. American Journal of Industrial Medicine 9: 459–475
Farrant W 1980 Stress after amniocentesis for high serum alpha-fetoprotein concentrations. British Medical Journal 1: 452
Ferguson D 1984 The 'new' industrial epidemic. Medical Journal of Australia, 17 March 1984, 318–319
Frank A L 1983 Effects on health following occupational exposure to video display terminals. Department of Preventive Medicine and Environmental Health. University of Kentucky. (Unpublished)
Frank A L 1985 Occupational Medicine. Journal of the American Medical Association 254: 2333–2334
Ghingirelli L 1982 Collection of subjective opinions on use of VDUs. In: Grandjean E, Vigliani (eds) Ergonomic aspects of visual display terminals. Taylor & Francis, London, p 227–231
Ghiselli E E, Brown C W 1948 Personnel and Industrial Psychology. New York. McGraw-Hill
Grandjean E 1984 Postures and the design of VDT workstations. Behaviour and Information Technology 3: 301–311
Grandjean E, Hunting W, Nishiyama K 1984 Preferred VDT workstation settings, body posture and physical impairments. Applied Ergonomics 15: 99–104
Graham H 1977 Smoking in pregnancy: The attitudes of expectant mothers. Social Science and Medicine 10: 399–405
Greenwald M J, Greenwald S L, Blake R 1983 Long-lasting visual after-effect from viewing a computer video display. New England Journal of Medicine 309, 315
Gould J D, Grischkowsky N 1984 Doing the same work with hard copy and with cathode-ray tube (CRT) computer terminals. Human Factors 26: 323–337
Gunnarsson E, Ostberg O 1977 The physical and physchological working environment in a terminal-based computer storage and retrieval system, Report No 35, National Board of Occupational Safety and Health, Stockholm
Gunnarsson E, Soderberg I 1983 Eye strain resulting from VDT work at the Swedish Telecommunications Administration. Applied Ergonomics 14: 61–69
Guy A W 1986 Health Hazard assessment of radio frequency electromagnetic fields emitted by video display terminals. International Scientific Conference: Work with Display Units. Stockholm, Sweden. 12–15 May 1986 (abstract)
Haider M, Kundi M, Weissenbock M 1980 Worker strain related to VDUs with differently coloured characters. In: Grandjean E, Vigliani E (eds) Ergonomic aspects of Visual Display Terminals. London. Taylor and Francis
Harvey S M, Electric-field exposure of persons using video display units. Bioelectromagnetics 1984, 5: 1–12
Harding G F A 1980 Photosensitive epilepsy. In: Oborne D J, Gruneberg M M, Eiser J R (eds) Research in Psychology and Medicine. London. Academic Press
Harding G 1986 Photosensitive epilepsy and employment. In Edwards F, Espir M, Oxley J (eds) Epilepsy and Employment — a medical symposium on current problems and best practices. London. Royal Society of Medicine Services Ltd
Hook E B 1971 Some general considerations concerning monitoring: Applications to utility of minor defects or markers. In: Hunt E B, Janerick D T, Porter I H (eds). Monitoring Birth Defects and Environment: The Problem of Surveillance. Academic Press. New York NY
Howarth P A, Istance H O 1985 The association between visual discomfort and the use of visual display units. Behaviour Information Technology 4: 131–149
Hytten F E 1984 The effect of work on placental function and fetal growth. In: Chamberlain G (ed) Pregnant Women at Work. Royal Society of Medicine/Macmillan Press. London and Basingstoke
Jaschinski-Kruza W 1984 Transient myopia after visual work. Ergonomics 27: 1181–1189
Jeavons P M, Harding G F A, Drasdo N, Furlong P L F, Bishop A I 1985 Visual display units and epilepsy. Lancet 2: 287
Johansson G 1979 Psychoneuroendocrine reactions to mechanised and computerised work routines. In: Mackay C J, Cox T (eds) Response to stress: occupational aspects
Johansson G, Aronsson G 1981 Stress reactions in computerised administrative work, Supplement 50, Reports from the Department of Psychology, The University of Stockholm
Juutilainen J, Saali K 1986 Effects of low frequency magnetic fields on the development of chick embryos. International Scientific Conference: Work with Display Units. Stockholm, Sweden 12–15 May 1986 (abstract).
Kaffman M, Elizur J, Harpazy L 1982 An epidemic of spontaneous abortion: Psychosocial factors. Israeli Journal of Psychiatry and Related Sciences 19: (3) 239–246

Kajiwara S 1984 Work and health in VDT workplaces (In Japanese) In-Service Training Institute for Safety and Health of Labour, Osaka, Japan, p 5–82

Kashiwazaki H, Uehara S, Koizumi, Mingkami H, Wada M, Suzuki T 1979 Factors affecting the incidence of miscarriage among the female employees. Japanese Journal of Industrial Health 21: 250–256

Knave B G, Wibom R I, Voss M, Hedstrom L D, Bergqvist U O V 1985a Work at video display terminals. An epidemiological health investigation of office employees. I. Subjective symptoms and discomforts. Scandinavian Journal of Work Environment and Health 11: 457–466

Knave B G, Wibom R I, Bergqvist U O V, Carlsson L W, Levin M I B, Nylen P R 1985b Work at video display terminals. An epidemiological health investigation of office employees. II. Physical exposure factors. Scandinavian Journal of Work Environment and Health 11: 467–474

Kuntz W D 1980 Pregnant Working Women: What advice should you give them? Contemporary Obstetrics and Gynaecology 15: 69–79

Kurppa K, Holmberg P C, Rantala K, Nurminen T 1984 Birth defects and video display terminals. Lancet ii: 1339

Kurppa K, Holmberg P C, Rantala K, Nurminen T, Saxen L 1985 Birth defects and exposure to video display terminals during pregnancy. Scandinavian Journal of Work Environment and Health II: 353–356

Landrigan P J, Melius J M, Rosenberg M J, Coye M J, Binkin N J 1983 Reproductive hazards in the workplace. Scandinavian Journal or Work Environment and Health 9: 83–88

Laubli Th, Hunting W, Grandjean E 1981 Postural and visual loads at VDT workplaces. II. Lighting conditions and visual impairments. Ergonomics 24: 933–944

Lee B V, McNamee R 1984 Reproduction and work with Visual Display Units: a pilot study. In: Allegations of Reproductive Hazards from Visual Display Units. Humane Technology, Loughborough

Lee W R 1985 Working with visual display units. British Medical Journal 291: 989–991

Lederman R P, Lederman A I, Kandall S R, Gertner L M 1976 Quality of care vs neonatal mortality rate. Journal of Pediatrics 89: 161–162

Lewis M J, Esterman A J, Dorsch M M 1982 A survey of the health consequences to females of operating visual display units. Community Health Studies 6: (2) 130–134

Licklider J C R 1960 Man-computer symbiosis. IRE Transactions on Human Factors in Electronics, HFE 1, March, p 4–11

Liden C, Wahlberg J E 1985a Work at video display terminals. An epidemiological health investigation of office employees. V. Dermatological examination. Scandinavian Journal of Work Environment and Health II: 489–494

Liden C, Wahlberg J E 1985b Does visual display terminal work provoke rosacea? Contact Dermatitis 12: 235–241

Linden V, Rolfsen S 1981 Video computer terminals and occupational dermatitis. Scandinavian Journal of Work Environment and Health, 1981 7: 62–67

McDonald A D, Cherry N M, Delorme C, McDonald J C 1984 Work and pregnancy in Montreal—preliminary findings on work with Visual Display Terminals. In: Allegations of Reproductive Hazards from VDUs. Humane Technology, Loughborough

McKeown T, Record R C 1960 Malformations in a population observed for five years after birth. In: Wolstenholme G E W, O'Connor C M (eds) CIBA Foundation Symposium on Congenital Malformations. London, Churchill

Maeda K, Hunting W, Grandjean E 1980 Localised fatigue in accounting machine operators. Journal of Occupational Medicine 22: 810–816

Mackay C J 1980 Human factors aspects of Visual Display Unit Operation. Health and Safety Executive Research Paper No 10. London, HMSO

Mackay C J 1984 Visual Display Units possible reproductive effects. In: Allegations of Reproductive Hazards from VDUs. Humane Technology, Loughborough

Mackay C J, Bishop C M 1984 Occupational Health of Women at Work: Some Human Factors Considerations. Ergonomics 27: (5) 489–498

Mackay C J, Cox T 1984 Occupational stress associated with visual display unit operation. In: Pearce B G (ed) Health Hazards of VDUs? Wiley. Chichester

Marriott I A, Stuchly M A 1986 Health aspects of work with video display terminals. Journal of Occupational Medicine (in press)

Mellner M, Moberg I 1983 Belastnings-och synbesvär vid arbete med dataterminal (Visual and muscular discomforts during VDT work, in Swedish). Oxens Företagshälsovard, Stockholm, Sweden

Michealson S M 1986 Health implications of exposure to emissions from video display terminals. International Scientific Meeting: Work with Display Units. Stockholm, Sweden. 12–15 May 1986 (abstract)

Miller W, Suther ThW 1983 Display station anthropometrics: Preferred height and angle settings of CRT and keyboard. Human Factors 25: 401–408

Mourant R R, Lakshmanan R, Chantadisai R 1981 Visual fatigue and cathode ray tube display terminals.

Human Factors 23: 520–540

Murray W E, Moss C E, Parr W H et al 1981 A radiation and industrial hygiene survey of video display terminal operations. Human Factors 23: 413

Nakaseko M, Grandjean E, Hunting W, Gieres R 1985 Studies on ergonomically designed alphanumeric keyboards. Human Factor 27: 175–188

National Institute for Occupational Safety and Health 1977 A report on electromagnetic radiation surveys of video display terminals. National Institute for Occupational Safety and Health, US Department of Health, Education and Welfare, NIOSH Technical Report, Publication 78–129, 1977

National Institute for Occupational Safety and Health 1981 Potential health hazards of video display terminals. National Institute for Occupational Safety and Health, US Department of Health and Human Services, NIOSH Research Report, Publication 81-129

National Research Council, Panel on Impact of Video Viewing on Vision of Workers 1983 Video displays, work and vision. National Academy Press, Washington DC

Newton R W, Hunt L P 1984 Psychosocial stress in pregnancy and its relation to low birth weight. British Medical Journal

Nilsen A 1982 Facial rashes in visual display unit operators Contact Dermatitis 8: 25–28

Office of Population Census and Surveys 1980 Hospital In-patient Enquiry. Maternity Tables. MB4 No 8. London, HMSO

Ong C N, Hoong B T, Phoon W O 1981 Visual and muscular fatigue in operators using visual display terminals. Journal of Human Ergology, 161–171

Ontario 1984 Video display units — characterization of electric and magnetic fields, Ontario Hydro, Research Division 1984, Report No 82-528-K

Ostberg I 1980 Accommodation and visual fatigue in display work. Displays, July 1980, 81–85

Ostberg O, Powell J, BlomKvist A C 1980 Laser optometry in assessment of visual fatigue. University of Lulea Technical Report. No 1980: 1T. University of Lulea, Sweden

Paulsson L E, Kristiansson I, Malmstrom I 1984 Stralning fran dataskarmar. Report No 84-08. Statens stralskyddsinstitut (in Swedish) ISSN 0281–1359

Pomroy C, Noel L 1984 Low-background radiation measurements on video display terminals. Health Physics 46: 413–417

Radl G W 1982 Experimental investigations for optimal presentation-mode and colours of symbols on the CRT-screen. In: Grandjean E, Vigliani E (eds) Ergonomic aspects of visual display terminals. Taylor and Francis, London p 127–135

Rey P, Meyer J J, Bousquet A 1982 Surveillance médicale des operateurs sur écran cathodique. Klin Mbl Augenheilk 180, 370–372

Roberts R S 1983 Adverse pregnancy outcomes associated with VDUs. Interpreting the evidence of case clusters. Unpublished Manuscript.

Rohmert W 1960 Ermittlung von er holungspousen fur statische arbeit des menschen. Int. Z angiew. Physio. Einschl. Arkeitsphysiol 18: 123–164

Roy C R, Joyner K H, Gies H P et al 1984 Measurement of electromagnetic radiation emitted from visual display terminals (VDTs). Radiation Protection in Australia 2: 26–30

Rupp B A 1981 Visual display standards: a review of issues. Proceedings of the Society for Information Display 22: 63–72

Rycroft R J G, Calnan C D 1984 Facial rashes among visual display unit (VDU) operators. In: Pearce B G (ed) Health hazards of VDTs? John Wiley & Sons, Chichester, pp 13–15

Sandstrom M, Hansson-Mild K, Lovtrup S 1986 Effects of weak pulsed magnetic fields on chick embryogenesis. International Scientific Meeting: Work with Display Units, Stockholm, Sweden 12–15 May 1986 (abstract)

Sauter S L, Gottlieb M S, Jones K C, Dodson V N, Rohrer K M 1983 Job and Health Implications of VDT Use: Initial Results of the Wisconsin-NIOSH Study. Communications of the ACM 26: 285–294

Sauter S 1984 Predictors of strain in VDT users and traditional office workers. In: Grandjean E (eds) Ergonomics and health in modern offices. Taylor & Francis, London, pp 129–135

Sauter S L, Chapman L J, Knutson S J, Anderson H A 1986 Wrist trauma in VDT Keyboard use: evidence mechanisms and implications for Keyboard and Wrist rest design. International Scientific Conference: Work with Display Units, Stockholm, Sweden, 12–15 May 1986 (abstract)

Shaw J A, Wheller P, Morgan D W 1970 Mother–infant relationship and weight gain during the first month of life. Journal of the American Academy of Child Psychiatry 9: 428–444

Smith A B, Tanaka S, Halperin W, Richards R D 1982 Report of a cross-sectional survey of video display terminal (VDT) users at the Baltimore Sun (National Institute for Occupational Safety and Health, Center for Disease Control, Cincinnati, Ohio) September 1982

Smith M J, Cohen B G F, Stammerjohn L W 1981 An investigation of health complaints and job stress in video display operations. Human Factors 23: 387–400

Smith M J, Stammerjohn L W, Cohen B G F, Lalich N R 1982 Job stress in video display operations.

In: Grandjean E, Vigliani E (eds) Ergonomics aspects of visual display terminals. Taylor & Francis, London, p 101–110

Snyder H 1983 Keyboard design. In: Ergonomic Principles in Office Automation. Ericsson Information Systems Stockholm

Snyder H L, Maddox M E 1978 Information transfer from computer-generated dot-matrix displays. Department of Industrial Engineering and Operations Research, Human Factors Laboratory Report No HFL-78-3/ARO-78-1. Blacksburg, Va: Virginia Polytechnic and State University

Sokil R J, Woolf R B, Rosen M G, Weingarden K 1980 Risk, Antepartum care and outcome. Impact of a maternity and intact care project. Obstetrics and Gynaecology 56: 150–156

Sosa R, Kennell J, Klaus M, Robertson S, Urruha J 1980 The effect of a supportive companion on perinatal problems, length of labour and mother-intact interaction. New England Journal of Medicine 303: 597–600

South Australian Health Commission 1984 Repetition strain symptoms and working conditions among keyboard workers engaged in data entry or word processing in the South Australian Public Service. Occupational Health Branch

Starr S J 1984 Effects of video display terminals in a business office. Human Factors 26: 347–356

Stuchly M AS, Lecuyer D W, Mann R D 1983a Extremely low frequency electromagnetic emissions from video display terminals and other devices. Health Physics 45: 713–722

Stuchly M A, Repacholi M H, Lecuyer D W et al 1983b Radiofrequency emissions from video display terminals. Health Physics 45: 772–775

Task H L 1979 An evaluation and comparison of several measures of image quality for television displays. Technical Report 79-7 (Wright-Patterson Air Force Base, Ohio; Aerospace Medical Research Laboratory).

Terrana T, Merkzzi F, Giudini E 1981 Electromagnetic radiations emitted by visual display units. In: Grandjean E, Vigliani E (eds) Ergonomic aspects of visual display terminals. Taylor and Francis, London, pp 25–38

Tjonn H H 1984 Report of facial rashes among VDU operators in Norway. In: Pearce B G (ed) Health hazards of VDTs? John Wiley & Sons, Chichester, pp 17–23

Tribukait B, Cekan E, Paulsson L E 1986 Effects of pulsed magnetic fields on embryonic development in mice. International Scientific Conference: Work with Display Units. Stockholm, Sweden 12–15 May 1986 (abstract)

Turner P J 1982 Visual requirements for VDU operators. Australian Journal of Optometry 65, 58–64. (Additional details in Coe J B, Cuttle K, McClellon W C, Warden N J, Turner P J, Visual Display Units. New Zealand Department of Health, Wellington 1980, Report W/1/180)

Ubeda A, Leal J, Tvillo M A, Jimenez M A, Delgado J M 1983 Pulse shape of magnetic fields influences chick embryogenesis. Journal of Anatomy 37: 513–536

Wallin L, Winkvist E, Svensson G 1983 Terminalanvandares arbetsmiljo — en enkatstudie vid Volvo i Goteborg (The work environment of terminal users — a questionnaire study at Volvo in Gothenburg, in Swedish), A B Volvo, Gothenburg, Sweden

Weate R A 1981 Eye tests and VDUs. In: Reading V (ed) Vision and Visual Display Units, Institute of Ophtholmology. London

Weiss M M 1983 The video display terminals — is there a radiation hazard? Journal of Occupational Medicine 25: 98–100

Weiss M M, Petersen R C 1979 Electromagnetic radiation emitted from video computer terminals. American Industrial Hygiene Association Journal 40: 300–309

Wilkins A J, Darby C E, Binnie C D 1979 Neurophysiological aspects of pattern-sensitive epilepsy. Brain 102: 1–6

Wilkins A J, Darby C E, Binnie C D, Stefansson S B, Jeavons P M, Harding G F A 1979 Television epilepsy — the role of pattern. Electroencephalography and Clinical Nuerophysiology 47: 163–171

Wolbarsht M L, O'Foghludha F A, Sliney D H et al 1979 Electromagnetic emission from visual display units: a non-hazard, non-ionising radiation — Proceedings ACGIH Topical Symposium 11, 1979, p 193–201

Wolkind S 1981 Prenatal emotional stress — effects on the feotus. In: Wolkind S, Zajicek E. (Ed) Pregnancy — a psychological and social study. London. Academic Press

World Health Organisation 1983 Women and Occupational Health Risks. EURO Reports and Studies, 76 WHO, Copenhagen

Yang R, Zweig A R, Porthitt T C, Federmann E J 1976 Successive relationships between maternal attitudes during pregnancy, analgesic medication during labour and delivery and newborn behaviour. Developmental Psychobiology 12: 6–14

Zaret M 1984 Cataracts and visual display units. In: Pearce B G (ed) Health hazards of VDTs? John Wiley & Sons, Chichester, p 47–54

Zipp P, Haider E, Halpern N, Rohmert W 1983 Keyboard design through physiological strain measurements. Applied Ergonomics 14: 117–122

Index

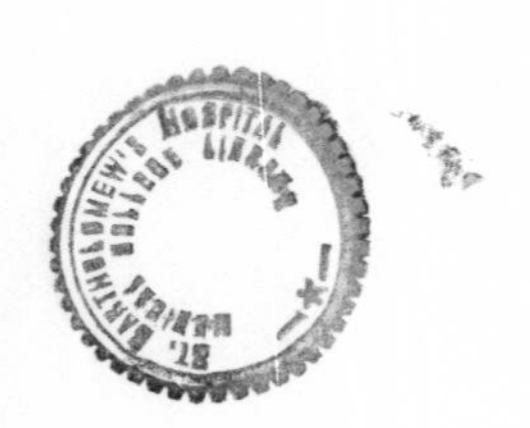